Springer Series in
Nuclear
and **Particle Physics**

Springer Series in **Nuclear** and **Particle Physics**

Editors: Mary K. Gaillard · J. Maxwell Irvine · Vera Lüth · Achim Richter

Hasse, R. W., Myers, W. D.
Geometrical Relationships of Macroscopic Nuclear Physics

Belyaev, V. B.
Lectures on the Theory of Few-Body Systems

Heyde, K. L. G.
The Nuclear Shell Model

Gitman, D. M., Tyutin I.V.
Quantization of Fields with Constraints

Sitenko, A. G.
Scattering Theory

Fradkin, E. S., Gitman, D. M., Shvartsman, S. M.
Quantum Electrodynamics with Unstable Vacuum

Brenner, M., Lönnroth, T., Malik, F. B. (Editors)
Clustering Phenomena in Atoms and Nuclei

Makhankov, V. G., Rybakov, Y. P., Sanyuk, V. I.
The Skyrme Model

Antonov, A. N., Hodgson, P. E., Petkov, I. Zh.
Nucleon Correlations in Nuclei

Akhiezer, A. I., Sitenko, A. G., Tartakovskii, V. K.
Nuclear Electrodynamics

V. G. Makhankov   Y. P. Rybakov
V. I. Sanyuk

The Skyrme Model

Fundamentals
Methods
Applications

With 22 Figures

Springer-Verlag
Berlin Heidelberg New York
London Paris Tokyo
Hong Kong Barcelona
Budapest

Professor Dr. Vladimir G. Makhankov
Joint Institute for Nuclear Research
141980 Dubna, Moscow Region, Russia

Professor Yurii P. Rybakov
Professor Valerii I. Sanyuk
People's Friendship University, 6, M. Maklai Str.
117198 GSP Moscow, Russia

ISBN-13:978-3-642-84672-4 e-ISBN-13:978-3-642-84670-0
DOI: 10.1007/978-3-642-84670-0

Library of Congress Cataloging-in-Publication Data. Makhankov, V. G. The Skyrme model: fundamentals, methods, applications/V. G. Makhankov, Y. P. Rybakov, V. I. Sanyuk. p. cm. – (Springer series in nuclear and particle physics) Includes bibliographical references and index. ISBN-13:978-3-642-84672-4
1. Skyrme model. I. Rybakov, Y. (Yurii), 1939-. II.
Sanyuk,V. (Valerii), 1950-. III. Title. IV. Series. QC793.3.S8M35 1993 539.7–dc20 92-28266

Softcover reprint of the hardcover 1st edition 1993

Typesetting: Data conversion by Springer-Verlag

56/3140-5 4 3 2 1 0 - Printed on acid-free paper

To our parents
and wives

Foreword

The December 1988 issue of the International Journal of Modern Physics A is dedicated to the memory of Tony Hilton Royle Skyrme. It contains an informative account of his life by Dalitz and Aitchison's reconstruction of a talk by Skyrme on the origin of the Skyrme model. From these pages, we learn that Tony Skyrme was born in England in December 1922. He grew up in that country during a period of increasing economic and political turbulence in Europe and elsewhere. In 1943, after Cambridge, he joined the British war effort in making the atomic bomb. He was associated with military projects throughout the war years and began his career as an academic theoretical physicist only in 1946. During 1946–61, he was associated with Cambridge, Birmingham and Harwell and was engaged in wide-ranging investigations in nuclear physics. It was this research which eventually culminated in his studies of nonlinear field theories and his remarkable proposals for the description of the nucleon as a chiral soliton.

In his talk, Skyrme described the reasons behind his extraordinary suggestions, which when first made must have seemed bizarre. According to him, ideas of this sort go back many decades and occur in the work of Sir William Thomson, who later became Lord Kelvin. Skyrme had heard of Kelvin in his youth. His maternal great grand-father Edward Robert knew and admired Kelvin, having been associated with the concentration of the Tidal Predictor under the direction of Kelvin and Tait. His admiration was so great that he named his son Herbert William Thomson Robert. This machine was in Tony's grandfather's house and greatly impressed Tony with its ingenuity.

Now Kelvin disliked hard point-like atoms. He was also not attracted to theories which cannot be mechanically visualized, although he came to accept Maxwell's electrodynamics in his later years. In 1867, he had proposed that atoms were vortices in a fluid and not the infinitely rigid points of chemists. Tony Skyrme seemed to have been vaguely aware of Kelvin's ideas, which to us would suggest a nonlocal structure for elementary particles.

In any event, somewhat like Kelvin, Tony too desired a visualizable description of the nucleon as an extended object. He was also led to this type of search by his intuition that fermions may emerge from self-interacting Bose fields just as bosons can arise as bound states of fermions.

By these emotional and imprecise means, and influenced as well by his research on nuclear matter and the fluid drop model, and by Pauli's work on meson theory, Tony was led to the sine-Gordon equation and to the proposal that nucleons are chiral solitons. His papers on these topics also contain seminal work on bosonization, vertex operators, quantization on multiply connected spaces and winding number as conserved charge.

Skyrme's ideas had to wait many, many years before they had a visible impact on the direction of nuclear and particle theory. This happened only after theoretical physicists generally became conscious of topological issues during the 1970s. This awareness was stimulated by the discovery of monopoles and instantons in nonabelian gauge theories and by the emergent classification theory of defects. Eventually, Skyrme's work was revived in the late 1970s and the first half of the 1980s by groups at Yale, Syracuse, and Princeton and soon thereafter rapidly found general acceptance in the particle and nuclear physics community.

Actually, however, even before their revival, interest in Skyrme's ideas was not limited to the aforementioned groups. There was for example the 1970 paper of Williams, a student of Skyrme, devoted to fundamental topological issues regarding the quantization of the chiral soliton. There were also the rigorous, sophisticated papers from the Soviet Union which in particular investigated the stability of soliton solutions in the Skyrme model. We were well aware of these Soviet papers and of their fundamental nature when we began our research on this model. Indeed they were among the formative influences leading us to take an interest in Skyrme's ideas.

This Soviet school of the past, and what is now a Russian school, has actively studied the chiral Lagrangian and nonlinear physics for well over a decade. Valery Sanyuk, a senior physicist from that group, was in Syracuse on a Fulbright Fellowship during 1990–91 when he initiated this book with his Russian colleagues. It is particularly fortunate that a group with long experience and scholarship undertook this task. The result is impressive. I feel honored that Valery invited me to write this Foreword.

There exist, of course, several reviews of the Skyrme model. There are first those articles with a strong phenomenological orientation which linger but briefly on its beautiful and important formal aspects. There are then also those reviews, including our own, with emphasis on formal issues, but their presentation is naturally less thorough than what could be accomplished in an entire book. A pedagogical and detailed treatment of the Skyrme model has thus been felt necessary, and the decision of our Russian colleagues was most timely.

This book is unique in several ways. It contains a comprehensive discussion of the topological aspects of the Skyrme soliton and its quantization, including an introductory and clear account of algebraic topology and its field theoretic applications. This material can be found elsewhere, but in a scattered and less accessible form. This is the only work I know with a systematic discussion of the stability theory of Skyrmions and nonlinear field theories. Its authors have themselves contributed extensively to this theory, and have written an account here which will attract all mathematically inclined physicists and mathematicians interested in physical applications of abstract methods. It is also the only book to be found with a fascinating history of Skyrme's ideas.

In summary, *The Skyrme Model: Fundamentals, Methods, Applications* is a comprehensive and user-friendly book on chiral solitons, nonlinear field theories and much else besides. It is to be recommended with enthusiasm to physicists with interest in the Skyrme model or topological ideas, and to mathematicians wishing to learn about the uses of their discipline in the physical sciences.

Syracuse
February 1993

A.P. Balachandran

Preface

This book is based on a lecture series given by the authors for undergraduate and postgraduate students at the People's Friendship University (Moscow), from 1983 to 1987 and for young scientists at the Joint Institute for Nuclear Research (Dubna) in 1987. In 1990 one of us (V.I. S.) delivered lectures at Syracuse University (U.S.) under the Fulbright Fellowship Program. The content of the courses varied in accordance with the requirements of the auditorium and our knowledge of the subject. During this period of time, interest in the Skyrme model has been renewed, mostly because of the efforts of two groups: the Syracuse group (A.P. Balachandran, F. Lizzi, V.P. Nair, S.G. Rajeev, V.J.G. Rodgers, J. Schechter, A.S. Stern) and the Princeton group (G.S. Adkins, C.R. Nappi, E. Witten). Many particle and nuclear physicists have become interested in the apparent possibility of understanding some hadron physics phenomena in the framework of the relatively simple Skyrme-model approach. The number of publications on the subject grows at an exponential rate; new reviews become out-of-date faster than they are published. In this situation it is more and more difficult for beginners to get a primary understanding of the fundamentals of the Skyrme approach.

This was one of our goals: to guide the initial steps of newcomers in the field and then to lead them quickly to an understanding of "the present state of the art" at a fairly elementary level. With this in mind we found it suitable to organize the book in the following manner. First of all we pay special tribute to the background ideas of the Skyrme approach, such as the hydrodynamical origin and vortex-like nature of Skyrmions. Second, we familiarize the reader with the deep topological content of the Skyrme model, with an explanation (where possible) of the topological concepts introduced from a physical standpoint. Third, instead of a hopeless attempt to collect all known facts on the Skyrme model and its applications, we try to understand which of Skyrme's conjectures can already be definitely affirmed.

Naturally, to achieve these goals the book must be written in as simple and concrete a language as possible. To understand the basic part of it the reader must have some previous knowledge of the standard courses in calculus and geometry relevant to classical physics (classical mechanics and field theory) and it is well to have an appreciation of the basic notions of particle physics. To confirm the validity of Skyrme's hypothesis and conjectures it appears necessary to use some methods used to study multi-dimensional solitons. Another equally important aim of the book is to present to the physics community ideas, such as the methods of extending the phase space and of spherical rearrangement, which proved to be useful in attaining the absolute minima of nonlinear functionals. The principle of symmetric criticality, in particular the Coleman-Palais theorem, is used in the search for the structure of invariant fields. The generalized Hobart-Derrick theorem, together with the Q-stability theorem, provides a strategy for studying the stability of Skyrmions. None of the methods listed are widely known even among practitioners in the field and we present them in a compromise between mathematical rigor and finite length.

We believe that this book can encourage some graduate and even undergraduate students to start their own research in the promising field of nonlinear physics. It can also serve as a supplementary text for university courses in nuclear and particle physics and, in general, in any other courses covering the soliton approach in many dimensions or applications of modern geometrical and functional analytic methods in field theory. We hope the book may be useful for physicists who are novices in the field and who want to get acquainted with the basic concepts of the soliton approach to hadron structure, as well as with some topological and variational methods for the investigation of nonlinear functionals. As the soliton approach has already led to a wide range of applications in condensed matter physics, the theory of ferromagnets, hydrodynamics and plasma physics, it is beyond any doubt that specialists in these fields will find some new applications for the methods presented in Part II of the book. Finally, mathematicians may also have fun in discovering how useful abstract methods in physical applications can be, and can use the book as a source of illustrative examples. They should not be too captious about the style of our mathematical expositions (none of the authors was ever educated as a mathematician), but all their helpful advice would be greatly appreciated.

Note that the present book is the outcome of reworking, reordering and extensively elaborating the lecture notes published in duplicated form in the series Lecture Notes for Young Scientists at the Joint Institute for Nuclear Research: V.G. Makhankov, Y.P. Rybakov, and V.I. Sanyuk: *The Skyrme Model and Solitons in Hadron Physics* (JINR Publ., Dubna, 1989) P4-89-568, 172 pp.

In conclusion we should like to express our deep gratitude to our colleagues at the People's Friendship University and the Joint Institute for Nuclear Research, whose valuable support and assistance made this project possible. One of us (V.I.S.) is very grateful to the Fulbright Fellowship Board for the award which provided him with excellent conditions in which to implement the project in its initial stage and to meet with many experts in the field (Professors E. Braaten, F. Gürsey, C. Sommerfield) during the lecturing program. It is a pleasant duty to acknowledge many stimulating talks with outstanding contributors to and initiators of the modern development of Skyrme's ideas. A special vote of thanks for their kind attention, warm hospitality and support goes to colleagues from the Physics Department of Syracuse University, and especially to Professors A.P. Balachandran, J. Schechter, K.C. Wali, and M. Goldberg. The project would never have been implemented in its final form without the incredible assistance of Dr. Theodore J. Allen, and Anand

Subbaraman, among others, who improved the English, instructed one of us (V.I. S.) in using TEX and much more. The authors greatly appreciate the good-will of Professor A.P. Balachandran in writing his excellent Foreword to the book. Thanks are due to our editors, Dr. Ernst F. Hefter and Prof. Maxwell Irvine, for their kind attention, encouragement and great patience, and also to Ms. Deborah Cliver for editorial assistance.

Moscow-Dubna
Russia
June, 1992

V.G. Makhankov
Y.P. Rybakov
V.I. Sanyuk

Contents

Introduction

"If I have seen further than others, it is because I have stood on the shoulders of giants."

Isaac Newton

It took more than two decades for the ideas and results of the distinguished English physicist Tony Hilton Royle Skyrme (1922–1987) to become widely recognized as the outstanding and far-reaching contributions to high-energy physics, mathematical physics and nuclear physics that they are. Until 1982 he was chiefly known among nuclear physicists as the author of the effective nucleon-nucleon potential, known also as the "Skyrme force" (Skyrme 1959), and as the co author of the Bethe-Rose-Elliott-Skyrme theorem (Elliott and Skyrme 1959), which is included in textbooks on nuclear physics. But Skyrme's main discovery (Skyrme 1961, 1962) – **the model of baryons as topological solitons** – was largely ignored till the early 1980's. The novelty of his approach in particle physics during the 1960's was so impressive that only a very few amateurs 'were splashing in the calm basin' of his theory which seemed to be rather far away from the mainstream of progress in particle physics.

The situation changed drastically in the early 1980's after the realizing that the Skyrme model could be considered as the possible low-energy limit of QCD. It became the time of the professionals who 'crowded the basin, showed the amateurs some good swimming styles' and then, after a few years of enormous activity, were carried away by the main stream of progress. To be straightforward let us note that some professionals, mostly nuclear physicists, are still keeping an eye on the flounderings of the amateurs in this not yet quiet basin with willingness to stretch out a helping hand on the occasion of a good catch. Nevertheless we are strongly convinced that, in spite of numerous efforts in the field during the 1970's, only elementary steps have been taken towards understanding all the possible advantages of Skyrme's approach. Real achievements are still ahead and merit the attention of new researchers. With this in mind we provide our readers with a historical survey, based on Skyrme's earliest papers and on his own published remembrances.

In the Introduction and Chapter 1 are found the motivations and the step-by-step evolution of Skyrme's model description of nuclear matter from the "Mesonic Fluid" model up to the final version, known as the baryon model. Not widely known even among specialists, this evolution provides an instructive peep into the workshop of one of the most original theorists of our century. We pay special tribute to some ideas, currently under investigation in physics including: Nuclear Democracy, the Solitonic Mechanism, the Nonlinear Realization of Chiral Symmetry, Topological Charges, Fermi-Bose Transmutations. All those ideas were mostly pioneered by and made their first appearence in the earlier papers of Skyrme (Skyrme 1954, 1955, 1958). The passage of more than three decades served to emphasize the originality and still actuality of their contents.

As stated in the biography by Dalitz (1988), after graduation from Trinity College, Cambridge, U.K. in 1943, Tony Skyrme spent some time in the U.S.A., where he took part in and made a notable contribution to the success of the Manhattan Project. He returned to Britain and in 1950 started his work in the Division of Theoretical Physics of the Atomic Center (A.E.R.E) in Harwell. It is difficult to determine when the earliest of Skyrme's papers were written. According to Dalitz: "it was common knowledge" among colleagues at A.E.R.E that the senior principal scientific officer, Tony Skyrme, "had a desk drawer full of manuscripts awaiting completion and submission for publication". In 1952 B. Flowers, on returning to A.E.R.E from Birmingham, to became Head of the Theoretical Division, learned of this drawer and "made it one of his priorities to move as many of these papers as possible into scientific journals." Because of the persistence of B. Flowers we may approximate the initial date of the Skyrme model as 1954, when Skyrme's first paper (Skyrme 1954) on the model description of nuclear matter was published. In this paper Skyrme proposed the "mesonic fluid" model in order to answer a question raised by experimental data of nuclear radii measurements. We will discuss this problem in more detail in Chapter 1. Here we would wish only to point out that already in the first of his pioneering papers he used some hydrodynamical ideas for the nonpointlike description of nucleus.

According to his own writings (Skyrme 1988), Skyrme never believed in the validity of any pointlike description of elementary particles in the framework of a linear field theory with any conceivable renormalization scheme. He considered the renormalization theory as "...just a very good and useful way of enabling us to live with our ignorance of what really goes on at short distances." For a possible alternative, Skyrme was looking for a nonlinear field theory, which admits a description of the aforementioned particles as extended objects. In that respect he agreed with W. Heisenberg that a real particle theory must be strongly nonlinear. But at the same time Skyrme deeply disagreed with another opinion of Heisenberg, who considered fermion fields as the most fundamental. The main motivation in Heisenberg's Unified Field Theory was to show that everything should be made from the so called "profield" - the fundamental self-interacting fermion field. Skyrme's point of view, on the contrary, was that fermions do not have any real physical meaning but rather are "...useful in some situations purely mathematical constructions." Despite his great intuition, which he demonstrated in a good number of physical problems, Skyrme "...always felt uneasy about quantum-mechanical concepts that do not have clear classical analogues." In particular this was the case with fermions. Already in the early 1950s Skyrme was of the firm opinion that there is no reason to consider bosonic fields as less fundamental than fermionic ones. One of the main motives, which led him to the now well-known baryon model, was: "...**it would be fun to see if I could get everything out of self-interacting boson field theory**" (Skyrme 1988).

Skyrme found unexpected support for the ideas mentioned above in comparatively ancient papers (Kelvin 1904 and 1910)[1]. He noticed that in the previous century Kelvin was deeply concerned with similar problems arising in the description of atomic structure and "...was very reluctant to accept the idea of infinitely rigid point-like atoms." As a more reasonable alternative Kelvin suggested the "Vortex Atom" model (Kelvin 1910). This peculiar "Vortex Atom" idea was inspired by Helmholtz's discovery of the so-called "Wirbelbewegung" or "vorticity" - an absolutely conserved quality in the motion of any portion of a perfect fluid[2]. Kelvin assumed this "Wirbelbewegung" as a desirable fluid-mechanical model of the atom, as he also could not understand any physical phenomenon without a simple mechanical model. He did not accept the usual way of explaining incomprehensible things by introducing conceptions that are even less understandable. On the basis of experiments with smoke-rings, conducted by his collaborator Tait, and some analytical results on vortex description, Kelvin developed the hypothesis that **all bodies are composed of vortex atoms in a perfect homogeneous liquid**. He thought that this approach had some advantages in comparison with the generally accepted ones at that time, "*... as vortex rings do not require any other property of the matter whose motion composes them than inertia and incompressible occupation of space.*" According to his writings, Kelvin had farreaching plans to develop a new kinetic theory of gases and to found a theory of elastic solids and liquids on the dynamics of vortex atoms. But as far as we know those plans were not realized by Kelvin or his collaborators and later were entirely rejected by most physicists, along with the ether theory.

The paper "On Vortex Atoms" demonstrates that Kelvin was a real pioneer in introducing topological concepts into physics: Different sorts of atoms were to differ from each other in his theory in accordance with the number of intersections of vortex rings. Let us quote this paper, dated 1867: "*It is to be remarked that two ring atoms linked together or one knotted in any manner with its ends meeting, constitute a system, which, however it may be altered in shape, can never deviate from its own peculiarity of "multiple continuity", it being impossible for the matter in any line of vortex motion to go through the line of any other matter in such motion or any other part of its line. In fact a closed line of vortex core is literally indivisible by any action result-*

1 It is curious that, again according to (Dalitz 1988), T. Skyrme was acquainted with, as he put it, the "eminent Victorian, Sir William Thomson", by literature as well as through the stories of his mother's grandfather Edward Roberts. His great-grandfather "...was in 1868 appointed Secretary to the Tidal Committee of the British Association for the Advancement of Science, being made responsible later for the construction of the first Tidal Predicter, which had been designed by Lord Kelvin for this Committee." The first model of this machine, which was held in Roberts's house at 7 Blessington Road, made a strong impression on the young Tony and greatly influenced the development of his later ideas, as Skyrme himself has recounted (Skyrme 1988).

2 Here the term "perfect fluid" means a fluid perfectly destitute of viscosity. Later on this concept was developed into the basic idea of the ether theory.

Fig. 1. Drawings of Kelvin's "Vortex Atoms"

ing from vortex motion." In Fig. 1, which is taken from Kelvin's next paper "On Vortex Motion", are displayed knotted or knitted vortex atoms "*... the endless variety of which is infinitely more than sufficient to explain the varieties and allotropies of known simple bodies and their mutual affinities.*" Now one can easily recognize that Kelvin was thinking about a concept similar to a topological invariant and, in particular, about the Hopf index, which is widely used in contemporary particle and condensed matter physics models. Kelvin complained in the same paper that he was not sufficiently acquainted with Riemann's "Lehrsätze aus der Analysis situs", which was known to him only through Helmholtz. But nevertheless he was one of the first physicists to understand that such topological properties as "multiple continuity" or, in present terminology "connectedness" may be important for a description of matter.

As a result of this historical excursion one may form an impression that it was quite natural for Skyrme to start with a hydrodynamical model of nuclear matter. Thus, it will not be a surprise if the images of extended particles appear in such a model as a nuclear "Wirbelbewegung."

In Chapter 1 we shall see how this natural idea has been implemented and what kind of conjectures Skyrme made on the basis of its realization. The central aim of this book is to demonstrate the validity of Skyrme's conjectures and hypotheses in the light of contemporary knowledge. Pursuing this aim and taking into account that the anticipated reader of this book is a graduate student in physics, a more phenomenological particle physicist or a novice in the field, we include the elementary exposition of some background information. Chapter 2 treats the classical field theory, using the language of differential geometry. We provide the reader with the physical meaning of the topological, or more precisely, homotopical classification of solutions, with elementary understanding of homotopy and cohomology groups in application to nonlinear field theory. Finally we formulate general principles of construction of topologically nontrivial field models and apply them to the construction of chiral invariant Lagrangians. In Chapter 3 we familiarize the reader with background information on the soliton stability in the Lyapunov sense, with restrictions resulting from the Hobart-Derrick "no-go" theorem and with the generalization of this theorem. To describe extended particle-like structures, the model equations are required to possess stable soliton-like solutions. For this reason we discuss in detail the concept of topological stability. These three chapters, composing Part I, are written at a fairly elementary level to help the beginner in a quick understanding of 'the present state of the art'. For completness two appendices supply a novice with necessary information on the chiral symmetry and basic concepts of the modern geometry.

There is a widespread persuasion that workable analytical methods for studying nonlinear models are limited mostly to the (1+1)-dimensional case and progress in the study of (3+1) models is mostly related with computer simulations. In Part II we attempt to convince the reader that this is really not 'a one-way situation' and that in addition to computer analysis, more or less rigorous examinations for a rather wide variety of (3+1) models are possible. In particular we present methods for studying extremals of the energy functionals, using the mathematical apparatus of functional analysis and direct methods of calculus of variations. In Chapter 4 we give an exposition of the group-theoretic methods in a search for the structure of invariant (equivariant) fields, known in the literature as *the Principle of Symmetric Criticality or the Coleman-Palais theorem.* This principle, in brief, asserts that *extremals of symmetric functionals (if any) are realizable in the class of invariant functions* and thus makes it possible to narrow our search for extremals to the class of invariant fields. We show how to find the maximal compact groups which possess equivariant fields with nontrivial topological charges in realistic (3+1)-dimensional models, and the corresponding structure of invariant fields. We also discuss an essential role of topological invariants in topologically nontrivial field models. It turns out that the value and the structure of the topological invariant has a decisive influence on the structure of equivariant field configurations. For example, for the Skyrme functional the absolute minima of energy in the first homotopy class is attainable by the spherically symmetric fields - the Skyrmion or the "hedgehog" configuration. For the second homotopy class the energy minimum is realizable in the class of the axisymmetric fields.

It is apparent that for a description of stable localized particle-like structures, configurations which correspond to the absolute minimum of the energy are of top-priority importance. In Chapter 5 we present *the method of extending the phase space* in order to seek the absolute minimum of the energy functional. This method consists of regarding the field functions and their derivatives as independent and then proving that the minimum found in this way is indeed a critical point of the original functional. The validity of this device is based on the fact that the minimum in the extended space will also be a true minimum since the imposition of constraints (between the functions and their derivatives, which reflect the fact of their dependence) can only raise the value of the energy functional. One can regard this method as a generalization of Gelfand's "valley method" used for a search for the absolute minimum point of a composite function. To establish the explicit forms of the ansätze as an appropriate tool we use *the spherical rearrangement method.*

In Chapter 6 we consider in detail the problem of existence of "hedgehog"-type solutions in the Skyrme model and expose the rigorous proof of Skyrme's conjecture. In Chapter 7 we consider the problem of multi-baryon states and that of a rotating Skyrmion. As was already mentioned, the minimal-energy configurations in higher homotopy classes are no longer spherically symmetric and one can observe the phenomena of breaking the symmetry of invariant fields or ansätze with an increase of the topological charge value. Optimal configurations of a rotating Skyrmion also have axial symmetry and that circumstance leads to variational problems of enormous complexity. To proceed in such cases one is usually forced to apply the method of trial functions. In Appendix D.2 we supply the main body text with a proof of existence of regular axisymmetric configurations in all higher homotopy classes, which can be easily extended to rotating Skyrmion configurations as well. The latter are the subject of the last section in Chapter 7, whereby we also pay attention to limitations of a semiclassical treatment of Skyrmions. The last chapter of Part II deals with the semiclassical quantization of Skyrmions. Using Bogolubov's collective variables method we demonstrate different quantization schemes, which are applicable to Skyrmions. In exposition we mostly rest upon the results of the Syracuse group. Chapter 8 ends with discussion of a possible scheme of "non-rigid" quantization, which from different points is preferable in application to Skyrmions.

A remarkable modern phenomenon is that with more steadfast attention to nonlinear effects we have already gain a rich spectrum of multidimensional localised structures, which have appeared under various names (topological defects, vortices, strings etc.) in particle physics, condensed matter physics, cosmology and so on. A glance at the functionals used for model descriptions of those structures convince one that the aformentioned methods might be efficacious in their study. With the aid of the presented methods it is possible to reduce a complicated multidimensional problem to the study of a dynamical system with 'one-and-a-half' degrees of freedom. Behavior of reduced systems is usually governed by ordinary differential equations, which in most cases are essentially nonlinear. Nevertheless, even with computer simulations it is easier to extract reliable results from those equations, than to get these answers in a frontal computer attack on the unreduced problem. This approach to study of multidimensional nonlinear structures in different branches of contemporary physics should be another benefit of reading this book by non-High Energy physicists. For easier understanding, Part II is written in a compromise between rigour and finite length. Here and throughout the book in the essential text we only formulate main ideas, assertions and results, transferring most proofs and technical details of derivations into Appendices, which constitute an integral part of the context.

Part III deals with applications of Skyrme's soliton approach to low-energy Hadron Physics. These are questions of extreme importance and attraction for particle and nuclear physicists. This field especially is under the most active development and as a result it brings a huge number of current publications, so that any overview of recent results and achievements becomes out-of-date incredibly quickly. For this reason we limit ourselves to a more moderate goal, namely, to provide the reader with a background of already universally adopted information. We start Chapter 9 with a pedagogical exposition of G. t'Hooft and E. Witten's results, showing in what sense QCD is simplified in the framework of $1/N_c$-decomposition and how it can be reduced to an effective mesonic theory. Then we try to understand what one can definitely assert on the form of the meson Lagrangian and to what extent Skyrme's Lagrangian can be regarded as a substitute for this unknown theory. Then we reproduce justifications of Skyrme's conjectures, which are derivable from the QCD first principles. In particular we consider arguments in support of the use of the chiral limit and that of the choice of field manifold in the Skyrme approach and the validity of topological charge interpretation as the baryon number. The topic of Chapter 10 is the validity of the most intriguing of Skyrme's suggestions, that Skyrmions could be regarded as fermionic states in a quantized theory. Following our retrospective exposition, we start with the formulation of Finkelstein's topological justification of this hypothesis and, step by step, come to its modern implementation. In Chapter 11 we give a sketch analysis of the modern ideology of derivation of the static properties of baryons in the $SU(3)$ generalized Skyrme model. It is supplied by a brief account of the Skyrmion phenomenology, starting from the Skyrmion-Skyrmion interactions up to the dense nuclear matter description. The latter issue one can regard as a modern stage of development of the initial Skyrme's "Mesonic Fluid" model. Those who are interested in more advanced information on applications of Skyrmions in particle and nuclear or condensed matter physics we refer to reviews on the subject (Pak and Tze 1979; Zahed and Brown 1986; Brown and Rho 1986; Holzwarth and Schwesinger 1986; Meissner and Zahed 1986; Meissner 1988; Makhankov, Rybakov and Sanyuk, 1992), lecture courses (Adkins 1987; Balachandran 1986; Makhankov, Rybakov and Sanyuk 1989; Sanyuk 1992) and proceedings of workshops (Chodos, Hadjimichael and Tze 1984; Jezabek and Praszalowicz 1987).

We hope that this book will find its readers, who will share our opinion that it is too optimistic to expect that the Skyrme model will be able to provide us with answers to all possible questions in hadron physics. Practitioners will probably say that Skyrme's approach does not give any advantage if compared with already well-tested methods of calculations of baryon properties and characteristics. And they are right, but only to some extent. From that point of view we may continue to consider our Earth as a plate, as we do not very often need in our common life to take into account the geoid form of the Earth. With the advent of the Skyrme model, the hadron structure and interactions can be conveniently treated in a more unified manner. This is the most obvious advantage of the approach. But if this is the only extraction after the reading of the book we can recommend: read it again and see how many interesting features, mostly originating in hydrodynamics and later endowed with deep topological context, possess this simple model of a nuclear "Wirbelbewegung". Make an attempt to suggest an even more original approach and, if it will not take too much time, we are going to be among the first readers of your future book.

Part I

Fundamentals

1. The Evolution of Skyrme's Approach

"I can never satisfy myself until I can make a mechanical model of a thing."

Lord Kelvin

1.1 The "Mesonic Fluid" Model

It was one of Tony Skyrme's characteristic features to work in close contact with experimental physicists, who interested him in different phenomenological problems. In this way Skyrme was involved in solving the experimental data puzzle, which arose as a result of the fast electron scattering experiments, conducted with the aim of studying the charge distribution within the nucleus (the results were published in 1953). As a consequence of those **electromagnetic type experiments**, as well as from μ-mesonic atoms spectroscopy data or from the isotopic shift data, the value for the radius of the charge distribution within the nucleus was suggested to be

$$\bar{R} \simeq 1.2\, A^{1/3}\, \mathrm{fm}, \tag{1.1}$$

where, as usual, A is the atomic number. At the same time it was already known from specifically **nuclear type experiments** such as α-decay, light nuclei reactions, from the penetrability factors calculated for the scattering of neutrons, or from the evidence of the difference in binding energies of mirror nuclei, that one can derive the value for the radius of the nucleus close to

$$R \simeq 1.5\, A^{1/3}\, \mathrm{fm}. \tag{1.2}$$

In combination, these results signified that the electrical charge of the nucleus lies within the radius $\bar{R}$, but that the specifically nucleonic interactions extend out to the radius R.

This picture conflicted with the description of the nucleonic interactions as short-range ones, centered on the constituent nucleons, and with the image of nucleus structure adopted at that time. It was difficult to understand how it could lead to the effective radius (1.2), proportional to $A^{1/3}$, if the electrically charged region of nucleus has the smaller radius $\bar{R}$. In fact, from the successful prediction of the magnetic moments on the basis of the shell model (Flowers 1952), it was clear that the charge of the nucleus must be quite closely confined to the neighborhood of the protons. This means that protons should occupy the region of radius $\bar{R}$ and, if we accept the charge independence of nuclear interactions, so must the neutrons. Thus there was no reasonable explanation in the framework of the adopted picture of nuclei for the value R, which was evident from nuclear experiments.

It is necessary to note, that the structure of nuclei was a subject of interest to Skyrme for a long time. He acknowleged in the paper (Skyrme 1955) the fruitful discussions on that topic and criticism of V. Weisskopf from M.I.T., where Skyrme spent the 1948-49 academic year. Already at that time he felt that substantially homogeneous nuclear matter should not be described by means of a discrete (i.e. shell-model) picture. As a more preferable description Skyrme considered a type of fluid drop model with the nucleons being some kind of local twists (or a nuclear "Wirbelbewegung") in the fluid. This puzzle appeared to be the problem where he found it possible to apply his ideas of continuum theory of nuclei, based on a semiclassical approach. Skyrme agreed with Yukawa's picture of nucleonic interactions most likely being due to the intermediate π-meson fields, so that π-mesons should be regarded as constituents of the nucleus. To bring the theory into agreement with the nuclear type experiments data, π-mesons should occupy the region with radius R. But as ordinary π-meson fields are electrically charged, the electromagnetic experiments are to give the same value R for the radius of charge distribution. To avoid this obstacle, Skyrme suggested in his initial paper (Skyrme 1954) that mesonic fields might be condensed into "Mesonic Fluid" so, that at any point of nucleus, mean densities of π^+ and π^- components would be equal to each other. Meantime spin and charge fluctuations would arise locally in this fluid and would lead to a much stronger coupling of nucleons to the mesonic fluid than to each other. As a result of these interactions between nucleons and mesonic fluid, the nucleons would occupy only a smaller interior region with radius $\bar{R}$. The main positions of the "Mesonic Fluid" model proposed by Skyrme in order to meet these requirements were as follows:

1. The nucleus is considered to be a classical, electrically neutral incompressible "Mesonic Fluid", which occupies the region with radius R. In the standard hydrodynamical manner the state of this fluid at any point can be described by some scalar density and a direction (a vector in isospace).
2. The nucleons are immersed into the "Mesonic Fluid", which saturates them, and they are free to move inside the region with the radius $\bar{R}$ with residual interactions that result from spin-charge fluctuations in the fluid. (As a very crude analog it is possible to represent the nucleonic source as a sort of "oscillatory pump" free to move within some region of the mesonic "liquid" drop.)

In the next paper (Skyrme 1955) he proposed a mathematical formulation of these ideas, starting from the usual expression for the Lagrangian density of nucleon fields interacting with a pseudo-scalar meson field (Pauli 1946):

$$\mathcal{L} = \frac{1}{2}\left[(\partial_\mu \underline{\phi})^2 - k^2 \underline{\phi}^2\right] + i\bar{\psi}(\gamma^\mu \partial_\mu + g\gamma_5 \underline{\tau} \cdot \underline{\phi})\psi. \tag{1.3}$$

Here the convention is accepted that underlined quantities as $\underline{\phi}$ are vectors in the isotopic spin space; k is a reciprocal length, $\underline{\phi}$ are mesonic fields; $\underline{\tau}$ are the Pauli matrices and spinor-isospinor fields ψ describe nucleons. Note that there is no "inertial" mass term of bare nucleons in the Lagrangian density (1.3). This is not an accident but a special assumption by Skyrme, that **all the nucleon's mass is of mesonic origin, i.e. it arises as a result of mesonic fluctuations.** This suggestion looks like a first mention of the idea of the **"Solitonic Mechanism"**, expressed in a clearer form later on (Faddeev 1975) as the following: *"The strong-interacting particles could be described in the framework of a nonlinear field theory as collective excitations in the system of weakly coupled fundamental fields. The corresponding Lagrangian should be expressed in terms of fundamental fields only and admits the existence of particlelike (solitonic) solutions. Those solitons are to describe the evident spectroscopy of heavy particles, as well as interactions among them."*

To realize the idea of the mesonic origin for the nucleon's mass, Skyrme proposed the following method: write down the mesonic field vector in polar coordinates $\underline{\phi} = \underline{\mathbf{n}}\phi$, where $\phi = |\underline{\phi}|$, so that $\underline{\mathbf{n}}$ is a unit pseudo-vector and ϕ is a scalar; then perform the unitary transformation

$$\psi \rightarrow \exp\left(\mathrm{i}\frac{\pi}{4}\gamma_5\underline{\tau}\cdot\underline{\mathbf{n}}\right)\psi = \frac{1}{\sqrt{2}}(1+\mathrm{i}\gamma_5\underline{\tau}\cdot\underline{\mathbf{n}})\,\psi. \tag{1.4}$$

In Skyrme's terminology this is "...a limiting form of the Foldy transformation, appropriate to zero inertial mass". In the transformed Lagrangian

$$\mathcal{L}' = \tfrac{1}{2}\left[(\partial_\mu\underline{\phi})^2 - k^2\underline{\phi}^2\right] + \overline{\psi}\left[\mathrm{i}\gamma^\mu\partial_\mu - g\phi + \tfrac{1}{2}(\gamma_5 + \mathrm{i}\underline{\tau}\cdot\underline{\mathbf{n}})\gamma^\mu\partial_\mu(\underline{\tau}\cdot\underline{\mathbf{n}})\right]\psi, \tag{1.5}$$

the additional "mass" term arises in the form $\overline{\psi}g\phi\psi$ adjustable for nucleons. Here g is the mesonic coupling constant. It is also possible to regard the transformation (1.4) as a chiral transformation. This is a remarkable fact, because this particular symmetry was chosen later by Skyrme in order to modify the initial version of his model so that it would be possible to take into account new motives emerging in particle physics after the discovery of parity violation in weak interactions. To conclude this section we note that the "Mesonic Fluid" model managed to explain the evident discrepancy in experimental data, and its consequences were in agreement with well established facts about the shell structure of nuclei, the collective motions in nuclei, etc. Those who are interested in more detailed information are referred to Skyrme's original papers (Skyrme 1954 and 1955).

1.2 The Chiral Modification

The real acknowledgement of the Chiral Symmetry ideas in particle physics can be dated to 1957, when it was discovered that the parity conservation law could be violated in weak interactions. The long-standing question, known as the "$\Theta - \tau$" puzzle, was at last resolved and "strange" Θ- and τ-mesons were unified into the no less "strange" K-meson. Those particles, we recall , were called "strange" because although they arose through the strong interactions, their decays were mediated through the weak channels. The Θ-mesons decayed, as a rule, into two π-mesons, and τ-mesons decayed into three π-mesons. These hereditary strange features of K-mesons demanded for their description an enlargement of the isotopic internal symmetry group $SU(2)$ in such a way that the parity-violating transformations were to be presented in a new internal symmetry group of strong interactions. Some of those possibilities were already studied by (Pais 1953, 1954) and it was shown how to join three $SU(2)$-generators of isotopic rotations $\mathbf{I}_k$, which do not change the parity of states, with three generators of chiral boosts $\mathbf{K}_j$, which do mix up the states with different parities. Thus the problem was to construct a realization of at least the 6-parameter symmetry group, but, as is well-known, there is no linear representation of such a group in the 3-dimensional isotopic space.

The possible solutions are twofold: one can either to extend the isotopic space by introducing additional components or to look for a nonlinear group realization. The first way is analogous to the extension of the $SO(3)$ rotation group to the homogenous Lorentz group: the superfluous coordinate x_0 could be introduced and in addition to the three generators $\mathbf{J}_k$ of rotation about space axes three Lorentz-boost generators in (x_0, x_i)-planes are also considered. In the same manner one can supply the 3-isovector $\underline{\phi}$ with the fourth component ϕ_0 and consider the resulting 4-vector as an element of 4-isospace. In that case generators $\mathbf{I}_k$ would mix components of $\underline{\phi}$ only and would not affect ϕ_0. At the same time the chiral boost generators $\mathbf{K}_j$ would mix ϕ_0 with components of $\underline{\phi}$. Those generators would form the following algebra:

$$[\mathbf{I}_i, \mathbf{I}_j] = i\epsilon_{ijk}\mathbf{I}_k; \quad [\mathbf{I}_i, \mathbf{K}_j] = i\epsilon_{ijk}\mathbf{K}_k; \quad [\mathbf{K}_i, \mathbf{K}_j] = i\epsilon_{ijk}\mathbf{I}_k, \tag{1.6}$$

which is locally isomorphic to the Lie algebra of $O(4)$ group. Introducing the left and right generators

$$\mathbf{L}_i = \tfrac{1}{2}(\mathbf{I}_i - \mathbf{K}_i); \quad \mathbf{R}_i = \tfrac{1}{2}(\mathbf{I}_i + \mathbf{K}_i), \tag{1.7}$$

the commutation relations (1.6) take the form

$$[\mathbf{L}_i, \mathbf{L}_j] = i\epsilon_{ijk}\mathbf{L}_k; \quad [\mathbf{R}_i, \mathbf{R}_j] = i\epsilon_{ijk}\mathbf{R}_k; \quad [\mathbf{L}_i, \mathbf{R}_j] = 0. \tag{1.8}$$

The algebra has splitted into two independent subalgebras, each isomorphic to a $su(2)$-algebra. That was the reason that for the corresponding chiral group the notation $SU(2)_L \otimes SU(2)_R$ [1] was accepted.

Skyrme in 1958 realized the necessity of a chiral modification of the "Mesonic Fluid" model and performed it in his work (Skyrme 1958), choosing from the aforementioned possibilities the second one – **the Nonlinear Realization of the Chiral Symmetry**. It seems that the problem of construction of a nonlinear σ-model was rather topical, as almost simultaneously there appeared other variants (Gürsey 1960, 1961; Gell-Mann and Lévy 1960). In his approach Skyrme drew much from already known results (Schwinger 1957);

[1] It is curious to note that the chiral terminology was introduced in physics by Lord Kelvin, who used it for the description of vortex orientations and for the description of light polarizations (Kelvin 1904, lecture 20). Kelvin's definition of "chiral" and his geometrical theory of chirality can be found in Appendix H to the refered Baltimore Lectures.

specifically, he used the possibility of constructing a desirable chiral generalization starting from the familiar $PS - PS$ theory of pions and nucleons[2] by replacing the nucleon mass by $g\phi_0$. Here g is the pseudo-scalar coupling constant and ϕ_0 is another meson field of similar character to the three ϕ_i. In this case the interaction would be really symmetric between four fields. Skyrme added to the isotriplet of meson fields $\underline{\phi}$ an additional field ϕ_0, and, in order to prevent the unphysical enlargement of isospace degrees of freedom, he imposed the following constraint on the fields $\phi_\rho = (\phi_0, \underline{\phi})$, with $\rho = 0, 1, 2, 3$:

$$\phi_0{}^2 + \phi_a\phi^a = 1; \quad a = 1, 2, 3. \tag{1.9}$$

But under restriction (1.9) the natural generalization of the meson mass term in (1.3), namely $\frac{1}{2}k^2 \sum_i \phi_i{}^2 \rightarrow \frac{1}{2}k^2 \sum_\rho \phi_\rho{}^2$, is no longer a mass term, so to avoid this obstacle Skyrme considered two possibilities: to introduce into the Lagrangian a fourth order term with respect to mesonic fields $\frac{1}{4}k^2 \sum_\rho \phi_\rho{}^4$, or to admit that "... **the meson mass originates from the nucleon coupling**". The latter proposal together with the previous hypothesis on the meson generation of the nucleon mass is perhaps one of the earliest suggestions of the **"Nuclear Democracy" Principle**[3]

In his paper (Skyrme 1958) Skyrme took the first possibility. Thus, after the chiral modification the Lagrangian may be written in the following form:

$$\mathcal{L}_{\mathrm{Ch}} = \frac{1}{2} \sum_\rho \left[(\partial_\mu \phi_\rho)^2 + \frac{1}{2}k^2 \phi_\rho{}^4 \right] + \overline{\psi} \left[\mathrm{i}\gamma^\mu \partial_\mu + g(\phi_0 + \mathrm{i}\gamma_5 \underline{\tau} \cdot \underline{\phi}) \right] \psi. \tag{1.10}$$

The detailed analysis of the symmetry properties and features of the modified theory with the Lagrangian (1.10), undertaken in papers (Skyrme 1958, 1959), in spite of some interesting and promising results, led him to the conclusion that for a further test of the model some estimation of particle masses and of interaction coupling constants are required. A deep understanding of the model necessitated such practicalities. As a first step toward this, Skyrme considered a simplified version of his model "... of the same general character..., in which the configuration space-time and the isotopic spin space have both been reduced to two dimensions" (Skyrme 1958).

---

[2] Here PS denotes the pseudo-scalar type of coupling.

[3] According to (Chew 1961; Jacob and Chew 1964) the term "Nuclear Democracy" was suggested by M. Gell-Mann. In its full extent it expresses the conjecture that each strongly interacting particle is a composite (bound) state of those S-matrix channels with which it communicates. It means that each nuclear particle generates other particles, which in turn generate it. Thus no nuclear particle is more fundamental than any other.

1.3 The Two-Dimensional Simplified (sine-Gordon) Model

Moving to a two-dimensional model was the most crucial step in the evolution of the baryon model. Here we pay special attention to this topic and also remind some facts, included already in modern textbooks on field theory, e.g. (Rajaraman 1982). If we consider the proposed simplification, where the only two components ϕ_0 and ϕ_1 are nontrivial and both are functions on the (1+1)-dimensional space-time with coordinates x and t, then it is possible to fulfill the constraint (1.9) by setting

$$\phi_0 = \cos\alpha(x,t); \quad \phi_1 = \sin\alpha(x,t). \tag{1.11}$$

The direct substitution of (1.11) into (1.10) after some redefinitions of variables gives rise to the following form of the mesonic part of the Lagrangian density

$$\mathcal{L}_{\rm SG} = \tfrac{1}{2}\left[(\partial_t\alpha)^2 - (\partial_x\alpha)^2\right] - k^2(1-\cos\alpha), \tag{1.12}$$

with the corresponding Euler-Lagrange equation:

$$\partial_x^2\alpha - \partial_t^2\alpha - k^2\sin\alpha = 0. \tag{1.13}$$

This equation, usually called the sine-Gordon equation, gained most of its popularity a bit later, after the discovery of the Inverse Scattering Transform Method in the mid 1960's. The Eq. (1.13) itself was known in the previous century and investigated in great detail by geometers in connection with the problem of an embedding of the Lobatschevskii plane into Euclidean space $\mathbb{R}^3$. Solutions of (1.13) correspond to different types of possible enclosures.

It looks as though Skyrme was not acquainted with these results, so he tried to get possible solutions through his own methods[4] in (Skyrme 1958, 1961b) and managed to find nontrivial solutions of the sine-Gordon equation. Those solutions are now known as one-soliton (2π-kink) solutions, and two-soliton solutions (in particular, bions and breathers) and were discovered again by different methods in (Hirota 1972; Caudrey et al. 1973; Ablowitz et al. 1973). In principle, Skyrme himself (Skyrme 1958, 1961b) and later with Perring (Perring and Skyrme 1962) were the first investigators in the new branch of mathematical physics, now called **Field Theoretic Integrable Models.** The complete integrability of the sine-Gordon model was proved later by Russian physicists (Zakharov, Takhtadzhian and Faddeev 1974) and this

[4] According to (Dalitz 1988) Tony Skyrme was consistently outstanding in Mathematics "...gaining the Hawtrey Prize in 1938, the Tomline Prize in 1939 and the Russell Prize in 1940 ..." during the years of study at Eton College. Later at Trinity College he gained "...1st Class Honours in Part II of the Mathematical Tripos in 1942, proceeding to Part III on Senior Scholarship in the following year". During his participation in the Manhattan Project he was known among colleagues as an expert in finding solutions of differential equations both analytically and numerically.

work became a point of departure for this new branch. Skyrme's central and the most inspired achievement in the study of the sine-Gordon model was the discovery of a new type of conserved quantity in physics – **"Topological Charges"**. In Section 1.4 we shall see in what sense this discovery strongly supported Skyrme's long-held hope of developing a **Unified Field Theory of Mesons and Baryons.** Here we just elucidate the idea of the Topological Charge, leaving the discussion of this topic in greater detail for Chapter 2.

1.3.1 The Idea of Topological Charge

Besides the travelling wave solutions, which Skyrme identified with mesonic excitations of small amplitude, the Eq. (1.13) possesses one-soliton or 2π-kink solutions of the form:

$$\alpha(x) = 4\tan^{-1}\left[\exp\{\pm k(x - x_0)\}\right], \tag{1.14}$$

which can be related to a "particle", localized at the point x_0 (see Fig. 1.1). Here we will, following (Skyrme 1961b), frequently use the term *particle* in the sense of *fermion.*

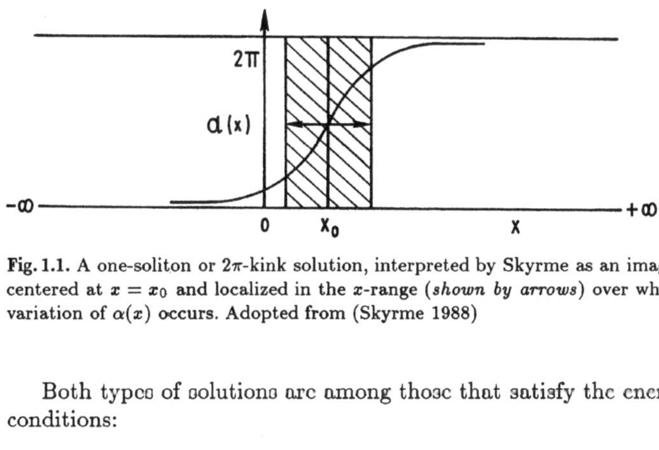

Fig. 1.1. A one-soliton or 2π-kink solution, interpreted by Skyrme as an image of "particle" centered at $x = x_0$ and localized in the x-range (*shown by arrows*) over which most of the variation of $\alpha(x)$ occurs. Adopted from (Skyrme 1988)

Both types of solutions are among those that satisfy the energy finiteness conditions:

$$\alpha(x) \to 0 \,(\mathrm{mod}\, 2\pi), \quad \text{as} \quad |x| \to \infty. \tag{1.15}$$

But in the case of mesonic type solutions the angular variable $\alpha(x)$ takes the zero value at both boundaries $x = \pm\infty$. On the other hand for (1.14) we have $\alpha(-\infty) = 0$ and $\alpha(+\infty) = 2\pi$. It means that $\alpha(x)$ interpolates two vacuum states. Those solutions are not transformable into each other by any continuous deformation, hence the space of solutions of Eq. (1.13) with boundary conditions (1.15) is splitted into distinct connected components. Since a continuous deformation could be regarded as an evolution in time of the classical dynamical system under study, it follows that two states, corresponding to the above mentioned solutions, are completely different, and there should be a characteristic, which differs on each type of solution but could not change its value under time evolution. Such characteristics were long ago used in mathematics as topological invariants of functional spaces (or spaces of maps) under different names: the winding number, the Brouwer degree of mapping, the Chern-Pontryagin number, etc.

Skyrme by himself found the explicit expression for a required topological invariant (Skyrme 1958). He paid attention to the existence in the sine-Gordon model of the conserved current J^μ, $\mu = 0, 1$; with components:

$$J^0 = \frac{1}{2\pi}\partial_x\alpha\,; \quad J^1 = -\frac{1}{2\pi}\partial_t\alpha\,; \quad J^\mu = \frac{1}{2\pi}\epsilon^{\mu\nu}\partial_\nu\alpha, \tag{1.16}$$

and with the usual conservation law $\partial_\mu J^\mu = 0$, which holds independent of the equation of motion (1.13) because of the continuity of the functions $\alpha(x,t)$. The corresponding integral conserved quantity is:

$$\mathbf{Q} = \int_{-\infty}^{\infty} dx\, J^0 = \frac{1}{2\pi}\int_{-\infty}^{\infty} dx\, \frac{\partial\alpha}{\partial x} = \frac{1}{2\pi}[\alpha(+\infty) - \alpha(-\infty)], \tag{1.17}$$

and it is easy to note, that the integrand in the second term of equality (1.17) is just the Jacobian of the change of the cartesian variable x into the angle variable α, thus determining the circle $\mathbf{S}^1$. The expression (1.17) is the explicit form of the winding number in the case of the sine-Gordon model, which takes only integer values in $\mathbb{Z}$. Each integer value can be ascribed to each class of solutions of Eq. (1.13), as is easily derived by substitution of a solution into (1.17).

Pursuing his initial motivations, which were mentioned in the Introduction, Skyrme tried to find out whether it is possible in the framework of the sine-Gordon field theory to treat the solutions (1.14), which are more massive in comparison with underlying fundamental fields, as *fermion states*. As it follows from Eq. (1.14), there are two types of kinks, corresponding to the $\pm$ signs, which can be considered as representing a particle and an antiparticle, respectively. Taking a number of x_0 values, which are sufficiently far apart from each other, the net function for this system of particles, say $\overline{\alpha}(x)$, may be well approximated by

$$\overline{\alpha}(x) = \sum_i \alpha^+(x - x_0^i) + \sum_j \alpha^-(x - x_0^j). \tag{1.18}$$

We denote here as $\alpha^\pm$, solutions corresponding to $\pm$ sign in (1.14). Each term of (1.18) contributes a kink or an antikink to the net function $\overline{\alpha}(x)$ and the net sum, which is

$$\overline{\alpha}(+\infty) - \overline{\alpha}(-\infty) = \sum_i n_i^+ - \sum_j n_j^-, \tag{1.19}$$

will remain constant throughout the motion, even though the kinks overlap. The last equation is identical with the statement that the number of particles minus that of antiparticles is a conserved quantity, usually seen only in the fermionic type field theories. Thus it is possible to say that Skyrme discovered in the purely bosonic sine-Gordon model the existence of conserved quantity, usually attributed to fermionic fields (one may be reminded of the electric charge, baryon and lepton numbers, etc.).

1.3.2 The Abelian Bosonization or Fermi-Bose Correspondence

To proceed further, Skyrme considered a possible quantization scheme for the sine-Gordon model. He managed to write down the fermion-boson correspondence relations and the explicit form for a creation operator of a fermion particle. This particle appeared to be neutrino-like (massless) and its field operator obeyed the anticommutation relations in the case where the energy-scale factor ε satisfied the condition $\varepsilon = \hbar c/2$ (Skyrme 1961b). That gave him cause to suggest that the quantum solitons of the sine-Gordon model may be equivalent to fermions interacting through a four-fermion interaction[5]. Skyrme's suggestion was proved only in (Coleman 1975), where it was shown that in the zero-charge sector the quantum sine-Gordon model is equivalent to the massive Thirring model in (1+1)-dimensions with the Lagrangian density

$$\mathcal{L}_{\mathrm{MT}} = i\overline{\psi}\gamma_\mu\partial^\mu\psi - \mathrm{m_f}\overline{\psi}\,\psi - \tfrac{1}{2}g(\overline{\psi}\gamma^\mu\psi)(\overline{\psi}\gamma_\mu\psi). \tag{1.20}$$

In the framework of perturbation theory Coleman compared the so-called n-point functions of the composite operator $[\overline{\psi}\,\psi]$ in the massive Thirring model (1.20) and those of the term $[\cos\alpha]$ in (1.12) and found that these two sets are equal for all n.

The Fermi-Bose transmutations are in the focus of interest at present in a variety of models in condensed matter and particle physics and there is an extensive literature on the subject. With this in mind it may be more useful to present the referred results in the modern language. More information on this can be found in the following reviews: (Balachandran, Gomm and Sorkin 1987; Tze 1987; Mackenzie and Wilczek 1988; Canright and Johnson 1990). The Thirring Lagrangian density (1.20) is invariant against $U(1)$ transformations of the type $\psi' = \exp(i\varphi)\cdot\psi$ and it provides the appearance of a conserved Noether current $j^\mu = \overline{\psi}\gamma^\mu\psi$, with $\mu = 0,1$. The second quantization of the system (Tze 1987) leads to the following equal time current algebra:

$$[j_0(x), j_0(y)] = [j_1(x), j_1(y)] = 0, \tag{1.21}$$

$$[j_0(x), j_1(y)] = i(\pi + g\hbar)^{-1}\partial_1\delta(x-y). \tag{1.22}$$

Here the term on the right hand side of (1.22) is the finite c-number "Schwinger term" with the interaction-dependent coefficient, which indicates a local "deformation" of the algebra of currents.

To get the equivalence relations between the massive Thirring model (1.20) and the quantum sine-Gordon system it is convenient to use the Lagrangian density (1.12) in the parameter-dependent form

[5] Recall that the Thirring model was proposed nearly at the same time (Thirring 1958) and it may be that Skyrme was not acquainted with this paper.

$$\mathcal{L}_{\mathrm{SG}} = \frac{1}{2}\left\{\partial_\mu\alpha\partial^\mu\alpha - \frac{\lambda}{\beta^2}[1-\cos(\beta\alpha)]\right\}. \tag{1.23}$$

The parameter λ is the "squared mass", associated with the spectrum of small oscillations about the minimum, and β measures the strength of the interactions between these small (mesonic) oscillations. The central conclusion of papers (Coleman 1975; Mandelstam 1975) may be formulated in terms of currents as follows: The quantum sine-Gordon system would in fact realize the same (Kac-Moody) current algebra (1.21–1.22) if the following operatorial equalities or Abelian bosonization rules hold:

$$j^\mu = -\frac{1}{2\pi}\hbar\beta\epsilon^{\mu\nu}\partial^\mu\alpha; \tag{1.24a}$$

$$m_f : \overline{\psi}\psi : = \lambda\beta^{-2} : [1 - \cos(\beta\alpha)] :; \tag{1.24b}$$

$$(\pi + g\hbar)^{-1} = \frac{1}{4\pi^2}\beta^2\hbar. \tag{1.24c}$$

It is easy to see that the energy-momentum tensor density $\mathbf{T}^{\mu\nu}$ has the same current-current form for both models:

$$\mathbf{T}^{\mu\nu} = \hbar(\pi + g\hbar)\left\{j^\mu j^\nu - \tfrac{1}{2}g^{\mu\nu}j^2\right\} + \sigma\cdot g^{\mu\nu}, \tag{1.25}$$

where $\sigma \equiv m_f : \overline{\psi}\psi :$ and $g^{\mu\nu}$ is the metric tensor. Coleman demonstrated that the strong coupling limit of the massive Thirring model ($g \gg 1$) corresponds to the weak coupling of the sine-Gordon fundamental fields ($\beta \ll 1$). This duality confirms the validity of Skyrme's conjecture that the elementary fermion, more precisely of the Thirring model, is identical with the quantum topological soliton of the sine-Gordon model. This is the only rigorous result on the bosonization to the moment. In subsequent chapters we will return to the discussion of Fermi-Bose correspondence in connection with an effective meson Lagrangian derivation and also with the treatment of Skyrmions as fermions.

It is widely recognized that this topic was one of the most original and attractive in Skyrme's approach. From Skyrme's subsequent papers (Perring and Skyrme 1962; Skyrme 1971) it may be deduced, that Skyrme was not satisfied with the result of the paper (Skyrme 1961b) and undertook different and rather resourceful attempts to improve it. Thus in (Skyrme 1971) he proposed the way of introducing collective coordinates and demonstrated that solitons of the sine-Gordon equations propogate according to one dimensional "neutrino-like" equations. Though the rigorous results in Fermi-Bose transmutations were gained in particle physics only recently, one can say that Skyrme's name ought to be listed among those of the first rank of physicists, who launched this very promising branch of modern investigation.

1.4 The Baryon Model – Topological Skyrmions

Encouraged by the nontrivial results obtained in the simplified (sine-Gordon) model, especially with the discovery of the conserved topological integral of motion, Skyrme came to his final modification of a (3+1)-dimensional model with the added specification, that the latter should also possess a conserved quantity of topological character. As we remember, the starting point of his constantly modified sequence of models was just the nuclear matter model, but in the process of evolution he gradually came to look more at the idea of a single nucleon as a nuclear "Wirbelbewegung" in some ether-like fluid. In that case an analogous topological conserved quantity, if any, could be interpreted as *the baryon number*. The nature of that conserved quantity, introduced into particle physics by E. Wigner and E. Stückelberg as a quite formal characteristic in order to restrict the possible types of reactions between baryons, was and remains the point of attraction for many investigators. The conservation law of the baryon number holds experimentaly with a mysterious accuracy to the present day (in spite of predictions of the Grand Unification Theory). It is clear that the nature and role of this number in hadron physics is quite different compared to that of the electric charge. The baryon number does not define in any sense the value of the coupling constant of strong interactions, contrary to the role of the electric charge in electrodynamics. That is why it seems unnatural to derive the baryon number conservation law from the invariance of the Lagrangian under gauge transformations, as is usually adopted, in close analogy with the derivation of the electric charge conservation law. It was Skyrme's rigorous conviction, shared by a number of thoughtful researchers, that this suggestion looks too artificial.

1.4.1 The (3+1)-dimensional "Angular" Variables

As could be learned from the simplified (sine-Gordon) lesson, to be able to construct a conserved quantity of the topological character in a more realistic physical model it is necessary to find a (3+1)-dimensional analog of the angular variable $\alpha(x)$. Skyrme managed to do this when he noticed that from a geometrical point of view the condition (1.9) is just the equation of the $\mathbb{S}^3$ - manifold in the isotopic spin space. The analogous condition in the (1+1)-dimensional case is the equation ${\phi_0}^2 + {\phi_1}^2 = 1$, which is just the $\mathbb{S}^1$-manifold (the circle) equation. Let us represent the complex-valued field $\phi(x) = \exp\{i\alpha(x)\}$ in the form $\phi = \phi_0 + i\phi_1$. Because of this one can write down the Jacobian for the change of variables (the integrand of the second term in Eq. (1.17)) in terms of the field ϕ as following

$$\frac{\partial\alpha}{\partial x} = -i\phi^* \frac{\partial\phi}{\partial x}, \tag{1.26}$$

where ϕ^* is the complex conjugate of ϕ, which coincides with ϕ^{-1}.

To perform the analogous change to "angle"-variables in the (3+1)-dimensional case it is suitable to use the corresponding generalization of complex numbers - the quaternionic representation of the group $SU(2)$, which is topologically isomorphic to the sphere $\mathbf{S}^3$. Let us choose a unit quaternion $\mathbf{U}$, parametrized by mesonic fields $\phi_\rho = (\phi_0, \underline{\phi})$:

$$\mathbf{U}(\boldsymbol{x},t) = \mathbb{1}\,\phi_0(\boldsymbol{x},t) + i\underline{\tau}\,\underline{\phi}(\boldsymbol{x},t), \tag{1.27}$$

where $\mathbb{1}$ is a unit 2×2 - matrix; $\underline{\tau}$ are isospin Pauli matricies and $\boldsymbol{x} \in \mathbb{R}^3$. The condition (1.9) in terms of $\mathbf{U}$ can be reexpressed as $\mathbf{U} \cdot \mathbf{U}^\dagger = \mathbb{1}$. The field $\mathbf{U}(\boldsymbol{x},t)$ is called *the chiral field* and at any fixed moment of time it is possible to think of this field as the map

$$\mathbf{U}(\boldsymbol{x}): \quad \mathbb{R}^3 \to \mathbf{S}^3. \tag{1.28}$$

The other useful parametrization of the field $\mathbf{U}$ is:

$$\mathbf{U}(\boldsymbol{x},t) = \exp\{i(\underline{\mathbf{n}} \cdot \underline{\tau})\Theta(\boldsymbol{x},t)\};\ \underline{\mathbf{n}} = \frac{\underline{\phi}}{|\underline{\phi}|};\ |\underline{\phi}| = \pm\sin\Theta;\ \phi_0 = \cos\Theta. \tag{1.29}$$

In what follows we will write down formulas for dynamical quantities of the Skyrme model using both of these parametrizations.

The next step is to find the form of the boundary conditions one should impose on the mesonic fields ϕ_ρ in order to provide a description of extended objects localized in space with finite dynamical characteristics. It is obvious that these conditions should be analogous to the Eq. (1.15) in the sine-Gordon model, and it means that we are to identify the points at spatial infinity, so that as $|\mathbf{x}| \to \infty$ the fields ϕ_ρ on $\mathbf{S}^3$ should tend to a fixed point on $\mathbf{S}^3$:

$$(\phi_0, \underline{\phi})(\boldsymbol{x}) \to (1,0), \quad \text{as} \quad |\boldsymbol{x}| \to \infty. \tag{1.30}$$

From definitions (1.27) and (1.29) it is clear that corresponding boundary conditions on $\mathbf{U}$ and on Θ look like the following:

$$\mathbf{U}(\boldsymbol{x}) \to \mathbb{1}; \quad \Theta(\boldsymbol{x}) \to 0, \quad \text{as} \quad |\boldsymbol{x}| \to \infty. \tag{1.31}$$

The conclusion follows easily that because of conditions (1.30) or (1.31) the space $\mathbb{R}^3$ is effectively compactified, so that $\mathbb{R}^3 \cup \{\infty\} \simeq \mathbf{S}^3$. This conclusion will be an important one for the geometrical definition of the topologically conserved quantity in Chapter 2.

Now we are in a position to write down the "angular" variable in (3+1)-model as the straightforward generalization of Eq. (1.26) in the case of quaternionic fields:

$$\mathbf{B}_\mu = \mathbf{U}^{-1}\partial_\mu \mathbf{U} = i\tau_a \mathbf{B}^a_\mu; \quad a = 1,2,3, \tag{1.32}$$

where in terms of $\mathbf{U}$, ϕ_ρ and alternatively in terms of $\mathbf{n}$ and Θ, the components of $\mathbf{B}^a_\mu$ are:

$$\begin{aligned} \mathbf{B}^a_\mu &= \frac{1}{2i}\mathrm{Tr}\,(\tau^a \mathbf{U}^{-1}\partial_\mu \mathbf{U}) \\ &= \phi_0\partial_\mu\phi^a - \phi^a\partial_\mu\phi_0 + \epsilon^{abc}\phi_b\partial_\mu\phi_c \\ &= n^a\partial_\mu\Theta + \sin\Theta\cos\Theta\partial_\mu n^a + \sin^2\Theta\epsilon^{abc}n_b\partial_\mu n_c. \end{aligned} \tag{1.33}$$

To be well defined the variable $\mathbf{B}^a_\mu$ should satisfy the compatibility conditions of the form:

$$\partial_\nu \mathbf{B}^a_\mu - \partial_\mu \mathbf{B}^a_\nu - 2\epsilon_{abc}\mathbf{B}^b_\mu \mathbf{B}^c_\nu = 0, \tag{1.34}$$

which are just another expression for the natural conditions $\partial_\mu\partial_\nu \mathbf{U} = \partial_\nu\partial_\mu \mathbf{U}$. The conditions (1.34), from the other side, mean that **the covariant curl of these "angular" variables $\mathbf{B}^a_\mu$ should be identical to zero.**

Later Skyrme's "angular" variables $\mathbf{B}_\mu$ were reintroduced in (Sugawara 1968; Sommerfield 1968) in connection with the field theory of currents. As a Lagrangian density in the referred papers they chose the simplest chiral invariant, which in terms of $\mathbf{B}_\mu$ reads:

$$\mathcal{L} = -\tfrac{1}{4}\,\mathrm{Tr}\,(\mathbf{B}_\mu \mathbf{B}^\mu), \tag{1.35}$$

and is a straightforward generalization of quadratic functionals in the linear field theory. Different aspects of those functionals have been studied by F. Gürsey, J. Cronin, S. Weinberg and other authors in the Phenomenological Lagrangian approach to strong interactions. The Lagrangian is manifestly invariant under the left transitions and $\mathbf{B}_\mu$ are conserved currents in the Sugawara-Sommerfield theory. Hence they were called *the left chiral currents*, given the notations $\mathbf{L}_\mu$ [6] in all recent papers. In what follows we use throughout the Sugawara-Sommerfield notation.

From a geometrical point of view the chiral currents $\mathbf{L}_\mu$ are vector fields defined on the $\mathbb{S}^3$-manifold with values in the $su(2)$-algebra. Then the conditions (1.34) are equivalent to

$$\partial_\mu \mathbf{L}_\nu - \partial_\nu \mathbf{L}_\mu + [\mathbf{L}_\mu, \mathbf{L}_\nu] = 0, \tag{1.36}$$

where [,] is the commutator of the Lie algebra. Equation (1.36) could be considered as the Maurer-Cartan structural equations or the zero curvature conditions. They are necessary and sufficient conditions for the reconstruction of the chiral field $\mathbf{U}$ (an element of the Lie group) in terms of the left chiral current $\mathbf{L}_\mu$ (an element of the Lie algebra). By this digression we have provided ourselves with the "building blocks" in terms of which we would construct in a moment the topological charge in the (3+1) field theory and formulate the dynamics of the Skyrme model.

1.4.2 The Topological Charge in the (3+1) Model

To write down an explicit form for the topologically conserved quantity Skyrme used the natural generalization of the expression for the Jacobian in (1.17) in the following form: $\partial\alpha/\partial x \to \det\{\mathbf{L}^a_i\}$ (Skyrme 1961a). The coefficient in (1.17), which is the reciprocal "volume" of the $\mathbb{S}^1$-sphere, he modified to the reciprocal "volume" of the $\mathbb{S}^3$-sphere. Then the following expressions for the (3+1) conserved quantity look straightforward:

[6] Let us hope that no confusion will occur with the notations for left chiral generators, denoted by the same letter in the Section 1.2.

$$\mathbf{Q} = \frac{1}{2\pi^2}\int d^3x\ \det\{\mathbf{L}_i^a\} = -\frac{\epsilon^{ijk}}{48\pi^2}\int d^3x \mathrm{Tr}\,(\mathbf{L}_i\cdot[\mathbf{L}_j,\,\mathbf{L}_k])$$
$$= -\frac{\epsilon^{ijk}\epsilon_{\alpha\beta\gamma\delta}}{12\pi^2}\int d^3x\,\phi^\alpha\partial_i\phi^\beta\partial_j\phi^\gamma\partial_k\phi^\delta$$
$$= -\frac{\epsilon^{ijk}\epsilon_{abc}}{4\pi^2}\int d^3x\sin^2\Theta\partial_i\Theta n^a\partial_j n^b\partial_k n^c. \tag{1.37}$$

There are many ways to check directly from one of the expressions (1.37) that $\mathbf{Q}$ is a conserved quantity, but the preferable way to do this is to write the corresponding conserved topological current:

$$\mathbf{J}^\mu = -\frac{\epsilon^{\mu\nu\lambda\rho}}{48\pi^2}\mathrm{Tr}\,(\mathbf{L}_\nu\cdot[\mathbf{L}_\lambda,\,\mathbf{L}_\rho]) = -\frac{\epsilon^{\mu\nu\lambda\rho}\epsilon_{abc}}{12\pi^2}\,\mathbf{L}_\nu^a\mathbf{L}_\lambda^b\mathbf{L}_\rho^c$$
$$= -\frac{\epsilon^{\mu\nu\lambda\rho}\epsilon_{\alpha\beta\gamma\delta}}{12\pi^2}\,\phi^\alpha\partial_\nu\phi^\beta\partial_\lambda\phi^\gamma\partial_\rho\phi^\delta$$
$$= -\frac{\epsilon_{abc}\epsilon^{\mu\nu\lambda\rho}}{4\pi^2}\,\sin^2\Theta\partial_\nu\Theta n^a\partial_\lambda n^b\partial_\rho n^c. \tag{1.38}$$

The conservation law of this current $\partial_\mu\mathbf{J}^\mu = 0$ is just a consequence of the Jacobi identity for Lie algebra commutators. Another easy way to justify $\mathbf{J}^\mu$ conservation comes from the third expression of (1.38), if one keeps in mind that fields ϕ_ρ are not independent because of the constraint (1.9). The conservation of the topological charge $\mathbf{Q}$, which has the current component $\mathbf{J}^0$ as the density, is then a straightforward consequence. It is clear that any derivation of this conservation law is independent of the dynamics of the model, which we are going to specify only now. By close analogy with the (1+1)-dimensional case one can expect that the topological charge $\mathbf{Q}$ will take integer values from $\mathbb{Z}$ on each topologically distinct class of solutions and can be considered as the winding number or the Brouwer degree of mapping $\mathbf{U} : \mathbb{R}^3 \to \mathbf{S}^3$. We will return to a more systematic exposition of those facts from the differential-geometric point of view in Chapter 2. In Chapter 3 we will see what a crucial role this topological integral of motion plays in the problem of stability of extended particle-like structures in the framework of nonlinear σ-models. And finally in Part III we discuss the validity of the interpretation of the quantity (1.37) as the baryon number. This Skyrme's hypothesis will then be examined on the basis of QCD first principles.

1.4.3 The Skyrme Model Dynamics

Skyrme suggested that the Lagrangian density for this modified model ought to be (Skyrme 1961a):

$$\mathcal{L} = \frac{\varepsilon}{4\pi^2}\left\{\kappa^2\,\mathbf{L}_\mu^a\,\mathbf{L}_\mu^a - \frac{1}{2}\left[(\mathbf{L}_\mu^a\,\mathbf{L}_\mu^a)^2 - (\mathbf{L}_\mu^a\,\mathbf{L}_\nu^a)^2\right]\right\}$$
$$= -\frac{1}{4\lambda^2}\mathrm{Tr}\,{\mathbf{L}_\mu}^2 + \frac{\epsilon^2}{16}\mathrm{Tr}\,[\mathbf{L}_\mu,\mathbf{L}_\nu]^2$$
$$= \frac{F_\pi^2}{16}\mathrm{Tr}\,(\partial_\mu\mathbf{U}\partial_\mu\mathbf{U}^+) + \frac{1}{32e^2}\mathrm{Tr}\,[\partial_\mu\mathbf{U}\mathbf{U}^+,\partial_\nu\mathbf{U}\mathbf{U}^+]^2. \tag{1.39}$$

Here, in addition to Skyrme's original Lagrangian density given in (Skyrme 1961a) (the first expression in (1.39)), we have also written the other forms found in recent papers in order to provide an easy passage from one form to another, as well as to get the relations between values of different constants. For example, from the second and the third lines in (1.39) one can find that $\lambda = 2/F_\pi$ and $\epsilon = (\sqrt{2}e)^{-1}$. Throughout this text we generally use the second expression in (1.39). The summation convention is the usual for Lorentz metric, so that:

$$\mathbf{L}_\mu^a \mathbf{L}_\mu^a = \mathbf{L}_0^a \mathbf{L}_0^a - \mathbf{L}_k^a \mathbf{L}_k^a; \quad k = 1, 2, 3. \tag{1.40}$$

The latter, quartic term in chiral currents in all variants of expression (1.39) for the Lagrangian density, is called in the recent literature *the Skyrme term* and there are many discussions of its nature.

A straightforward and elementary way to understand the origin of this term is to recall the roots of Skyrme's approach in Kelvin's work. From this position one can regard the left chiral currents $\mathbf{L}_\mu$ as *generalized velocities of the Mesonic Fluid,* then the Skyrme term may be considered as a squared *generalized vorticity.* In accordance with relations (1.36), an explicit expression for this quantity may be written in the form of an ordinary curl:

$$\mathbf{C}_{\mu\nu}^a \equiv \tfrac{1}{2}(\partial_\nu \mathbf{L}_\mu^a - \partial_\mu \mathbf{L}_\nu^a) = \epsilon^{abc} \mathbf{L}_\mu^b \mathbf{L}_\nu^c, \tag{1.41}$$

which justifies the aforementioned analogies. For some possible advantages of such a hydrodynamical treatment of the Skyrme model one may consult (Sanyuk 1990a,b).

Now we may write down the Hamiltonian of the model:

$$\mathbf{H} = -\int d^3x \left\{ \frac{1}{4\lambda^2} \mathrm{Tr}\,({\mathbf{L}_0}^2 + {\mathbf{L}_i}^2) + \frac{\epsilon^2}{8} \mathrm{Tr}\,([\mathbf{L}_0, \mathbf{L}_i]^2 + \frac{1}{2}[\mathbf{L}_i, \mathbf{L}_k]^2) \right\}, \tag{1.42}$$

and following (Skyrme 1961a) obtain the estimate of $\mathbf{H}$ from below through the topological charge $\mathbf{Q}$ from Eq. (1.37)

$$\begin{aligned} \mathbf{H} &\geq -\int d^3x \left\{ \frac{1}{4\lambda^2} \mathrm{Tr}\,{\mathbf{L}_i}^2 + \frac{\epsilon^2}{16} \mathrm{Tr}\,[\mathbf{L}_i, \mathbf{L}_k]^2 \right\} \\ &= -\int d^3x \left\{ \frac{1}{4\lambda^2} \mathrm{Tr}\,{\mathbf{L}_i}^2 + \frac{\epsilon^2}{32} \mathrm{Tr}\,(\epsilon_{ijk}[\mathbf{L}_j, \mathbf{L}_k])^2 \right\} \\ &> \frac{\epsilon}{4\sqrt{2}\lambda} \int d^3x \left| \mathrm{Tr}\,(\epsilon_{ijk}\, \mathbf{L}_i [\mathbf{L}_j, \mathbf{L}_k]) \right|, \end{aligned}$$

where we have used the standard inequality for vectors $\boldsymbol{a}^2 + \boldsymbol{b}^2 \geq 2|(\boldsymbol{a} \cdot \boldsymbol{b})|$. The presented elementary calculations persuade us that the following estimate holds:

$$\mathbf{H} > 6\sqrt{2}\pi^2 \frac{\epsilon}{\lambda} |\mathbf{Q}|. \tag{1.43}$$

On behalf of this estimate Skyrme suggested, that the Euler-Lagrange equation of the model

$$\partial_\mu \left(\mathbf{L}^\mu - \frac{\epsilon^2 \lambda^2}{2} [\mathbf{L}_\nu, [\mathbf{L}^\mu, \mathbf{L}^\nu]] \right) = 0 \tag{1.44}$$

possesses stable solutions with finite dynamical characteristics. One may easily derive Eq. (1.44) from the standard variational principle for the Lagrangian (1.39), taking into account the left chiral currents definition (1.32) together with an obvious relation $\delta \mathbf{U}^{-1} = -\mathbf{U}^{-1} \cdot \delta \mathbf{U} \cdot \mathbf{U}^{-1}$ for chiral fields $\mathbf{U}$, so that

$$\delta S = \delta \int d^4x \mathcal{L} = \int d^4x \mathrm{Tr} \left\{ \mathbf{U}^{-1} \delta \mathbf{U} \left(\frac{1}{2\lambda^2} \partial_\mu \mathbf{L}^\mu - \frac{\epsilon^2}{4} \partial_\mu [\mathbf{L}_\nu, [\mathbf{L}^\mu, \mathbf{L}^\nu]] \right) \right\} = 0.$$

The Eq. (1.44) has the form of a local conservation law and to find the corresponding conserved current one has to remember that the Lagrangian (1.39) is invariant against the chiral $SU(2)_L \otimes SU(2)_R$ transformations and because of this we have at least two conserved Noether currents:

$$\mathbf{N}^k_{\mu,R} = \mathrm{i}\,\mathrm{Tr} \left\{ \tau^k \left(\frac{1}{2\lambda^2} \mathbf{L}_\mu - \frac{\epsilon^2}{4} [\mathbf{L}^\nu, [\mathbf{L}_\mu, \mathbf{L}_\nu]] \right) \right\} ; \tag{1.45}$$

which corresponds to right generators in (1.7), and the analogous current for the left transformations

$$\mathbf{N}^k_{\mu,L} = \mathrm{i}\,\mathrm{Tr} \left\{ \tau^k \left(\frac{1}{2\lambda^2} \mathbf{R}_\mu - \frac{\epsilon^2}{4} [\mathbf{R}^\nu, [\mathbf{R}_\mu, \mathbf{R}_\nu]] \right) \right\} , \tag{1.46}$$

where $\mathbf{R}_\mu = \mathbf{U} \partial_\mu \mathbf{U}^{-1}$ are the right chiral currents by definition. In general in the framework of the Lagrangian formalism all Noether currents can be derived from the formula:

$$\mathbf{N}^k_\mu = \mathrm{Tr} \left(\frac{\partial \mathcal{L}}{\partial(\partial_\mu \mathbf{U})} \cdot \delta^k \mathbf{U} + \frac{\partial \mathcal{L}}{\partial(\partial_\mu \mathbf{U}^{-1})} \cdot \delta^k \mathbf{U}^{-1} \right) , \tag{1.47}$$

where by $\delta^k \mathbf{U}$ we denote the variations of the chiral field under respective transformations. For example, the variations of $\mathbf{U}$ under isotopic rotations with generators $\mathbf{I}^k$ are as follows

$$\delta^k_V \mathbf{U} = \mathrm{i} \left[\frac{\tau^k}{2}, \mathbf{U} \right] , \tag{1.48}$$

and the conserved isovector (or isospin) current can be obtained by substitution (1.48) into (1.47) in the form

$$\mathbf{V}^k_\mu = \mathrm{i}\,\mathrm{Tr} \left\{ \tau^k \left(-\frac{1}{2\lambda^2} \mathbf{L}_\mu + \frac{\epsilon^2}{4} [\mathbf{L}^\nu, [\mathbf{L}_\mu, \mathbf{L}_\nu]] + (\mathbf{L} \to \mathbf{R}) \right) \right\} , \tag{1.49}$$

so that the expressions in parentheses of Eq. (1.44) and of Eq. (1.49) are identical. Summarizing one can say that the Euler-Lagrange equations of the Skyrme model express the conservation of the isotopic spin current.

For the sake of completeness and for the convenience of the reader let us express the standard dynamical characteristics for the Skyrme model. By definition of the energy-momentum tensor density:

$$\mathbf{T}_{\mu\nu} = \mathbf{P}^a_\mu \mathbf{L}^a_\nu - \delta_{\mu\nu}\mathcal{L}, \tag{1.50}$$

and the canonical momentum

$$\mathbf{P}^a_\mu = \frac{\delta\mathcal{L}}{\delta\mathbf{L}^a_\mu} = \left\{2\epsilon^2(\mathbf{L}^a_\nu\mathbf{L}^b_\nu\mathbf{L}^b_\mu - \mathbf{L}^a_\mu\mathbf{L}^b_\nu\mathbf{L}^b_\nu) - \frac{1}{2\lambda^2}\mathbf{L}^a_\mu\right\}; \tag{1.51}$$

it is readily seen that all these quantities may be presented as polynomials in $\mathbf{L}^a_\mu$. Substituting (1.51) into (1.50) one obtains

$$\mathbf{T}_{0k} = \mathbf{L}^a_0\left\{2\epsilon^2(\mathbf{L}^a_i\mathbf{L}^b_i\mathbf{L}^b_k - \mathbf{L}^a_k\mathbf{L}^b_i\mathbf{L}^b_i) - \frac{1}{2\lambda^2}\mathbf{L}^a_k\right\} \tag{1.52}$$

and one can write down the following expression for components of the angular momentum

$$\mathcal{J}_\mu = \epsilon_{\mu lk}\int d^3x\, x_l \mathbf{T}_{0k}\,. \tag{1.53}$$

The energy density of the system

$$\mathcal{E} = \mathbf{T}_{00} = \mathcal{E}_{\text{stat}} + \mathcal{E}_{\text{rot}}, \tag{1.54}$$

is also composed of two parts: the static part

$$\mathcal{E}_{\text{stat}} = \left\{\frac{1}{2\lambda^2}\mathbf{L}^a_k\mathbf{L}^a_k + \frac{\epsilon^2}{2}\left[(\mathbf{L}^a_i)^2(\mathbf{L}^b_k)^2 - (\mathbf{L}^a_i\mathbf{L}^a_k)^2\right]\right\}, \tag{1.55}$$

which determines the mass of configuration, and the rotational part

$$\mathcal{E}_{\text{rot}} = \left\{\frac{1}{2\lambda^2}\mathbf{L}^a_0\mathbf{L}^a_0 + \epsilon^2\left[(\mathbf{L}^a_0)^2(\mathbf{L}^a_i)^2 - (\mathbf{L}^a_0\mathbf{L}^a_k)^2\right]\right\}, \tag{1.56}$$

on which we focus in the subsequent discussion of the rotating Skyrmion and in quantization procedure.

1.5 Skyrme's Results and Conjectures

Leaving a more detailed treatment of the final version of the Skyrme model for subsequent chapters, let us outline to what extent Skyrme managed to study the proposed theory by himself. The equally important aim of this section is to list the main suggestions and hypotheses of Skyrme, which were expressed in his papers (Skyrme 1961a, 1962, 1971) and became the subject of great interest among particle physicists during the 1980's. Briefly, the story of the revival of interest in Skyrme's ideas is as follows. The first substantial step in that direction was by (t'Hooft 1974). Looking for an expansion parameter in QCD he found suitable the so called $1/N_c$-expansion, where N_c is the number of colors. t'Hooft's next discovery was that QCD is drastically simplified in the framework of this expansion in the limit $N_c \to \infty$. If one assumes confinement, then QCD gives an effective description of the low-energy events (the Hadron Physics region) in terms of meson fields and glueballs, with the meson coupling constant of order $1/N_c$. The further development of these ideas in (Witten 1979) showed that for large N_c, baryon masses are of order N_c, while the baryon size, baryon-baryon and baryon-meson cross sections are of order one. On the basis of those results Witten suggested that the effective meson theory should be of the σ-model type, with a spontaneously broken internal symmetry, which admits soliton solutions. It happened that the Skyrme model, which substantially predates the formulation of the main principles of QCD, is one of the simplest variants of such a theory, endowed with most of the aforementioned features.

After this digression, let us return to Skyrme's results and proposals.

1. In order to bring his initial "Mesonic Fluid" model into the chiral invariant form Skyrme suggested that the mesonic fields could take their values on $\mathbf{S}^3$ as *the field manifold.* This assumption was crucial for the subsequent discovery of the conserved quantity of a new type - the topological charge. The study of the features of this novel quantity drew him to the hypothesis that **the Topological Charge might be interpreted as the Baryon Number.**
2. Skyrme's next prominent suggestion was, that topologically nontrivial solutions of the field equations (1.44) can be obtained in the form of **the "hedgehog" ansatz**[7]:

$$\phi_o = \cos\Theta(r)\,; \quad \phi_i = \frac{x_i}{r}\cdot\sin\Theta(r)\,; \quad i = 1, 2, 3. \tag{1.57}$$

3. Due to the availability of the already demonstrated estimate (1.43) of the energy from below Skyrme presupposed that solutions (1.57) could describe **a stable extended particle with the unit topological charge and all finite dynamical characteristics.**

To proceed further Skyrme undertook (with the help of a vacation student at A.E.R.E., A.J. Leggatt) the computer calculations in search of numerical solutions of the Eq. (1.44) on the "hedgehog" ansatz (1.57) and found of the profile function $\Theta(r)$, reported in (Skyrme 1962). In the same article Skyrme checked the possibility of obtaining a description of bound states in the framework of his model. From the one side, he proved that the topological charge $\mathbf{Q}$ is the additive conserved number described, when the two-particle state is considered, by the field $\mathbf{U}_{12}(\mathbf{x})$ which Skyrme suggested to take as a product

$$\mathbf{U}_{12}(\boldsymbol{x}) = \mathbf{U}(\boldsymbol{x} - \boldsymbol{x}_1)\cdot\mathbf{U}(\boldsymbol{x} - \boldsymbol{x}_2). \tag{1.58}$$

[7] Skyrme mentioned, that this form of solutions was inspired by the great influence of Pauli's semiclassical picture of spin and isospin locked together (Skyrme 1961a, 1988), but as one can find in the obituary on the occasion of Professor T.H.R. Skyrme's death in 1987, one of his characteristic features was to find a person to whom it was possible to ascribe his own findings.

Here $\mathbf{U}(\boldsymbol{x} - \boldsymbol{x}_i)$ are chiral fields, describing single particles with locations $\boldsymbol{x}_1$ and $\boldsymbol{x}_2$. But from the numerical analysis it was found that the total energy of the state (1.58) with $\mathbf{Q} = 2$ was about three times larger than that of the energy of one-particle configurations: $E^{(2)} \simeq 3E^{(1)}$. Those results Skyrme interpreted as the nonexistence of a stable particle with $\mathbf{Q} = 2$ or, from an alternative point, of the existence of a short-range repulsion between particles in the energy scale equal in order of magnitude to a particle mass.

4. As a possible description of the interaction picture among two particles Skyrme suggested the so-called **"product-ansatz"** (1.58), which he considered as a good approximation when the particles are far apart from each other.
5. The last, but not the least important suggestion was that configurations described by **solutions of the form (1.57) with the unit topological charge should be quantized as fermions,** so that it would be possible to identify the states with equal isotopic I and total J spins with the nucleon doublet in the case $I = J = \frac{1}{2}$, and with the Δ resonance in the case $I = J = \frac{3}{2}$.

It is interesting to note that Skyrme was one of the first to adopt Finkelstein's ideas (Finkelstein and Misner, 1959; Finkelstein 1966) of a topological treatment of fermion states in nonlinear field theories. As the theme of the thesis for his graduate student J. Williams he suggested a rigorous proof of the existence of Finkelstein's spinorial structures in the (3+1)-dimensional model (Williams 1970). Later the particle states with all foregoing features were called *Skyrmions*[8] and nowadays this term has acquired a nominal meaning and is widely used for description of solitonic states with a nontrivial topological index, which arise in purely bosonic theories but obey the Fermi-Dirac statistics. To conclude this section let us take one more quotation from Skyrme's talk: **"I still entertain the hope that some type of nonlinear theory will yield an explanation of elementary particles that can be visualised in a semiclassical way, and that quarks or leptons introduced as sources in most theories will be seen to be mathematical constructs helpful in the understanding, rather than fundamental constituents, just as the idea of vortices in a fluid is an indispensable way of talking about certain types of fluid motion".**

[8] Thus after E. Fermi and S. Bose, T. Skyrme became the third person in Particle Physics history whose last name by adding the suffix -"on" was used for a designation of a special type of particles, which are fermions composed from bosons.

2. Elements of Field Theory with Topological Charges

"...the steady progress in physics requires for its theoretical formulation a mathematics that gets continually more advanced. This is only natural and to be expected."

Paul Dirac

2.1 Geometric Viewpoint on the Classical Field Theory

Along with nonabelian gauge theories in the 1960s, the Skyrme model came to catalyze the application of homotopy theory and other machinery of algebraic topology to the study of nonlinear field theory problems. Topology and differential geometry have become a common language for particle physicists and condensed matter physicists and now geometrical concepts conquer the nuclear physics community. There exists a firm opinion that classical field theory is differential geometry endowed with physical meaning, while quantum field theory one could expect to be related to random geometry. We will not either support, nor deny this view, but rather formulate some familiar field theoretic concepts using geometrical language.

The geometrical approach to the Skyrme model was pioneered in two remarkable articles (Finkelstein and Misner 1959, Finkelstein 1966), where it was found suitable to treat the new type of conservation laws, discovered by Skyrme, from the homotopy viewpoint. In what follows we greatly rely on these papers, but it is apparent that in this chapter we are able to provide the reader with only an elementary account of the subject. For a more thorough account, which would be sufficient to understand the topological content of the Skyrme model, the reader should consult Appendix B. As a guide to further reading we must mention the classical textbooks and monographs on homotopy theory (Steenrod 1951; Hu 1959), which are more suitable for mathematically-minded readers. For physicists, more appropriate would be the expositions in (Dubrovin, Novikov and Fomenko 1984a,b, 1985; Choquet-Bruhat, De Witt-Morette and Dillard-Bleick 1977; Felsager 1981; Bott and Tu 1982; Nash and Sen 1983). There are also some reviews and lecture notes, written on the so-called physical level of the mathematical rigor (Boya, Cariñena and Mateos 1978; Eguchi, Gilkey and Hanson 1980, Olshanetskiĭ 1982; Goddard and Mansfield 1986), which might be recommended for rapid acquisition of basic concepts. Anyone with a superficial knowledge of the elements of algebraic topology or of modern geometry of nonlinear field theory may well ignore this brief introductory chapter.

2.1.1 The Configuration Space

To specify any mathematical model for physical phenomena it is necessary first to define *a "configuration space"* for the system or, in other words, to set up a kinematical framework for the theory. For this purpose, let us take for the central object of study – classical fields – the following geometrical definition.

Definition 2.1. *A smooth map $\phi(x)$ from the space-time $X = \{\boldsymbol{x}, t\}$ into the field manifold $\varPhi$:*

$$\phi(x) \; : \; X \; \to \; \varPhi \tag{2.1}$$

is called **a classical field.**

Throughout this text we take the domain of classical fields to be the usual Minkowskian space-time. The choice of the range of the classical fields depends on the field theory. It is clear that if one takes $\varPhi = \mathbb{R}$, one will get a scalar field theory, $\varPhi = \mathbb{R}^n$, gives a vector field theory, and so on. Those field manifolds may be covered by a single coordinate system, e.g. formed by field components $\{\phi^a\}$; $a = 1, .., n$, and are trivial from topological point of view. It is apparent that the field manifold of the Skyrme model $\varPhi = \mathbb{S}^3$ cannot be covered globally by any single coordinate system. Following (Finkelstein 1966) let us note that the application of standard perturbation methods in the case of topologically nontrivial field theories effectively replaces $\varPhi$ by a vector space (the tangent space to a fixed point on $\varPhi$), than one cannot obtain any information on the global structure of nontrivial field configurations through those methods. We recall Skyrme's opinion on perturbation methods, cited in the Introduction. Here we will deal only with topologically nontrivial field theories, using nonperturbative methods to study them.

One can regard as the configuration space of a field theory an infinite dimensional functional space of mappings (2.1), usually denoted by $\mathrm{Map}(X, \varPhi)$. For physical applications, especially when one pursues the aim of describing extended objects which are localized in space and have all dynamical characteristics finite, one must restrict himself to some subspace of $\mathrm{Map}(X, \varPhi)$. The natural way is to impose on the fields, $\phi(x)$, boundary condition analogous to Eq. (1.15) and Eq. (1.30):

$$\phi(x) \to \phi_0, \quad \text{as} \quad |\boldsymbol{x}| \to \infty, \tag{2.2}$$

where ϕ_0 is a fixed point on the field manifold $\varPhi$. The rate of approach $\phi \to \phi_0$ should be fast enough to guarantee the finiteness of energy and other dynamical quantities. Now we are in a position to define the configuration space.

Definition 2.2. *The space of smooth maps (2.1), which satisfy boundary conditions of the type (2.2), is called* **the configuration space** *for the topologically nontrivial field theory, with adopted notation* $\mathrm{Map}^0(X, \varPhi)$.

As was already discussed in Section 1.4.1, one can easily reach the conclusion that under boundary conditions (2.2) at any fixed moment of time the domain of the classical fields (2.1) – the Euclidean space $\mathbb{R}^3$ – is effectively compactified. Under maps from $\mathrm{Map}^0(X, \Phi)$ all points at spatial infinity, usually denoted by $\{\infty\}$, are sent to one and the same fixed point $\phi_0 \in \Phi$. In this situation one may regard the pre-image of the maps (2.1) as the topological equivalent of the sphere $\mathbb{S}^3$, so that $\mathbb{R}^3 \cup \{\infty\} \simeq \mathbb{S}^3$. This equivalence relation reflects the nontriviality of the manifold $\mathbb{S}^3$, in the sence that at least two coordinate systems (or charts) are needed to cover the sphere globally (to form its atlas).

To illustrate this point, let us take a well-studied example of the maps of the type (2.1) – the stereographic projection:

$$\boldsymbol{x} = \frac{\boldsymbol{\xi}}{1+\xi_0}; \quad \boldsymbol{x} \in \mathbb{R}^3 \tag{2.3}$$

together with its inverse mappings

$$\xi_0 = \frac{1-r^2}{1+r^2}; \quad \boldsymbol{\xi} = \frac{2\boldsymbol{x}}{1+r^2}; \quad r = |\boldsymbol{x}|, \tag{2.4}$$

where $(\xi_0, \boldsymbol{\xi})$ are stereographic coordinates on $\mathbb{S}^3$, satisfying the equation ${\xi_0}^2 + \boldsymbol{\xi}^2 = 1$. From Eq. (2.4), taking $r \to \infty$, one may readily see that all points at $\{\infty\}$ of $\mathbb{R}^3$ are mapped to the fixed point $\xi = (-1, \mathbf{0})$ on $\mathbb{S}^3$, namely to the south pole of $\mathbb{S}^3$. In this way we find that two charts are necessary to cover $\mathbb{S}^3$ globally and it is in this sense that we are to understand the above equivalence. As the next step it is possible to see that all functions, well-defined on $\mathbb{R}^3$, which satisfy the boundary condition (2.2), stay well-defined under this change of domain, which one can perform by a rule analogous to Eq. (2.3). Thus, instead of mappings (2.1) with boundary condition (2.2) one may consider equivalent maps:

$$\tilde{\phi} : \mathbb{S}^3 \to \Phi \tag{2.5}$$

and, correspondingly, the equivalent configuration space $\mathrm{Map}^0(\mathbb{S}^3, \Phi)$. In Section 2.2.2 we shall see that this will be important for the correct definition of the topological invariants.

Only after this brief set-up of the kinematical framework, does it become possible to represent the dynamics in the form of a variational principle on the configuration space. Because we are interested in field theory with topological charges, we would like to describe field configurations which are stable against continuous deformations for topological reasons. These deformations may result from the time evolution of the physical system and, if they may be taken small enough, there is a well-known procedure for establishing the Euler-Lagrange equations of motion. In the next section we are going to see how these common facts can be reformulated in terms of homotopy concepts and what benefits one may receive from this reformulation.

2.1.2 Homotopy as a Formalization of Dynamical Evolution

Let us describe the state of a dynamical system at a fixed moment of time t_1 by the function $\phi_1(\boldsymbol{x}) \equiv \phi(\boldsymbol{x}, t_1)$ and at the subsequent moment t_2 – by another function $\phi_2(\boldsymbol{x}) \equiv \phi(\boldsymbol{x}, t_2)$. The state functions of the system may be considered as mappings from $\mathbb{R}^3 = X_{|t=\text{const}}$ into the field manifold Φ:

$$\phi(\boldsymbol{x}) : \mathbb{R}^3 \rightarrow \Phi. \tag{2.6}$$

If the dynamics of the system is governed by the Euler-Lagrange (or any other reasonable) equation of motion, then a continuous solution $\phi(\boldsymbol{x}, t)$ of this equation has to link the state $\phi_1(\boldsymbol{x})$ with the subsequent one $\phi_2(\boldsymbol{x})$. Formally this evolution process can be expressed in the form:

$$\phi(\boldsymbol{x}, t) = \begin{cases} \phi(\boldsymbol{x}, t_1) \equiv \phi_1(\boldsymbol{x}), \\ \phi(\boldsymbol{x}, t_2) \equiv \phi_2(\boldsymbol{x}). \end{cases} \tag{2.7}$$

In the language of homotopy the same facts are expressed in the following way:

Definition 2.3. *Two continuous maps $\phi_1(\boldsymbol{x})$ and $\phi_2(\boldsymbol{x})$ are* **homotopic**, *written $\phi_1 \sim \phi_2$, when there exists a continuous function in the joint variables $\phi(\boldsymbol{x}, t)$ –* **the homotopy**, *which interpolates between the two maps $\phi_1(\boldsymbol{x})$ and $\phi_2(\boldsymbol{x})$.*

In other words, there exists a continuous deformation, which transforms $\phi_1(\boldsymbol{x})$ into $\phi_2(\boldsymbol{x})$ and can be given as a map

$$\phi(\boldsymbol{x}, t) : \mathbb{R}^3 \otimes T \rightarrow \Phi, \tag{2.8}$$

for which the condition (2.7) holds. Here $T = [t_1, t_2]$ is the time interval and it is convenient to transform it into the unit interval $I = [0, 1]$.

It is important (Steenrod 1951, Hu 1959) that the homotopy relation possesses such properties as symmetry, reflexivity and transitivity and, therefore, it is *an equivalence relation.* The first two properties follow straightforwardly from the definition of homotopy (2.7) and (2.8). As for transitivity, we have to consider three maps (2.6): $\phi_1(\boldsymbol{x})$, $\phi_2(\boldsymbol{x})$ and $\phi_3(\boldsymbol{x})$, so that if $\phi_1 \sim \phi_2$ and $\phi_2 \sim \phi_3$, then there exists $\phi^{(1)}(\boldsymbol{x}, t) : \mathbb{R}^3 \otimes I \rightarrow \Phi$ and $\phi^{(2)}(\boldsymbol{x}, t) : \mathbb{R}^3 \otimes I \rightarrow \Phi$, such that

$$\begin{aligned} &\phi^{(1)}(\boldsymbol{x}, 0) = \phi_1(\boldsymbol{x}); \\ &\phi^{(1)}(\boldsymbol{x}, 1) = \phi_2(\boldsymbol{x}) = \phi^{(2)}(\boldsymbol{x}, 0); \\ &\phi^{(2)}(\boldsymbol{x}, 1) = \phi_3(\boldsymbol{x}). \end{aligned} \tag{2.9}$$

Then the transitivity follows from the fact that we can construct the continuous map

$$\phi(\boldsymbol{x},t) = \begin{cases} \phi^{(1)}(\boldsymbol{x},2t)\,; & 0 \le t \le \frac{1}{2}\,; \\ \phi^{(2)}(\boldsymbol{x},2t-1)\,; & \frac{1}{2} \le t \le 1\,, \end{cases} \tag{2.10}$$

which is a homotopy between ϕ_1 and ϕ_3, thus $\phi_1 \sim \phi_3$.

As does any other equivalence relation, the homotopy enables us to divide the collection of all continuous maps, e.g. defined by (2.6), into disjoint classes of equivalent or mutually homotopic maps. The set of all maps may be turned into topological space $\mathrm{Map}(\mathbb{R}^3, \Phi)$ by endowing it with the compact-open topology and further into the configuration space of the field theory $\mathrm{Map}^0(\mathbb{S}^3, \Phi)$, after imposing the boundary conditions discussed in the previous subsection. Thus the formalization of the evolution process provides us with a tool for dividing our configuration space into disjoint classes or connected components. But where are the benefits and advantages of introducing all this heavy machinery and what does all this stuff have to do with the classical field theory? To get a notion of that and to keep the reader from a desire to close this book and reading no further let us take an illustrative example from classical mechanics (Finkelstein 1966).

Example 2.1. *Particle Number of Classical Mechanics*
Denoting by $\mathcal{Q}_n$ the configuration space of n particles it is apparent that the configuration space of an unspecified number of particles may be considered as the union

$$\mathcal{Q} = \bigcup_n \mathcal{Q}_n. \tag{2.11}$$

In standard classical mechanics the number of particles, n, does not change under any evolution process and it is possible to reexpress this trivial fact using the obvious disconnectedness of the configuration space (2.11). All continuous trajectories of a system with any fixed number of particles $n = m$ belong to the connected component of the configuration space (2.11), namely to $\mathcal{Q}_m$. There is no continuous trajectory which joins two distinct components $\mathcal{Q}_m$ and $\mathcal{Q}_k$, with $m \neq k$. Thus the conservation law of the number of particles in classical mechanics may be considered as a consequence of the continuity of classical trajectories.

On the basis of this observation Finkelstein formulated the following assumption: *Whenever the configuration space $\mathcal{Q}$ admits a resolution (2.11) into connected components $\mathcal{Q}_n$, there exists a strong conservation law, independent of symmetries of the classical theory and dependent only on continuity.* This is one of the central ideas of the proposed geometrical approach. Finkelstein's assumption can be regarded as a criterion to obtain a classical field theory with conserved for topological reasons quantity. One can do this whenever the configuration space of the theory admits a resolution (2.11) into connected components. In this way Skyrme's discovery of topological charge acquires the geometrical background and may be "derived" in the framework of a regular procedure. It is clear that homotopy theory, which was specificially developed for study of connectedness problems in functional spaces, is the right tool to use for formulating general principles of construction of field theories with topological charges. It can provide us with the answer to the question in what cases the configuration space of the field theory under study $\mathrm{Map}(\mathbb{R}^3, \Phi)$ admits splitting into the equivalence classes in analogy with (2.11):

$$\mathrm{Map}(\mathbb{R}^3, \Phi) = \bigcup_i [\mathbb{R}^3, \Phi]_i . \qquad (2.12)$$

These equivalence classes, denoted by $[\mathbb{R}^3, \Phi]_i \equiv \mathcal{Q}_i$, are called **homotopy or Chern-Pontryagin classes**. The existence of the decomposition (2.12) means, in close analogy with Example 2.1, that if the state function of some system belongs to any one homotopy class and the time evolution of the system is supposed to be continuous, then at any subsequent moment of time this function has to stay in the same class.

The continuous deformation of the system from one state to another (homotopy) is commonly performed by the continuous solution of the equation of motion. Thus the problem of finding the homotopy becomes equivalent to that of seeking for solutions of the field equations. The homotopy theory will not provide us with the solution of the equation of motion, but it does tell us what kind of solution we can expect. In each homotopy class $\mathcal{Q}_i$ of decomposition (2.12) there should be a continuous solution, but it is useless to look for continuous solutions which interpolate between two distinct classes.

Fields (or solutions) from different homotopy classes in (2.12) can be distinguished by the value of the homotopic characteristics or topological index **Q**. To each homotopy-related class of fields one can assign a concrete value of **Q**. This homotopic dynamical variable, in which we readily recognize the **topological charge**, would be strictly conserved by the reason of the continuity of classical fields only. The conservation law of this quantity holds independent of the dynamical symmetries of the model. The natural appearance of the topological charge is another advantage of the geometrical approach.

To summarize, one may say that the homotopy theory provides us with the scheme of *topological classification of solutions* in nontrivial field theoretic models. On the other hand it explains the origin of new types of conserved quantities in physics. For some mathematicians that would be the end of the matter. However, for physicists it is rather the starting point, as the most important task for them is to find an actual construction procedure and then to make sure that it really works. With this in mind, we turn to the implementation of those rather sketchily developed ideas and start, for the sake of clarity, from the topological classification of the sine-Gordon solutions.

2.2 The Topological Classification of Solutions

2.2.1 Homotopy Classes of the sine-Gordon Model

It was already mentioned in Section 1.3, that the field manifold of the sine-Gordon model is the circle $\mathbb{S}^1$. The classical fields $\phi_0(x,t)$ and $\phi_1(x,t)$ in (1.11) may be considered, at any fixed moment of time, as a map

$$\phi(x) = \exp[i\alpha(x)] \; : \; \mathbb{R}^1 \; \rightarrow \; \mathbb{S}^1, \tag{2.13}$$

which is the exponential representation of complex valued field $\phi = \phi_0 + i\phi_1$ with a unit norm. As we are interested only in field configurations with a finite energy

$$\mathcal{E}_{\text{S-G}} = \int_{-\infty}^{\infty} dx \, \left[\tfrac{1}{2}(\partial_t\alpha)^2 + \tfrac{1}{2}(\partial_x\alpha)^2 + k^2(1-\cos\alpha)\right], \tag{2.14}$$

we have to impose the boundary conditions

$$\alpha(x) \rightarrow 0 (\operatorname{mod} 2\pi), \quad \text{as} \quad |x| \rightarrow \infty. \tag{2.15}$$

The form of conditions (2.15) means that the angle-like field variable $\alpha(x)$ is a multivalued function and to obviate it one should identify angles, which differ by $2\pi n$ in value, where $n \in \mathbb{Z}$. Under conditions (2.15) the real axis $\mathbb{R}^1$ is evidently compactified, as the points at infinity ($x = \pm\infty$) are mapped into one and the same point, say the "north pole" of the circle $\mathbb{S}^1$. Thus $\mathbb{R}^1 \cup \{\infty\} = \mathbb{S}^1$ and the maps (2.13) are effectively the maps: $\mathbb{S}^1 \rightarrow \mathbb{S}^1$.

In this example it is easy to obtain a descriptive picture of splitting the configuration space of the model, $\text{Map}^0(\mathbb{R}^1, \mathbb{S}^1)$, or, equivalently, $\text{Map}(\mathbb{S}^1, \mathbb{S}^1)$, into disjoint homotopy classes. If the angle variable $\alpha(x)$ takes the value zero at both boundaries (one may say, at both vacua), $x = \pm\infty$, then the corresponding solutions are the travelling waves or the breathers and have the corresponding profile shown in Fig. 2.1 a. The image of the axis $\mathbb{R}^1$ on the field manifold $\mathbb{S}^1$ under the map (2.13) would be a closed loop on $\mathbb{S}^1$, which does not cover the whole circle $\mathbb{S}^1$. Such a loop could be shrunk in a continuous way to a point on $\mathbb{S}^1$. These maps are called *topologically trivial.*

But we can as well have a situation in which a time independent $\alpha(x)$ interpolates from one vacuum at $x = -\infty$ ($\alpha(-\infty) = 0$) to another at $x = +\infty$ ($\alpha(+\infty) = 2\pi$). The corresponding class of solutions of Eq. (1.13) was written in the explicit form in Section 1.3.1 and is displayed on Fig. 2.1 b. In this case the image of $\mathbb{R}^1$ on $\mathbb{S}^1$ would be the loop, which covers the whole circle $\mathbb{S}^1$ and cannot be shrunk in any continuous way to a point on $\mathbb{S}^1$. These solutions are called 2π-kinks and they are *topologically nontrivial.*

There exists one more class of solutions, known as two-kink solutions, which satisfy the boundary conditions in the form: $\alpha(-\infty) = 0$, $\alpha(+\infty) = 4\pi$. The closed loop, still the image of axis $\mathbb{R}^1$, in that case covers the circle twice and the corresponding maps are topologically different from the previous two (see Fig. 2.1 c). The two-kink solution can never be continuously deformed

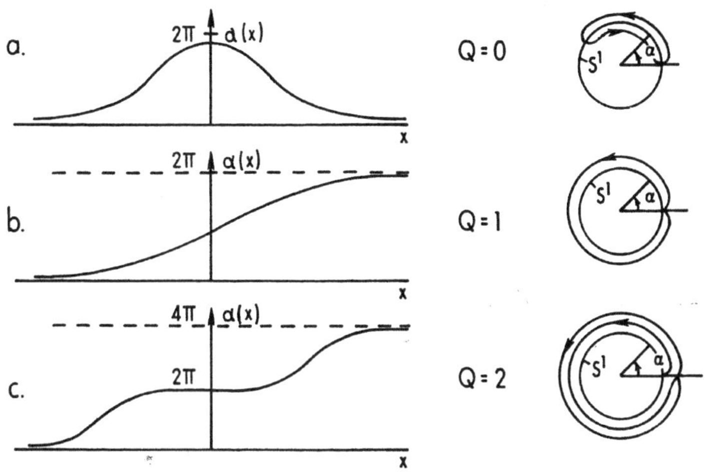

Fig. 2.1 a–c. Descriptive Topological Classification of the sine-Gordon Solutions. Left hand side profiles represent behavior of the angular variable $\alpha(x)$, as the right hand side pictures display the images of the domain $\mathbb{R}^1$ on the field manifold $\mathbb{S}^1$ under the maps (2.13). Situations: (a) corresponds to travelling wave or breather solutions with $Q = 0$; (b) represents 2π-kink solutions with $Q = 1$; (c) represents two-kink solutions with $Q = 2$

into breathers or 2π-kinks. All other possible cases just differ by the number of times the image of the axis $\mathbb{R}^1$ winds around the circle $\mathbb{S}^1$. The used terminology comes from the close analogy with the process of winding a "thread" ($\mathbb{R}^1$) on a "bobbin" ($\mathbb{S}^1$). Maps with the same number of closed loops (no matter how long those loops are nor what form they have, except that they should not contain knots) around the field manifold $\mathbb{S}^1$ could be considered as belonging to the same class of solutions. This is the vivid implementation of the homotopic variable or topological index **Q** as the **winding number**. This image is so natural, that the name winding number is usually extrapolated in physical papers to all other cases where we have the Brouwer degree of a mapping as a topological charge.

The same results can be obtained in a more formal way if, instead of the maps (2.13), one uses equivalent mappings of circles:

$$\phi[\xi(x)] \;=\; \exp\{i\alpha[\xi(x)]\} \;:\; \mathbb{S}^1 \;\rightarrow\; \mathbb{S}^1, \tag{2.16}$$

where $\xi(x)$ is the inverse stereographic projection $\mathbb{R}^1 \;\rightarrow\; \mathbb{S}^1$, defined as one-dimensional variant of Eq. (2.4), which parametrizes the domain circle $\mathbb{S}^1$. The maps (2.16) become continuous after imposing the condition

$$\alpha(2\pi) \;=\; \alpha(0) \;+\; 2\pi n, \tag{2.17}$$

where $n \in \mathbb{Z}$ is the number of times that the field α winds about the range circle $\mathbb{S}^1$ as ξ spans over the domain circle $\mathbb{S}^1$. This number n is called **the**

degree of mapping and is the well-defined homotopic invariant for continuous maps between sphere manifolds[1]. Maps with the same degree of mapping are mutually homotopic, those with different degrees belong to different homotopy classes. This is another implementation of the central idea of *the topological* (or, more precisely, *homotopical*) *classification* of maps. Notice that in this formal way we have got at once the integer-valued homotopic invariant.

Let us note that in fact we have come to the idea of algebraization of topological problems, which led H. Poincaré to the foundation of algebraic topology itself. Actually, we have got the notion how to set up a correspondence between purely geometric objects – classes of maps, and integer numbers n with well-known algebra. This indicates the possibility of setting up an algebraic structure in the set of homotopy classes and, after doing that, of using a powerful and well-developed calculus for solving complicated geometrical problems.

2.2.2 The Fundamental and Higher Homotopy Groups

To elucidate the latter statement of the previous section, let us look at the problem of topological classification of mappings from another point of view. Consider, for example, the set of homotopy classes $\{\mathcal{Q}_i\} \equiv \{[\mathbb{R}^1, \Phi]_i\}$ for mappings of the type (2.13). It is possible to put a composition law on this set, which turns the set into a group from the algebraic point of view. The idea of how to induce, for example, the abelian group structure comes in a natural way (Finkelstein and Misner 1959; Finkelstein 1966).

First of all we have to define the composition law on the set $\{\mathcal{Q}_i\}$ e.g.

$$\mathcal{Q}_1 \oplus \mathcal{Q}_2 = \mathcal{Q}_3, \tag{2.18}$$

so that homotopy classes $\mathcal{Q}_i$ are themselves the elements of an additive group. To achieve this we must take special representative fields in each homotopy class $\mathcal{Q}_i$, which will take on fixed value $\phi_0 \in \Phi$ not only for $|x| \to \infty$, but also for the half-axis e.g. $x \geq 0$. To obtain such representatives in each class we can use homotopy. Taking $\phi_1(x) \in \mathcal{Q}_1$ such that $\phi_1(x) = \phi_0$ at $x \geq 0$ and $\phi_2(x) \in \mathcal{Q}_2$ such that $\phi_2(x) = \phi_0$ at $x \leq 0$ we may define their composition under the law (2.18) to be the class $\mathcal{Q}_3$ with the representative field $\phi_3(x) \in \mathcal{Q}_3$, defined as:

$$\begin{aligned} \phi_3(x) &= \phi_1(x) \quad \text{at} \quad x \leq 0; \\ &= \phi_2(x) \quad \text{at} \quad x \geq 0. \end{aligned}$$

This means that the nontrivial parts of $\mathcal{Q}_1$ and $\mathcal{Q}_2$ are pushed to opposite sides of the x-axis with respect to the point $x = 0$ and then matched at this point. The role of the identity element in this group structure can be ascribed to a constant map $\overline{\phi}(x)$, which sends all domain points to the fixed point ϕ_0. It is a simple exercise to see that for each group element there is an inverse, so that

[1] For a strict definition of this invariant see Appendix B.

their composition (2.18) is equal to the identity element. In Appendix B we give in more detail an explicit method of finding the appropriate composition law on the set of homotopy classes $\{[\mathbb{S}^1, \mathbb{S}^1]_i\}$ for sine-Gordon theory, which shows the associativity property and provides the presence of inverse and identity elements[2].

Definition 2.4. *The set of homotopy classes of maps (2.13) with fulfilled boundary conditions (2.15) and endowed with the composition law (2.18) is called* **the Poincaré fundamental or the first homotopy group**. *It is denoted by* $\pi_1(\Phi)$.

To compute the fundamental group for any manifold Φ means to establish an isomorphism between $\pi_1(\Phi)$ and the group (or any subgroup) of integers $\mathbb{Z}$. The results of calculation – established isomorphisms – tell us what arithmetic can be applied when one adds homotopy classes. If $\pi_1(\Phi) = \mathbb{Z}$, it means that homotopy classes possess the addition operation just as ordinary integer numbers. In the case $\pi_1(\Phi) = \mathbb{Z}_2$ we have to apply the addition by modulo 2 and so on. From the other point we get information of how many pathwise-connected components are contained in the corresponding functional space of maps. In this algebraic approach the homotopic variable $\mathbf{Q}$ defines the abovementioned isomorphism between groups, e.g. for the maps (2.16) we have:

$$\mathbf{Q} \, : \, \pi_1(\mathbb{S}^1) \, \to \, \mathbb{Z}. \tag{2.19}$$

In that case it is easy to find the explicit form for this variable

$$\mathbf{Q} = \frac{1}{2\pi}\int_0^{2\pi} \frac{\partial \alpha}{\partial \xi} d\xi, \tag{2.20}$$

which coincides with that from Eq. (1.17). But in general it is not easy to establish such an explicit form as will be clear from the discussion in the next section.

Now we are more or less prepared for a passage to higher dimensional field theories. From the previous consideration it follows that, for the most part, we should be interested in the set of homotopy classes $\{[\mathbb{S}^3, \Phi]_i\}$ corresponding to classical fields (mappings) from Eq. (2.5). But the presented in section scheme is quite general and valid for classical field domains of any dimension.

[2] In general, the set of homotopy classes does not form a group under operation (2.18), but a groupoid. The only difference from the properties of a group is that the composition law (2.18) is not defined for every pair of classes, but only for pairs whose representative maps from the unit interval $[0, 1]$ satisfy the following conditions: $\phi_1(1) = \phi_2(0)$. In our cases those conditions may be considered as fulfilled because of imposed boundary conditions (2.2) and the effective compactification of the domain into $\mathbb{S}^1$. The same reasoning should be taken into account in passing to higher-dimensional homotopy groups. In this sense the conditions (2.2) are important for the proper definition of homotopy groups and consequently for that of topological charges.

It should be already apparent that with the correct composition law the collection of homotopy classes $\{[\mathbb{S}^3, \Phi]_i\}$ would be endowed with the structure of the third homotopy group $\pi_3(\Phi)$. Note that the set of connected components of the configuration space itself $\mathcal{M} \equiv \mathrm{Map}^0(\mathbb{R}^3, \Phi)$ is usually regarded as the zeroeth homotopy group $\pi_0(\mathcal{M})$, which can be identified with $\pi_3(\Phi)$. We are no longer going to overload the reader with mathematical details about the homotopy groups, but rather sketch the information that is really of use for physics nowadays. There exist methods to compute higher homotopy groups (see Bott and Tu 1982, and references therein), but this is a rather difficult problem. Fortunately, for such manifolds as spheres $\mathbb{S}^n$, classical Lie groups $\mathbf{G}$, and projective spaces $\mathbb{R}\mathbf{P}^n$ and $\mathbb{C}\mathbf{P}^n$, those homotopy groups which are commonly applied in physical models are already computed. Some of the known homotopy groups are listed in Table B.1. In Appendix B, mostly for curious readers, we give some simple rules for calculating unknown homotopy groups, starting from the known results for the manifolds most popular in physics.

Actually, in construction of physical models with nontrivial topological charges the so-called "inverse approach" is usually exploited. Field manifolds whose homotopy groups are already known are used to construct theories having the desired group $\pi_n(\Phi) = \mathbb{Z}$. The homotopic variable $\mathbf{Q}$, which realizes that isomorphism, plays the role of the topological charge of the theory and can be endowed with a physical meaning, according to the assumed interpretation. The actual problem in this "inverse approach" is to obtain an explicit expression for the topological quantity in terms of the underlying classical fields of the physical model. There are various methods available to solve this problem, but we have found that the most simple and powerful algorithm for construction of topological charges had been proposed in (Isham 1977). In what follows we develop this algorithm in application to the Skyrme model (Rybakov and Sanyuk 1981) and derive the explicit forms of the topological current and topological charge, which coincide with expressions (1.37) and (1.38).

2.3 Isham's Construction of Topological Charges

The key point of Isham's algorithm is to exploit the existing connections between the homotopy and cohomology groups. Thus, we must first give a brief account of the relationship between homotopy and cohomology groups.

2.3.1 Cohomology – Homotopy Relationship in Brief

Both the concepts of cohomology and homotopy arose as higher-dimensional analogs of the most intuitively evident topological invariant of a space, which is connectedness. Furthermore both theories are aimed at studying the same problem of connectedness of functional spaces. From their definitions, homotopy and cohomology groups in some sense are dual to each other. The homotopy groups are very easily and naturally defined, but the great drawback of the invariants serving as generators of these groups is that they are very hard to compute (Bott and Tu 1982). On the other hand the existing definitions of cohomology groups are not so evident, but they are comparatively easily computed. For proper definitions of cohomology groups and for fundamentals of exterior differential forms calculus we refer the reader to already listed literature, and for a brief account of needed information, to Appendix B. Here we only mention some features, which are essential to understanding Isham's algorithm.

Roughly speaking, the primitive idea of the homotopy theory is to test the connectedness of a space by mapping into it "unit intervals" or n-cubes, further transforming their images into "loops" and then to understand whether or not these "loops" are shrinkable. Two maps are considered equivalent if they could be continuously deformed into each other. The cohomology theory[3] deals in a similar manner with the generalization of integrands – exterior differential forms. In that sense it is possible to say that cohomology theory consider maps of the space Φ itself, where those integrands are defined, into the real axis $\mathbb{R}$, where the corresponding integrals take their values. Instead of "loops", in cohomology theory we have cocycles or closed differential forms. Two cocycles are considered equivalent – *(co)homological* – if they differ only by an exact form. (This is generalization of the fact that the integrand is defined up to a complete differential). In the same manner, as we have seen in the previous section for homotopy classes, closed forms are collected into elements of cohomology groups $H^n(\Phi;\mathbb{R})$ - the cohomology classes. Thus the nth de Rham cohomology group could be defined as

$$H^n(\Phi;\mathbb{R}) = \{\text{closed } n\text{-forms on } \Phi\}/\{\text{exact } n\text{-forms on } \Phi\}.$$

When Φ is chosen to be a Lie group $\mathbf{G}$, the cohomological information on the connectedness of the functional space of maps is the same as that obtained from the homotopy theory. Then, instead of direct construction of the isomorphism $\pi_n(\Phi) = \mathbb{Z}$ we can reduce this problem to an equivalent one in terms of cohomology groups. This is the idea, which was realized by Isham.

Let us turn back to the Skyrme model. The classical fields in the model, as emphasised in Section 1.4, take their values on the sphere $\mathbb{S}^3$. For the above reasons it would be convenient to use, instead of maps (2.5), the so-called *principal chiral fields* $\mathrm{U}(\boldsymbol{x},t)$ with values on the $SU(2)$-group manifold, which is isomorphic to $\mathbb{S}^3$-manifold. These chiral fields we consider as maps:

$$\mathrm{U}(\phi(\boldsymbol{x})): \quad \mathbb{S}^3 \to SU(2), \tag{2.21}$$

where the fields $\phi(\boldsymbol{x})$, defined in turn as maps from Eq. (2.6), parametrize the $SU(2)$-group elements $\mathbf{U}$ in accordance with Eq. (1.27). The map (2.21) induces the homomorphism between cohomology groups with real coefficients:

$$\mathrm{U}^* : H^3(SU(2);\mathbb{R}) \to H^3(\mathbb{S}^3;\mathbb{R}). \tag{2.22}$$

[3] Here and elsewhere, when we refer to the cohomology theory we will mean the de Rham theory, which is the easiest to handle and is equivalent to other variants of cohomology theory when one restricts oneself to differentiable manifolds.

Let **U** and **V** be two maps of the type (2.21). If they belong to the same homotopy class ($\mathbf{U} \sim \mathbf{V}$) then by the basic theorem of algebraic topology (see e.g. Dubrovin, Novikov and Fomenko 1984b, Chap. 1), the corresponding induced maps (2.22) are cohomologic $\mathbf{U}^* = \mathbf{V}^*$. As $SU(2)$ is isomorphic to $\mathbb{S}^3$, the Brouwer degree theorem states that the converse $\mathbf{U}^* = \mathbf{V}^* \Rightarrow \mathbf{U} \sim \mathbf{V}$, also holds. Thus instead of studying homotopy properties of maps (2.21) it is sufficient to investigate the induced maps (2.22). Trivial computations, given for convenience in Appendix B, show that $H^3(\mathbb{S}^3;\mathbb{R}) \approx \mathbb{R}$. From the general theory of Lie group cohomology the same result holds for the group $H^3(SU(2);\mathbb{R}) \approx \mathbb{R}$. According to the definition of cohomology groups, $H^3(SU(2);\mathbb{R})$ is the space of closed differential 3-forms on the $SU(2)$ manifold modulo exact 3-forms. The generator of this group can be obtained as a closed 3-form on $SU(2)$, which should not be at the same time exact. The plan of attack could be the following. After constructing an explicit form of the $H^3(\mathbb{S}^3)$ generator, we may repeat this sequence of arguments in reverse, to be sure that we have the desired homotopic invariant Q.

2.3.2 Derivation of Topological Charge in the Skyrme Model

We start with the setting of the left-invariant 1-form on $SU(2)$, which may be defined as

$$\Lambda(\mathbf{U}, d\mathbf{U}) = \mathbf{U}^{-1} \cdot d\mathbf{U}, \tag{2.23}$$

and, after the substitution of parametrization (1.27), can be written as follows

$$\begin{aligned} \Lambda &= \mathrm{i}\tau_a \omega^a = \mathrm{i}\tau_a \xi_i^a d\phi^i \\ &= \mathrm{i}\tau_a \left\{ \left(\delta_i^a + \frac{\phi^a \phi_i}{{\phi_0}^2} \right) \phi_0 + \delta^{ab} \epsilon_{bci} \phi^c \right\} d\phi^i . \end{aligned} \tag{2.24}$$

Here the 1-forms $\omega^a = \xi_i^a d\phi^i$ are the basic left-invariant 1-forms, which realize the left-invariant basis on the $SU(2)$-manifold. Note that in this representation we may treat the fields ϕ^i, $i = 1,2,3$ as a set of independent coordinates on $SU(2)$. Basic 1-forms ω^a satisfy the Maurer-Cartan structure equations

$$d\omega^a = -\tfrac{1}{2} \delta^{ab} \epsilon_{bdc} \omega^d \wedge \omega^c, \tag{2.25}$$

where $\wedge$ and d denote the exterior product and derivative respectively.

Using 1-forms ω^a as "building blocks", we can construct the 3-form on $SU(2)$:

$$\begin{aligned} \Omega &= \epsilon_{abc} \omega^a \wedge \omega^b \wedge \omega^c \\ &= \epsilon_{abc} \xi_i^a \xi_j^b \xi_k^c d\phi^i \wedge d\phi^j \wedge d\phi^k \\ &= (1 + 2{\phi_0}^2) \phi_0 \epsilon_{ijk} d\phi^i \wedge d\phi^j \wedge d\phi^k , \end{aligned} \tag{2.26}$$

where the latter expression has been obtained after substitution of Eq. (2.24). Making use of the Maurer-Cartan equations (2.25) it is straightforward to convince oneself that 3-form Ω is closed. In fact it can be presented as

$$\Omega = -2\, d\omega^a \wedge \omega_a .$$

Then taking the exterior derivative of this expression and making use of the Poincaré lemma we find $d\Omega = 0$. Thus this 3-form satisfies all the conditions necessary to be a generator of the cohomology group $H^3(SU(2);\mathbb{R})$. The standard way to obtain the generator of the group $H^3(\mathbf{S}^3;\mathbb{R})$ is to use the pullback of the generator (2.26) by the map $\mathbf{U}^*$ from Eq. (2.22), so that

$$\begin{aligned} \mathbf{U}^*\Omega &= \epsilon_{abc}\xi^a_i\xi^b_j\xi^c_k\frac{\partial\phi^i}{\partial x^l}\frac{\partial\phi^j}{\partial x^n}\frac{\partial\phi^k}{\partial x^m}dx^l\wedge dx^n\wedge dx^m \\ &= (1+2\,{\phi_0}^2)\phi_0\epsilon_{ijk}\frac{\partial\phi^i}{\partial x^l}\frac{\partial\phi^j}{\partial x^n}\frac{\partial\phi^k}{\partial x^m}dx^l\wedge dx^n\wedge dx^m. \end{aligned} \tag{2.27}$$

In terms of independent fields ϕ^i and using Eq. (2.23), it is not difficult to see that the components of left chiral currents $\mathbf{L}^a_\mu$, defined in Section 1.4.1 (see Eq. (1.32) and Eq. (1.33)), can be written as

$$\mathbf{L}^a_l = \xi^a_k\frac{\partial\phi^k}{\partial x^l}. \tag{2.28}$$

This makes it possible to rewrite the pullback of Eq. (2.27) in the more familiar form

$$\begin{aligned} \mathbf{U}^*\Omega &= \epsilon_{abc}\,\mathbf{L}^a_l\mathbf{L}^b_n\mathbf{L}^c_m\,dx^l\wedge dx^n\wedge dx^m \\ &= \epsilon_{abc}\epsilon^{lnm}\,\mathbf{L}^a_l\mathbf{L}^b_n\mathbf{L}^c_m d^3x. \end{aligned} \tag{2.29}$$

For two chiral fields from Eq. (2.21) $\mathbf{U}$ and $\mathbf{V}$, the corresponding 3-forms $\mathbf{U}^*\Omega$ and $\mathbf{V}^*\Omega$ are cohomologous on $\mathbf{S}^3$ if and only if their cocycles coincide

$$\int_{\mathbf{S}^3}\mathbf{U}^*\Omega = \int_{\mathbf{S}^3}\mathbf{V}^*\Omega, \tag{2.30}$$

which means that $\mathbf{U}^* = \mathbf{V}^*$ and, consequently, $\mathbf{U}\sim\mathbf{V}$. Thus $\mathbf{U}$ and $\mathbf{V}$ would belong to the same homotopy class with homotopy invariant

$$\tilde{\mathbf{Q}} = \int_{\mathbf{S}^3}\mathbf{U}^*\Omega = \epsilon_{abc}\epsilon^{ijk}\int d^3x\,\mathbf{L}^a_i\mathbf{L}^b_j\mathbf{L}^c_k. \tag{2.31}$$

Using the result that $H^3(\mathbf{S}^3;\mathbb{R})\approx\mathbb{R}$ we find that invariant (2.31) takes its values on the real axis $\mathbb{R}$. Since our aim is to realize isomorphism $\pi_3(SU(2)) = \mathbb{Z}$, we need an invariant which takes on only integer values. Here there comes into play another useful result from cohomology theory, known as *the universal coefficients theorem*:

$$H^3(\mathbf{G};\mathbb{R}) = \mathbb{R}\otimes H^3(\mathbf{G};\mathbb{Z}), \tag{2.32}$$

which tell us that there exists a real number, say $\rho_0\in\mathbb{R}$, such that the 3-form $\rho_0\Omega$ is the generator of the cohomology group with integer coefficients $H^3(SU(2);\mathbb{Z})\approx\mathbb{Z}$. After the required modification of the Eq. (2.31) we come to the result that

$$\mathbf{Q} = \int_{\mathbf{S}^3}\mathbf{U}^*(\rho_0\Omega) = \rho_0\epsilon_{abc}\epsilon^{ijk}\int d^3x\,\mathbf{L}^a_i\mathbf{L}^b_j\mathbf{L}^c_k \tag{2.33}$$

is the explicit form sought for topological charge. The value of ρ_0 can be easily calculated either by derivation of the integral, or by purely topological methods, as is demonstrated in (Isham 1977). Comparing with (1.37), we take $\rho_0 = -(12\pi^2)^{-1}$.

The formula for the topological current (1.38) can be derived in the same manner; instead of maps (2.21) we consider time-dependent maps

$$\mathbf{W}\left(\phi(\boldsymbol{x},t)\right): \quad \mathbb{S}^3 \otimes \mathbb{R}^1 \rightarrow SU(2), \tag{2.34}$$

where $\mathbb{R}^1$ is the time axis. In this case the generator of the group $H^3(\mathbb{S}^3;\mathbb{Z})$ takes the form

$$\mathbf{W}^*(\rho_0\Omega) = \epsilon_{abc}\rho_0\mathbf{L}_\nu^a\mathbf{L}_\lambda^b\mathbf{L}_\rho^c dx^\nu \wedge dx^\lambda \wedge dx^\rho. \tag{2.35}$$

With the help of the Hodge duality operation we obtain from the generator $\mathbf{W}^*$ above the dual 1-form

$$\mathbf{J} = *\mathbf{W}^*(\rho_0\Omega) = \mathbf{J}^\mu dx^\mu, \tag{2.36}$$

or explicitly,

$$\mathbf{J}^\mu = \rho_0\epsilon^{\mu\nu\lambda\rho}\epsilon_{abc}\mathbf{L}_\nu^a/\mathbf{76}\mathbf{L}_\lambda^b\mathbf{L}_\rho^c. \tag{2.37}$$

This is the explicit form for the conserved topological current with the conservation law, which can readily be seen from the fact of closedness of the 3-form $\rho_0\Omega$.

Some remarks are in order. By no means can one regard the above algorithm as superfluous in application to the problem, although it is true that it was practically solved intuitively by Skyrme in 1958, without all the heavy machinery. But we must note that, first of all, Skyrme's derivation of the topological charge was largely due to his great intuition - a subject which can never be taught by any textbook. Secondly, it is doubtful that this intuitive way is applicable in the gauge generalization of the Skyrme model (Faddeev 1976a; Kundu, Rybakov and Sanyuk 1979). For those readers who would like to check this statement we pose the problem in a clearer form.

In the gauge generalization of the Skyrme model, in addition to the chiral fields (2.21) we consider the triplet of vector fields $\mathbf{A}_\mu^a$, which might be conveniently represented by the matrix $\mathbf{A}_\mu = \frac{i}{2}\tau^a \cdot \mathbf{A}_\mu^a$. Introducing the standard covariant derivative $D_\mu = \partial_\mu \cdot + [\mathbf{A}_\mu, \cdot \quad]$, we have the straightforward gauge generalizations for the left $\tilde{\mathbf{L}}_\mu = U^{-1}D_\mu U$ and right $\tilde{\mathbf{R}}_\mu = -U \cdot D_\mu U^{-1}$ chiral currents. For the topological current of the gauge model Isham's procedure gives the expression

$$\mathbf{J}_G^\mu = -\frac{\epsilon^{\mu\nu\lambda\rho}}{48\pi^2}\,\mathrm{Tr}\left\{\tilde{\mathbf{L}}_\nu \cdot [\tilde{\mathbf{L}}_\lambda, \tilde{\mathbf{L}}_\rho] + 3\mathbf{F}_{\nu\lambda} \cdot (\tilde{\mathbf{L}}_\rho + \tilde{\mathbf{R}}_\rho)\right\}, \tag{2.38}$$

where $\mathbf{F}_{\mu\nu}$ is the Yang-Mills field strength tensor $\mathbf{F}_{\mu\nu} = \partial_\mu\mathbf{A}_\nu - \partial_\nu\mathbf{A}_\mu + [\mathbf{A}_\mu, \mathbf{A}_\nu]$. It might be that the expression (2.38) can be obtained in a simpler way, but there is no doubt that a regular procedure is always preferable. Still others are of the notorious opinion, that good intuition could be even greater, if supported by the right mathematical tool.

2.4 Guiding Principles in the Choice of Model Lagrangians

Through all preceding sections of this chapter we touched only briefly on the problem of dynamics. All the information we needed about the model was the choice of the field manifold Φ. Now it is time to pay tribute to dynamics itself and to understand what guiding principles in the choice of the Lagrangian of the model are to be used. First of all we must consider any appropriate symmetries, which originate either from geometry, or from the physics of the phenomena we are intending to describe in our model. In the spirit of this chapter we will take into account the symmetries of the domain X, where classical fields are defined, and those of the field manifold Φ. As for symmetries of the domain (geometrical symmetries) it is hardly negotiable that any reasonable variational principle should have manifest Lorentz-invariance. This aspect is described in any physics textbook from the last century and thus needs not be discussed here. Under symmetries of the field manifold we should understand the so-called dynamical symmetries of the model, and they are dictated in general by available experimental data. A reasonable approximation for the low-energy sector in hadron physics is the so-called chiral limit, when one can ignore the quark masses. We have already discussed physical motivations for the chiral symmetric approach in Section 1.2. In Part III we intend to look for its justification from QCD first principles. Thus in this section we will concentrate our attention on the geometrical aspects of chiral symmetries with respect to construction of chiral symmetric Lagrangians.

On the other hand we have to learn what general requirements should be imposed on a Lagrangian to guarantee a description of particles as extended objects. From the discussion in Chapter 1, we know part of the answer. The model equations ought to possess regular particle-like solutions with finite dynamical characteristics. At present there is the special popular name for such solutions – **solitons**. Thus we can formulate our goal as **to outline general principles of the construction of field theoretic models which admit multi-dimensional solitonic solutions with nontrivial topological charges.** The problems of searching for multidimensional solitons had been discussed in the literature in the 1970s and some general principles had been stated (see e.g. Faddeev 1976b; Coleman 1977). Taking all that into account, the main requirements of the model Lagrangian are:

1. Lorentz-invariance, as manifestation of space-time symmetries.
2. Chiral Invariance, justified by symmetries of strong interactions.
3. The stability condition and, in particular, topological stability requirements.

The latter principle of topological stability merits exposition in greater detail, which we shall do in the next chapter. In what follows we discuss implementation of the chiral symmetry requirements, namely the construction of chiral invariant Lagrangians. This problem can be solved by geometrical methods and to ward that end we need some elementary information on the Riemannian geometry of the curved isospin space, which one can find in (Meetz 1969; Ebrahim and Gürsey 1974; Ebrahim and Savci 1984) and in Appendix A.2.

2.4.1 Chirally Invariant Lagrangians

It is apparent that chiral invariant Lagrangians must be composed from invariants of the chiral group $SU(N)_L \otimes SU(N)_R$. Here we use the simplest geometrical approach in the construction of chiral invariants as algebraic functions of left chiral currents $\mathbf{L}_\mu$, or, equivalently, of the coefficients ξ_a^i, defined by Eq. (2.24). The currents $\mathbf{L}_\mu$, or what is the same their coefficients in the concrete decomposition $\mathbf{L}_\mu = \mathrm{i}\,\mathbf{L}_\mu^a \tau_a$, entering into the expressions for the chiral invariants, have two types of indices: space-time and internal indices. To construct invariant functions from those building blocks we must be able to saturate those indices. The space-time indices can be saturated with the Minkowski metric and with the Levi-Civita symbol $\epsilon^{\mu\nu\lambda\rho}$, and the internal (isospace) indices by the corresponding Cartan-Killing metric on the field manifold and structure constants or other covariants, which define an algebraic operation on vector fields. This is the general idea. Now we turn to its implementation for the $SU(2)_L \otimes SU(2)_R$ chiral group. The action of this group on the $SU(2)$-manifold is described in Appendix A.2 where, in particular, it is shown that the metric tensor on the $SU(2)$-manifold can be written in terms of fields ϕ_i as

$$g_{ij}(\phi) = \xi_i^a \xi_j^a = \delta_{ij} + \frac{\phi_i \phi_j}{{\phi_0}^2}. \tag{2.39}$$

There is a unique chiral invariant quadratic in the currents $\mathbf{L}_\mu$ or, equivalently, in the ξ_i^a:

$$\begin{aligned}\mathcal{L}_{(2)} &= \mathrm{Tr}(\mathbf{L}_\mu \mathbf{L}^\mu) = -2\delta_{ab}\mathbf{L}_\mu^a \mathbf{L}^{b\mu} = -2(\mathbf{L}_\mu^a)^2 \\ &= -2\xi_i^a \xi_j^a \partial_\mu \phi^i \partial^\mu \phi^j = -2g_{ij}(\phi)\partial_\mu \phi^i \partial^\mu \phi^j \\ &= -2\big\{\partial_\mu \Theta \partial^\mu \Theta + \sin^2\Theta\, \partial_\mu n^a \partial^\mu n^a\big\}. \end{aligned} \tag{2.40}$$

The last line corresponds to the parametrization of chiral fields (1.29), with $\mathbf{L}_\mu^a$ given by the last expression in Eq. (1.33). This invariant has a clear geometrical meaning as the line element of the spherical surface in the isotopic space. The corresponding simplest chiral invariant Lagrangian is known in mathematics under the name of the Dirichlet functional. From the Weinberg point of view (Weinberg 1969), **the Lagrangian density (2.40) can be regarded as the first term in the decomposition of the true mesonic Lagrangian in powers of chiral invariants.** As we shall see in Chapter 3 by means of the Hobart-Derrick theorem, the Lagrangian corresponding to (2.40) admits stable solutions only for one-dimensional theories. Realistic three dimensional field configurations are unstable against collapse. One of the possible ways to avoid this and to get a realistic theory with stable soliton solutions is to involve chiral invariants of higher order in $\mathbf{L}_\mu$.

The fourth order chiral invariants may be obtained from the antisymmetric combinations $C_{abc}\mathbf{L}_\mu^a\mathbf{L}_\nu^b$, formed with the help of structure constants C_{abc} of the Lie algebra of vector fields on $SU(2)$. For the sake of simplicity, we assume that the appropriate choice of basis has been made, so that these structure constants coincide with the Levi-Civita symbols ϵ_{abc}. Saturating the free indices in two antisymmetric combinations and using, where it is needed, the Maurer-Cartan structure equations in the form (1.36), we can obtain three linearly independent chiral invariants of the fourth order:

$$\mathcal{L}_{(4)} = \alpha \mathrm{Tr}\big[(\mathbf{L}_\mu\mathbf{L}^\mu)(\mathbf{L}_\nu\mathbf{L}^\nu)\big] + \beta \mathrm{Tr}\big[\mathbf{L}_\mu\mathbf{L}_\nu\mathbf{L}^\mu\mathbf{L}^\nu\big] + \gamma \mathrm{Tr}\big[\partial_\mu\mathbf{L}_\nu\partial^\mu\mathbf{L}^\nu\big]. \quad (2.41)$$

Here α, β, γ are some arbitrary real-valued constants. All three invariants satisfy the Lorentz-invariance requirements, together with parity and G-parity requirements. From the symmetry point of view there is no compelling argument to prefer one or another invariant. But there are some indirect arguments, as the only combination of invariants (2.41) which contains time derivatives up to the second power is the following:

$$\begin{aligned}\mathcal{L}_{(4)} &= \mathrm{Tr}\{(\mathbf{L}_\mu\mathbf{L}^\mu)(\mathbf{L}_\nu\mathbf{L}^\nu) - (\mathbf{L}_\mu\mathbf{L}_\nu)(\mathbf{L}^\mu\mathbf{L}^\nu)\} = -\tfrac{1}{2}\mathrm{Tr}\big[\mathbf{L}_\mu, \mathbf{L}_\nu\big]^2 \\ &= 4\,(g_{ij}g_{kl} - g_{il}g_{kj})\partial_\mu\phi^i\partial_\nu\phi^j\partial^\mu\phi^k\partial^\nu\phi^l \\ &= 4\,\sin^4\Theta\,\{(\partial_\mu n^a)^2(\partial_\nu n^b)^2 - (\partial_\mu n^a\partial_\nu n^a)^2\} \\ &\quad + 8\,\sin^2\Theta\,\{(\partial_\mu\Theta)^2(\partial_\nu n^a)^2 - \partial_\mu\Theta\partial_\nu\Theta\partial^\mu n^a\partial^\nu n^a\}. \end{aligned} \quad (2.42)$$

This combination provides the positive definiteness of the corresponding energy functional and proves to be most convenient for the semiclassical quantization procedure. It is this combination which has been added by Skyrme (1958, 1961) to the pion Lagrangian (2.40) or (1.35). In that sense it is possible to regard the Skyrme term (2.42) as the distinguished one among fourth-order chiral invariants. In the next section we will clarify the crucial role of this term in the stability of realistic soliton structures.

The next order chiral invariants can be obtained with the help of the following combination

$$\epsilon^{\mu\nu\lambda\rho}\epsilon_{abc}\mathbf{L}_\nu^a\mathbf{L}_\lambda^b\mathbf{L}_\rho^c, \quad (2.43)$$

where we make use of both space-time and "isospin" Levi-Civita symbols. It is not difficult to recognize in (2.43) the structure of the topological current $\mathbf{J}^\mu$ from Eq. (1.38). Among the sixth-order chiral invariants there is only one,

$$\mathcal{L}_{(6)} = \mathrm{Tr}(\mathbf{J}_\mu\mathbf{J}^\mu), \quad (2.44)$$

which depends on the time derivatives of fields up to the second power and thus shares the benefits with the previous Skyrme term. This term has been studied with the aim of possible improvements of the Skyrme model predictions by taking into account ω-meson exchanges in (Jackson et al. 1985).

In the same manner it is possible to continue the construction of higher order chiral invariants, as was done, for example, in (Marleau 1990), where for the hedgehog configurations (1.57) a systematic approach was suggested and terms of order 8, 10, 12,... have been constructed. For our purpose it is enough to understand the procedure of the construction in principle, as we are not going to use chiral invariants of order higher than four. To conclude the chapter and provide the reader with a guide for further reading on the subject some remarks might be useful. The construction of chiral invariants was intensively discussed in the literature in the 1960s in relation to the problem of nonlinear realizations of chiral symmetry. Here we should mention Gürsey's papers (1960 and 1961), where the simplest chiral invariant (σ-model) Lagrangian (1.35) was introduced and studied, and especially the paper (Chang and Gürsey 1967), where the so-called Gürsey matrix first appeared. This matrix has been found useful in construction of chiral invariants of high order later on (Barnes et al. 1977). There also should be mentioned the remarkable paper of Weinberg (1967), who constructed effective Lagrangians for soft pion interactions, which reproduced the current algebra results. Those results were developed in subsequent papers in the new approach to strong interactions, named the effective or phenomenological Lagrangian approach (for review see e.g. Gasiorowicz and Geffen 1969). This approach was very popular from the late 1960s till the end of 1970s and the development of Skyrme's approach after 1983 can be regarded as a fresh and promising inspiration in that field.

3. Topological Stability

"Correctness and Truth are points so fine that our instruments are too blunt to touch them exactly."

Blaise Pascal

3.1 Some General Remarks

One of the remarkable features of the soliton approach in particle and nuclear physics is that in the context of a relativistic field theory the soliton solution already exists on a classical level. Meanwhile, in traditional approaches, not only particles and nuclei, but even larger objects such as atoms and molecules exist only in the quantum theory. Already at an early stage of development of the soliton approach it was widely accepted (Goldstone and Jackiw 1975; Christ and Lee 1975; Friedberg, Lee and Sirlin 1976, Makhankov 1978) that a necessary condition for the applicability of solutions of classical field theory to the construction of extended quantum-mechanical objects is the stability of such solutions with respect to small initial perturbations. In this connection the stability of solitons acquires special importance. In view of this, and in order to give a more detailed exposition of basic methods and background concepts, we have put the stability problems of multidimensional solitons in general, and of Skyrmions in particular, in a separate chapter. Our exposition is largely based on some previous publications (Makhankov 1978, 1980, 1983, 1990; Rybakov 1985a,b, 1991).

We start with the simplest stability criterion, known in the literature as the Hobart-Derrick theorem. This virial-type theorem considers energy variations under scale perturbations. The presence of a scale parameter λ makes it possible to deal with energy functionals $\Phi[\phi_\lambda]$ as with ordinary functions of one variable $\Phi(\lambda)$ by means of classical analysis. To check the scale stability of solitons one must check the fulfillment of standard sufficient conditions for the function of one variable to have a minimum: $\Phi'_\lambda = 0, \Phi''_{\lambda\lambda} > 0$. In spite of its simplicity this very effective and popular "no-go" test cancels the claims of some models to describe stable static structures[1]. Nevertheless, it should be stressed here and will be evident from what follows, that if a model passes this test it means no more than the admittance of scale-stable solutions. Those solutions might well be unstable with respect to more general perturbations. The general formulation and particular applications of what is usually called the Hobart-Derrick theorem are given in Section 3.2.

[1] Remark, that it happens as a rule but not always. There are exceptions, e.g. models with noncompact internal symmetries (Makhankov, 1990).

Unfortunately (or, perhaps, fortunately) Nature is more ingenious and produces a much wider variety of perturbations, which affect physical systems. To deal with those perturbations, and to establish stability criteria, first of all we must decide what concept of stability is most suitable for our problem. As for soliton stability, the Lyapunov concept of stability with respect to shape proves to be most appropriate. The definition of stability in the Lyapunov sense is located in Section 3.3. Although the methods for the investigation of stability appear to be well developed, their practical application meets with serious mathematical difficulties. Roughly speaking, to meet the Lyapunov stability requirements we need to check that the second variation $\delta^2 V$ is positive definite, where V is the appropriate Lyapunov functional. In this preliminary section we attempt to present the main ideas and tools used in the study of stability problems, without cloaking them in mathematical complexity. To begin with, it may be useful to remind the reader of some textbook facts about the calculus of variations (see e.g. Gelfand and Fomin 1963).

The second variation of a given functional in the simplest one-dimensional variational problem

$$F[y] = \int_a^b dx\, \mathcal{F}(x, y, y'), \qquad y' \equiv \frac{dy}{dx}, \tag{3.1}$$

under an increment of the function $y(x) \to y(x) + \xi(x)$, satisfying the trivial boundary conditions: $\xi(a) = 0, \xi(b) = 0$ can be presented (through Taylor's expansion with the subsequent integration by parts) as the quadratic functional

$$\delta^2 F[\xi] = \int_a^b dx\, \left(P(x)\xi'^2 + Q(x)\xi^2\right), \tag{3.2}$$

where the following notations are used:

$$P(x) = \frac{1}{2}\mathcal{F}_{y'y'}, \qquad Q(x) = \frac{1}{2}\left(\mathcal{F}_{yy} - \frac{d}{dx}\mathcal{F}_{yy'}\right).$$

Ignoring for a moment its origin, it is possible to treat the analysis of the functional (3.2) as an independent problem. Then it is possible to find conditions, which are both necessary and sufficient for the functional (3.2) to be greater than zero for all admissible $\xi(x) \not\equiv 0$, i.e. to be positive definite. This requirement is analogous to the necessary and sufficient conditions for a function of one variable to have a minimum since another condition $\delta F[\xi] = 0$, which is the Euler-Lagrange equation, is assumed to be already fulfilled.

There are two main approaches to the solution of this problem. One of them is based on the presentation of $\delta^2 F[\xi]$ as a scalar product in a Hilbert space $L_2(\mathbb{R})$:

$$\delta^2 F[\xi] = (\xi, \hat{K}(x)\xi), \tag{3.3}$$

where $\hat{K}$ is the Jacobi operator of the functional (3.1) and at the same time the Euler-Lagrange operator for the functional (3.2)

$$\hat{K}(x)\xi \equiv \left[-\frac{d}{dx}\left(P\cdot\frac{d}{dx}\right)+Q\right]\xi. \tag{3.4}$$

Thus one can study the positivity of $\delta^2 F[\xi]$ as the eigenvalue problem for the Jacobi operator. The further analysis can be carried on as the analysis of the spectrum of the linear operator (3.4). To this end Courant's theorem about the positivity of the first eigenfunction (Courant and Hilbert 1953) appears to be useful. In this way the general statement due to (Rybakov 1984) was obtained, that in field theories with translationally invariant functionals there could not be any absolutely stable stationary solitons in the Lyapunov sense and that the only possibility is conditional stability or, the so-called $\mathcal{Q}$-stability (Makhankov 1978), as a special case. This means that stable solitons could arise in such kinds of theories only under some restrictions on the form of initial perturbations. This statement, which generalizes the aformentioned Hobart-Derrick theorem, will be discussed together with the $\mathcal{Q}$-stability concept in Section 3.3.

Another approach is based on the following simple observation. The positiveness of the functional (3.2) would be proved if under an assumption $P(x) > 0$ we could reduce Eq. (3.2) to the form

$$\int_a^b dx P(x)\Psi^2(\ldots),$$

where $\Psi^2(\ldots)$ is some expression which cannot be identically zero unless $\xi = 0$ (Gelfand and Fomin 1963). Since the integrand is defined up to a complete differential, to achieve this it is possible to use *the method of reduction of a quadratic form to a perfect square.* We can add to the integrand in Eq. (3.2) a quantity of the form $d(w\xi^2)$ and we have to select a function $w(x)$ such that the resulting expression

$$P\xi'^2 + Q\xi^2 + \frac{d}{dx}(w\xi^2) = P\xi'^2 + 2w\xi\xi' + (Q+w')\xi^2$$

is a perfect square. Straightforward calculations show that this will be the case if $w(x)$ is a regular solution of a Riccati equation

$$P(Q+w') = w^2. \tag{3.5}$$

Thus the problem is reduced to the solution of a nonlinear first order equation and if such a solution exists, it proves that the second variation of the functional (3.1) is positive definite.

A generalization of this trick has proved to be useful in topologically nontrivial models, with energy functionals estimated from below by a positive definite increasing function of a topological charge $\mathbf{Q}$. As was demonstrated by Bogomol'nyi (1976), when the above mentioned estimate is exact, then it is possible to reduce the Euler-Lagrange equation for a given energy functional to a first order (self-duality) equation. In this case the corresponding solutions are stable for topological reasons and this kind of stability has gained the special name *Topological Stability.* But as become apparent later on, solutions of

Bogomol'nyi's equations satisfy the boundary conditions of the original problem mostly for scale-invariant models. For the Skyrme model functional this is not the case as we shall see in Section 3.4.

3.2 The Hobart-Derrick Theorem

The scale instability arguments for multi-dimensional solitons were formulated in (Hobart 1963; Derrick 1964) and give necessary conditions of the virial type on the energy functional claiming to possess stable solutions. Let $\phi(\boldsymbol{x})$ be a static solution of the field equation, which one may arrive at, for example, from a variational principle

$$\delta\Phi[\phi] = 0. \tag{3.6}$$

Consider the simplest type of perturbations - scale transformations, performed in the manner

$$\boldsymbol{x} \to \lambda\boldsymbol{x}; \quad \phi(\boldsymbol{x}) \to \phi_\lambda = \phi(\lambda\boldsymbol{x}), \tag{3.7}$$

so that the variational principle (3.6) takes the form:

$$\delta\Phi[\phi_\lambda] = \left.\frac{\partial\Phi[\phi(\lambda\boldsymbol{x})]}{\partial\lambda}\right|_{\lambda=1} \cdot \delta\lambda = 0 \tag{3.8}$$

Treating the functional $\Phi[\phi(\lambda\boldsymbol{x})]$ as a function $\Phi(\lambda)$ and assuming that it is a sum of homogeneous functions, one can present it as

$$\Phi(\lambda) = \sum_{\nu=-n_1}^{n_2} \Phi^{(\nu)}(\lambda), \tag{3.9}$$

where $\Phi^{(\nu)}(\lambda)$ is a function with the degree of homogeniety ν. Substituting function (3.9) into Eq. (3.8) we obtain the identity of virial type

$$\left.\frac{d\Phi(\lambda)}{d\lambda}\right|_{\lambda=1} = \sum_{\nu=-n_1}^{n_2} \nu\Phi^{(\nu)}(1) = 0, \tag{3.10}$$

which is known as **the Hobart-Derrick necessity condition** for existence of static solutions to the variational equation (3.6) If one restricts the admissible perturbations of fields to scale deformations (3.7), then it is possible to treat the Hobart-Derrick condition (3.10) as a consequence of the equations of motion, derivable from (3.6).

To study applications of this theorem we start from the simplest chiral (σ-model) Lagrangian (1.35) and (2.40) and write the corresponding energy functional for space with arbitrary dimension D:

$$\mathbf{E}_{(2)} = \int d^D x g_{ij}(\phi)\partial_k\phi^i\partial_k\phi^j. \tag{3.11}$$

Under scale transformations (3.7) the functional (3.11) changes to

$$\mathbf{E}_{(2)}[\phi_\lambda] = \int \frac{d^D(\lambda \boldsymbol{x})}{\lambda^D} g_{ij}(\phi_\lambda)\cdot\lambda^2 \frac{\partial\phi^i(\lambda\boldsymbol{x})}{\partial(\lambda x^k)}\frac{\partial\phi^j(\lambda\boldsymbol{x})}{\partial(\lambda x^k)} = \lambda^{2-D}\mathbf{E}_{(2)}[\phi], \quad (3.12)$$

which means that $\mathbf{E}(\lambda)$ is the homogeneous function with the degree $\nu = 2-D$. From (3.12) it follows that the functional (3.11) is scale-invariant only for $D = 2$ and thus there could exist static solutions. For $D \geq 3$ **there is no minimum** (even extremum), as we always have the possibility of lowering the value of the functional by a scale deformation. The minimal energy corresponds to the limit $\lambda \to \infty$ when the particle collapses to zero radius.

It is not difficult to see that addition of an arbitrary potential energy

$$\mathbf{V}[\phi] = \int d^D x\, \mathcal{V}(\phi(\boldsymbol{x}))$$

to Eq. (3.11) will not improve this result, as under scale transformations (3.7)

$$\mathbf{V}[\phi] \to \mathbf{V}[\phi_\lambda] = \lambda^{-D}\cdot\int d^D(\lambda\boldsymbol{x})\,\mathcal{V}(\phi_\lambda) = \lambda^{-D}\cdot\mathbf{V}[\phi], \quad (3.13)$$

so that we can write the total energy as a function of λ

$$\mathbf{E}(\lambda) = \lambda^{2-D}\mathbf{E}_{(2)} + \lambda^{-D}\mathbf{V}, \quad (3.14)$$

in support of assumption (3.9). Using the virial statement (3.10) we get

$$\left.\frac{\partial\mathbf{E}(\lambda)}{\partial\lambda}\right|_{\lambda=1} = (2-D)\cdot\mathbf{E}_{(2)} - D\cdot\mathbf{V} = 0. \quad (3.15)$$

For the potential energy Eq. (3.15) implies the equality

$$\mathbf{V} = D^{-1}(2-D)\cdot\mathbf{E}_{(2)}, \quad (3.16)$$

with two evident possibilities either $D = 2$ and $\mathbf{V} = 0$, or for $D \geq 3$ we have $\mathbf{V} < 0$. Once again this means that for $D \geq 3$ there is no static regular solution if $\mathbf{V} > 0$. But one can claim that those solutions could exist for a sign changing potential $\mathbf{V}$. Let us show that if any soliton solutions exist under such an assumption, they would be unstable. To do that we must, as usual, examine the positive definiteness of the second variation of the energy functional: $\delta^2\mathbf{E}[\phi] \geq 0$, or, equivalently in our case, that of the second derivative

$$\left.\frac{\partial^2\mathbf{E}(\lambda)}{\partial\lambda^2}\right|_{\lambda=1} \geq 0.$$

Differentiating Eq. (3.14) twice

$$\left.\frac{\partial^2\mathbf{E}(\lambda)}{\partial\lambda^2}\right|_{\lambda=1} = (2-D)(1-D)\cdot\mathbf{E}_{(2)} + D(D+1)\cdot\mathbf{V}, \quad (3.17)$$

and taking into account Eq. (3.16), we get

$$\left.\frac{\partial^2 \mathbf{E}(\lambda)}{\partial \lambda^2}\right|_{\lambda=1} = 2(2-D) \cdot \mathbf{E}_{(2)} < 0, \quad \text{when } D \geq 3. \tag{3.18}$$

Thus all static soliton solutions in $D \geq 3$ models with any sign changing potentials should be unstable. Further examples on applications of the Hobart-Derrick criterion can be found in (Makhankov 1977, 1978; Hietarinta 1977).

As pointed out already by Hobart in (1963, 1964), one of the real ways to improve the situation is to add terms of higher power in field derivatives. This possibility, in fact, is realized in the case of the Skyrme model, with Lagrangian

$$L = \int d^3x \{\mathcal{L}_{(2)} + \mathcal{L}_{(4)}\}, \tag{3.19}$$

and the corresponding energy, presented as a function of λ:

$$\mathcal{E}(\lambda) = \lambda^{2-D}\mathcal{E}_{(2)} + \lambda^{4-D}\mathcal{E}_{(4)}. \tag{3.20}$$

Actually, from calculations of the first derivative

$$\left.\frac{\partial \mathcal{E}(\lambda)}{\partial \lambda}\right|_{\lambda=1} = (2-D) \cdot \mathcal{E}_{(2)} + (4-D) \cdot \mathcal{E}_{(4)} = 0, \tag{3.21}$$

and that of the second one

$$\left.\frac{\partial^2 \mathcal{E}(\lambda)}{\partial \lambda^2}\right|_{\lambda=1} = (2-D)(1-D) \cdot \mathcal{E}_{(2)} + (4-D)(3-D) \cdot \mathcal{E}_{(4)}, \tag{3.22}$$

we find from Eq. (3.21) the necessary condition for a minimum as

$$\mathcal{E}_{(4)} = -(4-D)^{-1}(2-D)\mathcal{E}_{(2)}. \tag{3.23}$$

Substituting (3.23) into (3.22) the value for the second derivative is

$$\left.\frac{\partial^2 \mathcal{E}(\lambda)}{\partial \lambda^2}\right|_{\lambda=1} = 2(D-2) \cdot \mathcal{E}_{(2)}. \tag{3.24}$$

For $D \geq 3$ the last expression provides

$$\left.\frac{\partial^2 \mathcal{E}(\lambda)}{\partial \lambda^2}\right|_{\lambda=1} \geq 0$$

and thus these types of models admit solutions which are stable against scale transformations. Note that addition of the sixth-order chiral invariant (2.44) to the Lagrangian (3.19) also leads to a model with solutions stable against scale transformations. Besides the Skyrme-type Lagrangian (3.19), the scale-invariant solutions are available in models with the so-called "exotic" Lagrangian densities

$$\mathcal{L}_{3/2} = \left\{\mathrm{Tr}(\mathbf{L}_\mu \mathbf{L}^\mu)\right\}^{3/2}. \tag{3.25}$$

This fact can be easily checked by calculations similar to those we performed in (3.11). The value of power $\frac{3}{2}$ is chosen to meet the Hobart-Derrick requirements (Duff and Isham 1976), but there are problems with a physical interpretation for such "exotic" Lagrangians.

3.3 Soliton Stability and the Second Variation Structure of the Lyapunov Functional

3.3.1 The Lyapunov Stability of Solitons

The study of the second variation structure is the crucial point in the stability analysis itself and in particular, as applied to stationary soliton solutions. More general nonstatic soliton solutions are of the greatest importance for the problem of soliton quantization. Actually, the most developed method of soliton quantization is the quasiclassical method. This method is based on the consideration of small fluctuations in the neighbourhood of a given classical soliton solution or that of collective excitations of the rotational type. To have a well-defined quantization procedure, we must be sure of the stability of underlying classical solitons. Stability problems arise in the description of solitons interactions, as well.

We begin by introducing soliton solutions of stationary type which include as a special case solutions with harmonic dependence on time. It will henceforth be assumed that we are dealing with n-component fields φ, so that $\varphi = (\varphi_1, \varphi_2 \dots \varphi_n)$.

Definition 3.1. *A soliton solution* $\varphi_0(\boldsymbol{x}, t)$ *is called* **stationary**, *if it satisfies the following variational equations*

$$\frac{\delta \Phi}{\delta \dot{\varphi}} = 0; \qquad \frac{\delta \Phi}{\delta \varphi} = 0, \tag{3.26}$$

where $\dot{\varphi}$ *denotes the time derivative and* $\Phi[\varphi]$ *is additive translationally invariant functional of the form*

$$\Phi = \int d^3x\, F(\varphi, \dot{\varphi}, \partial_i \varphi), \quad i = 1, 2, 3. \tag{3.27}$$

To elucidate the Definition 3.1 let us consider an illustrative example.

Example 3.1. *Complex Scalar Fields with Harmonic Time-Dependence*
We take the complex scalar field $\varphi_0(\boldsymbol{x}, t)$ which is a solution of the field equation

$$\left(\frac{\partial^2}{\partial t^2} - \Delta\right) \varphi = f'(|\varphi|^2)\varphi \tag{3.28}$$

and has a structure with evident harmonic dependence on time:

$$\varphi_0(\boldsymbol{x}, t) = u(\boldsymbol{x}) \exp\{-\mathrm{i}\,\omega t\}. \tag{3.29}$$

It is not difficult to see that the function (3.29) is a critical point of the stationary type for the functional $\Phi = \mathcal{E} - \omega Q$, where $\mathcal{E}$ and Q are the energy and the charge for the field $\varphi(\boldsymbol{x}, t)$. By writing this functional in the explicit form

$$\Phi = \int d^3x \{|\dot{\varphi}|^2 + |\nabla\varphi|^2 + f(|\varphi|^2) - \mathrm{i}\,\omega(\varphi^*\dot{\varphi} - \dot{\varphi}^*\varphi)\}, \tag{3.30}$$

one can easily check that (3.30) is an additive, translationally invariant functional of the type (3.27). The Eqs. (3.26) for the functional (3.30) are

$$\begin{aligned} \frac{\delta\Phi}{\delta\dot{\varphi}^*} &= \dot{\varphi} + \mathrm{i}\,\omega\varphi = 0, \\ \frac{\delta\Phi}{\delta\varphi^*} &= -\Delta\varphi - \mathrm{i}\,\omega\dot{\varphi} + f'(|\varphi|^2)\varphi = 0, \end{aligned} \tag{3.31}$$

which really express standard relations between φ and $\dot{\varphi}$ for fields with a harmonic dependence on time.

As was emphasised in the previous section, the Hobart-Derrick theorem deals with static soliton solutions $\varphi_0 = u(\boldsymbol{x})$ for field equations of the type

$$\left(\frac{\partial^2}{\partial t^2} - \Delta\right)\varphi = f(\varphi)$$

and proves their instability. The proof was based on the apparent negativeness of the second variation (derivative) in some neighbourhood of static soliton solutions under scale perturbations $\delta\varphi = x^i\partial_i u$. To extend this criterion to more general classes of solutions we have to examine the structure of the second variation of the energy functional in the neighbourhood of stationary solitons. Technically this is a more difficult problem, as we have to deal with more general than scale perturbations and cannot use the scale parameter trick any more. Taking this obstacle into account and trying to keep the exposition as elementary as possible, here we only formulate central statements and results, together with main ideas of how those results may be obtained. For clarity, the main statements are given in the form of theorems but instead of rigorous proofs we prefer to illustrate them with explanatory simple examples. All the necessary rigorous proofs of theorems and technical details of derivations can be found in Appendix D.1.

It may be useful to give here the definition of soliton stability in the Lyapunov sense, the concept of stability with respect to shape which proves to be most suitable for soliton problems. Let $\varphi_0 = u$ be a stationary (periodic in time) soliton solution whose size is supposed to be localized in the domain of unity. This signifies that u itself and its derivatives are decreasing fast enough as $r = |\boldsymbol{x}| \to \infty$, e.g. so that

$$|\nabla u| = \mathcal{O}(r^{-3/2+\alpha}), \quad \alpha > 0. \tag{3.32}$$

As a preparatory step, following the papers (Movchan 1960, Kumar, Nisichenko and Rybakov 1979, Rybakov and Chakrabarti 1984; Makhankov 1990) we introduce two functional metrics $\rho_0[\xi^0(\boldsymbol{x})]$ and $\rho[\xi(\boldsymbol{x},t)]$ for the characterization of the initial perturbations $\xi^0 \equiv \varphi(\boldsymbol{x},0) - u(\boldsymbol{x},0)$ and the current perturbations $\xi \equiv \varphi(\boldsymbol{x},t) - u(\boldsymbol{x},t)$, respectively. Putting $\xi = \xi_1 + \mathrm{i}\,\xi_2$, $\xi_i^* = \xi_i$, and denoting the norms in Hilbert space $L_2(\mathbb{R}^3)$ and Sobolev space $W_2^1(\mathbb{R}^3)$ by $\|\cdot\|$ and $\|\cdot\|'$ respectively[2], we choose the metrics ρ_0, ρ in the form

$$\rho_0[\xi^0] = \sum_{i=1}^{2}\{\|\dot{\xi}_i^0\| + \|\xi_i^0\|'\}, \quad \rho[\xi] = \inf_{u\in U}\sum_{i=1}^{2}\{\|\dot{\xi}_i\| + \|\xi_i\|\}, \tag{3.33}$$

where $\dot{\xi}_i \equiv \partial_0\xi_i$ and U denotes the set of functions obtained from the stationary solution u of the type e.g. (3.29) by means of 3-translations, 3-rotations and gauge transformations $u \to u \exp i\alpha$, $\alpha = \text{const}$. The function $\varphi \notin U$, by definition, describes the perturbed soliton. We see that the metric ρ is continuous with respect to the metric ρ_0 and $\rho[\xi] = 0$ if $\varphi \in U$.

Definition 3.2. *The regular soliton solution u is said to be* **stable in the Lyapunov sense** *with respect to the metrics ρ_0, ρ, if for each $\epsilon > 0$ there exists a number $\delta(\epsilon) > 0$ such that from $\rho_0[\xi^0] < \delta$ it follows that $\rho[\xi] < \epsilon$ for any $t > 0$.*

According to the general theorem in the theory of stability (Zubov 1957; Malkin 1962) for stability of the soliton solutions with respect to the metrics ρ_0, ρ, it is necessary and sufficient that there exists a Lyapunov's functional $V[\varphi]$ with the following properties:

1. The soliton solution u is a stationary point of $V[\varphi]$.
2. $V[\varphi]$ is continuous with respect to the metric ρ_0.
3. $V[\varphi]$ is positive definite with respect to the metric ρ in some neighbourhood of u.
4. $V[\varphi]$ does not increase along the trajectories of motion.

Now we are equipped for the study of more delicate questions in the stability of stationary solitons. The following material can be ommitted in the first reading by those readers who are already convinced of the validity of statements in Section 3.1. They may pass directly to Section 3.4, where we continue remarks of a general character on the topological stability of Skyrmions.

3.3.2 The Generalized Hobart-Derrick Theorem

The study of stability of stationary solitons in scalar and spinor models (Rybakov 1966, 1979) shows that *absolutely stable stationary solitons cannot exist.* In $D > 1$ case those results can be readily seen from the following lemma of variational calculus.

Lemma 3.1. Let the functional

$$V[\phi] = \int d^3x\, \mathcal{V}(\phi, \nabla\phi) \tag{3.34}$$

[2] Recall, that by definition $\|\xi\|' = \|\nabla\xi\| + \|\xi\|$.

be defined on the class of sectionally smooth functions $\phi(\boldsymbol{x}) : \mathbb{R}^3 \to \mathbb{R}^n$ with $\phi(\infty) = 0$. Let $u(\alpha; \boldsymbol{x})$ be the family of its extremal fields given by the parameters $\alpha = \{\alpha_i\}$. If there exist constants C_i, not all equal to zero, such that

$$f \equiv \sum_i C_i \left(\frac{\partial u}{\partial \alpha_i}\right)\Big|_{\alpha=0} = 0 \tag{3.35}$$

on some surface S separating in $\mathbb{R}^3$ a domain Ω of nonzero measure, then $\delta^2 V$ is sign changing in the neighbourhood of $u(0; \boldsymbol{x})$.

Proof. Consider the perturbations $\delta\phi$ such that $\delta\phi = 0$ for $\boldsymbol{x} \in \Omega$ and $\delta\phi = f$ for $\boldsymbol{x} \notin \Omega$. Then due to properties of extremal fields $\delta^2 V[\delta\phi] = 0$. However, the perturbation $\delta\phi$ is not the extremal point of the functional $\delta^2 V$, as it violates the Weierstrass-Erdmann matching condition (see e.g. Gelfand and Fomin 1963). Hence, $\delta^2 V$ is a sign changing. □

Thus, if we suppose a scalar soliton solution u to be stable, then in accordance with Definition 3.2 and general stability requirements there must exist a positive definite functional $V[\phi]$ for which u is the extremal field. If the model is invariant under 3-translations $\boldsymbol{x} \to \boldsymbol{x} + \boldsymbol{a}$, where $\boldsymbol{a} = \text{const}$ then $u(\boldsymbol{x} + \boldsymbol{a}, t)$ is also the extremal field for $V[\phi]$. However, the equation $\partial_x u = 0$ can be satisfied on some surface S because $u(\boldsymbol{x}, t)$ is by assumption a regular solution and $u(\infty, t) = 0$. Hence according to the Lemma 3.1 $\delta^2 V$ is sign changing, which contradicts the stability of φ and proves the following theorem (Rybakov 1979, 1991):

Theorem 3.1. The regular soliton solutions are unstable, in the Lyapunov sense with respect to the metric $\rho = \rho_0$, in any model with the Lagrangian density

$$\mathcal{L} = -F(p, s); \qquad p = -\partial_\mu \varphi^* \partial^\mu \varphi; \qquad s = |\varphi|^2. \tag{3.36}$$

Therefore it is reasonable to consider *the conditional stability of solitons*, that is, to impose some subsidiary conditions on initial perturbations ξ_0, or equivalently, to choose some special second metric $\rho[\xi]$. This procedure seems to be inevitable to provide the stability of solitons. The general character of the fact that in any realistic three-dimensional model only conditionally stable solitons can exist is emphasised by **the Generalized Hobart-Derrick Theorem** (Rybakov 1984, 1991; Rybakov and Chakrabarti 1984), with the proof given in Appendix D.1.

Theorem 3.2. If a stationary soliton solution $\varphi_0 = u$ is the critical point of the functional (3.27), then its second variation $\delta^2 \Phi$ is sign changing at the point φ_0.

The idea of the proof in brief is as follows. We assume the stability of the solution u and then demonstrate that the additive Lyapunov functional with the properties mentioned above (requirements 1. – 4.) does not exist. In fact in a translationally invariant theory, given by Φ from Eq. (3.27), the additive Lyapunov functional should have the form

$$V[\varphi] = \int d^3x\, \mathcal{V}(\varphi, \dot{\varphi}, \partial_i\varphi), \quad i = 1, 2, 3, .$$

In order to check whether this functional meets the requirements, in close analogy with the Hobart-Derrick theorem, it is enough to examine the positive definiteness of the second variation $\delta^2 V[\varphi]$. Already this test proves that $\delta^2 V[\varphi]$ is sign changing and thus contradicts the previous assumption on stability of the soliton solution u.

Corollary 3.1. In a translationally invariant theory the stationary solitons cannot be absolutely stable in the Lyapunov sense with respect to the metric $\rho = \rho_0$. There could only be conditionally stable solitons with some restrictions on the form of the initial perturbations.

3.3.3 The $\mathcal{Q}$-stability of Solitons

One of the simplest conditions that can be imposed on the initial perturbations ξ^0 is the condition of charge fixation, which for the model (3.36) can be taken in the form:

$$\mathcal{Q} = \frac{\mathrm{i}}{2} \int d^3x\, F_p(\varphi^* \partial_0 \varphi - \partial_0 \varphi^* \varphi) = \mathcal{Q}[\varphi_0] \equiv \mathcal{Q}_0, \tag{3.37}$$

where $F_p = \partial F / \partial p$. The stability under the condition (3.37) is known under the name of **$\mathcal{Q}$-stability** (Makhankov 1977, 1978).

The simplest way to establish sufficient criteria for $\mathcal{Q}$-stability is to consider non-nodal regular solutions for the model (3.36) in the form

$$\varphi_0(\boldsymbol{x}, t) = u(\boldsymbol{x}, t) \exp\{-\mathrm{i}\,\omega(t)\}, \quad u^* = u. \tag{3.38}$$

Another consequence, which one may obtain from Lemma 3.1 is that corresponding nodal solutions are $\mathcal{Q}$-unstable. The Lyapunov functional might be chosen in the form

$$V = E - \omega\, \mathcal{Q}, \tag{3.39}$$

where E is the field energy. The second variation of V can be written as

$$\delta^2 V = \left(\dot{\xi}_1, F_p \dot{\xi}_1\right) + \left(\dot{\xi}_2, (F_p - 2\, F_{pp} \omega^2 s) \dot{\xi}_2\right) + \sum_{i=1}^{2} \left(\xi_i, \hat{L}_i \xi_i\right), \tag{3.40}$$

where $(\cdot, \cdot)$ denotes the scalar product in $L_2(\mathbb{R}^3)$ and the Hermitian operators $\hat{L}_i$, $i = 1, 2$, have the form

$$\hat{L}_1 = \hat{L}_2 - 2\mathrm{div}\left[F_{pp}\nabla u(\nabla u\nabla)\right] + \mathrm{div}\left(F_{pp}\omega^2 - F_{ps}\right)\nabla s + 2s\left(F_{ss} - 2\omega^2 F_{ps} + \omega^4 F_{pp}\right),$$

$$\hat{L}_2 = -\,\mathrm{div}F_p\nabla - \omega^2 F_p + F_s.$$

It is clear from (3.40) that if $\delta^2 V$ is positive definite, then

$$F_p > 0, \qquad h \equiv F_p - 2F_{pp}\omega^2 s > 0, \qquad F_p + 2F_{pp}(\nabla u)^2 > 0. \tag{3.41}$$

In the approximation linear with respect to ξ, the fixed charge condition (3.37) can be written as

$$\left(hu, \dot{\xi}_2\right) = 2\omega\left(g, \xi_1\right) \tag{3.42}$$

where

$$g \equiv -\mathrm{div}\left(F_{pp}s\nabla u\right) + u\left(F_p - \omega^2 sF_{pp} + sF_{ps}\right).$$

The usual method to study the structure of the second variation in the case when particular solutions are unknown is the method of variational estimates. Therefore from Eq. (3.42), using Schwartz's inequality, we obtain

$$\left(\dot{\xi}_2, h\dot{\xi}_2\right) \geq 4\omega^2\left(g, \xi_1\right)^2\left(u, hu\right)^{-1}. \tag{3.43}$$

From Eq. (3.40) and Eq. (3.43) we deduce the estimate for $\delta^2 V$:

$$\delta^2 V \geq \left(\dot{\xi}_1, F_p\dot{\xi}_1\right) + \left(\xi_2, \hat{L}_2\xi_2\right) + \left(\xi_1, \hat{K}\xi_1\right) \equiv W\left[\dot{\xi}, \xi\right], \tag{3.44}$$

where

$$\hat{K}\xi_1 = \hat{L}_1\xi_1 + 4\omega^2 g\left(g, \xi_1\right)\left(u, hu\right)^{-1}. \tag{3.45}$$

Now we have to find the conditions on W to be positive definite with respect to the metric ρ, which can be enlarged by including $\| \dot{\xi}_1 \|$. Note that due to the field equations we have $\hat{L}_2 u = 0$ and, therefore can apply Courant's theorem (Courant and Hilbert 1953)

Theorem 3.3. The first eigenfunction of a self-conjugate elliptic operator of the second order is a non-nodal function and the corresponding eigenvalue is a non-degenerate one.

The spectrum of $\hat{L}_2$ will be positive since $u > 0$, and the zero mode is excluded here according to the definition of ρ, because for $\xi_2 = u \in U, \rho[u] = 0$. Further, for the spectrum of $\hat{K}$ to be positive, it is necessary that $\hat{L}_1$ have no more than one negative eigenvalue, because in the opposite case $\left(g, \xi_1\right) = 0$ can always be attained for $\left(\xi_1, \hat{L}_1\xi_1\right) < 0$.

Let $\lambda(\omega)$ be the first eigenvalue of $\hat{K}$. According to Theorem 3.1, $\lambda(0)$ is always negative. Let us find the critical frequency ω_0 for which $\lambda(\omega_0) = 0$ and which defines the boundary of the domain of Q-stability $\omega > \omega_0$, if $\lambda(\omega) \geq 0$.

Owing to the evident symmetry $\omega \to -\omega$, it is sufficient to consider only positive frequencies $\omega > 0$. As

$$\operatorname{sgn} \min_{\rho=\epsilon} \delta^2 V = \operatorname{sgn}(\omega - \omega_0),$$

it is possible to say that $u(\omega_0)$ is the saddle point of V with the curve of descent $u(\omega)$. Thus, for the point $\omega = \omega_0$ we have

$$\min_{\rho=\epsilon} \delta^2 V = \min_{\rho=\epsilon}(\xi_1, \hat{K}\xi_1) = 0$$

and this minimal point is achieved when $\xi_1 = u_\omega \equiv \partial u/\partial\omega$. Hence $\hat{K}(u_\omega) = 0$, which along with the equation $\hat{L}_2 u = 0$ leads to the following equation for ω_0 (Zastavenko 1965, Rybakov 1966):

$$\mathcal{Q}_{0\omega} \equiv \frac{d}{d\omega}\left(\omega \int d^3x\, F_p s\right) = 0. \tag{3.46}$$

If $\omega > \omega_0$, then $(u_\omega, \hat{K}u_\omega) > 0$ or $\mathcal{Q}_{0\omega}(\mathcal{Q}_{0\omega} - (u, hu)) > 0$, whence with the help of Eq. (3.46), we get the inequality to determine the domain of $\mathcal{Q}$-stability (Vakhitov and Kolokolov 1973; Friedberg, Lee and Sirlin 1976, Makhankov 1977):

$$\mathcal{Q}_{0\omega} < 0. \tag{3.47}$$

Note, once again, that the zero modes of the type $\xi_i = C_i\partial_i u$ are excluded according to the definition of the metric ρ. The obtained results can be summarized in **the $\mathcal{Q}$-stability theorem:**

Theorem 3.4. Non-nodal regular solutions (3.38), in the model (3.36), are $\mathcal{Q}$-stable and the domain of $\mathcal{Q}$-stability is determined by the inequality (3.47), if the following conditions hold:

a. The first eigenfunction ψ of operator $\hat{L}_1$ satisfies the condition $(g, \psi) \neq 0$;
b. The operator $\hat{L}_1$ has at most one negative eigenvalue.

The generalization of this theorem for multiple-charged solitons is given in (Rybakov and Chakrabarti 1984). Here we would rather take an illustrative example to explain in a simpler way some concepts and methods, used in the previous derivation, and also to make the reader familiar with their practical implementation. One can find more systematic expansion of the Lyapunov direct method and its applications to solitons' stability analysis in the recent review (Rybakov, 1991) with special emphasize to this topic.

Example 3.2. *The $\mathcal{Q}$-Stable Solitons in Synge's Model*
This model of the type (3.36) is specified by the choice of F in the form

$$F = p + s - \frac{s^n}{\ }. \tag{3.48}$$

For $1 < n < 3$ and $|\omega| < 1$, there exist a spherically symmetric regular solution (3.38). Changing the variables

$$|\boldsymbol{x}| = \rho(1-\omega^2)^{1/2}, \qquad u = v(1-\omega^2)^{1/2(n-1)}, \tag{3.49}$$

we get

$$\mathcal{Q}_0(\omega) = \text{const}\,\omega(1-\omega^2)^{(5-3n)/2(n-1)}. \tag{3.50}$$

Therefore the $\mathcal{Q}$-stability condition (3.47) is fulfilled only if $n < 5/3$. Let us take $n = 3/2$, which corresponds to the so-called Synge model (Synge 1961). The operator $\hat{K}$ has the following structure:

$$\hat{K} = \hat{L}_1 + 4\omega^2 P_u, \qquad \hat{L}_1 = -\Delta - \omega^2 + 1 - 2u,$$

with P_u being the projector on $u/\| u \|$.

Now let us verify that the operator $\hat{L}_1$ has only one negative eigenvalue. As from (3.49),

$$\hat{L}_1 = (1-\omega^2)(-\Delta + 1 - 2v)$$

consider the equation

$$\left[-\Delta + 1 - 2v(\rho)\right]\chi = \nu\chi. \tag{3.51}$$

Separating the angular variable in Eq. (3.51): $\chi_{lk} = R_l(\rho)\cdot Y_{lk}$, where Y_{lk} are spherical harmonics, we reach the conclusion that for $l = 1, R_1 = dv/d\rho$ and $\nu = 0$. Further, as $v(\rho)$ is monotonic, R_1 has no internal zeros and hence, according to Theorem 3.3, $\nu = 0$ is the lowest eigenvalue for $l \neq 0$. Therefore ν is negative only for s-states. According to Sturm's theorem (Tricomi 1961) the number of s-states with $\nu < 0$ is equal to the number of internal zeros of the solution to the equation

$$\left[-\frac{d^2}{d\rho^2} + 1 - 2v(\rho)\right] y(\rho) = 0,$$

if $y(0) = 0, y'(0) = 1$. Numerical calculations show that $y(\rho)$ has only one internal zero for $\rho \approx 1.32$. Thus we see that all conditions of Theorem 3.4 are fulfilled. The domain of $\mathcal{Q}$-stability is $1 > |\omega| > 2^{-1/2}$.

3.4 The Topological Stability of Skyrmions

Chiral models with Lagrangians composed at least from chiral invariants of the second and of the fourth order in currents $\mathbf{L}_\mu$ (see Eq. (2.40) and Eq. (2.41)) have one more attractive feature. The energy of a static field configuration in such models can be estimated from below by a positive definite function of a topological charge $\mathbf{Q}$ from Eq. (1.37), which is the third order combination of the chiral currents $\mathbf{L}_\mu$ themselves. In Section 1.4.3 we have seen how this type of estimate Eq. (1.43) can be deduced on the basis of the triangle inequality for vectors. The other way to obtain such an estimate is based on the observation that $\mathbf{L}_i$ and $\epsilon_{ijk}[\mathbf{L}_j, \mathbf{L}_k]$ are antihermitian matrices. The well-known property of any antihermitian matrix $\mathbf{A}$ is that $\operatorname{Tr}\mathbf{A}^2 \leq 0$. Presenting the energy of the model $\mathcal{E}[\phi] = \mathcal{E}_{(2)}[\phi] + \mathcal{E}_{(4)}[\phi]$ in the form of a perfect square plus the topological charge

$$\mathcal{E} > -\int d^3x \operatorname{Tr}\left(\frac{1}{2\lambda}\mathbf{L}_i - \frac{\epsilon}{4\sqrt{2}}\epsilon_{ijk}[\mathbf{L}_j, \mathbf{L}_k]\right)^2 + 6\sqrt{2}\pi^2\frac{\epsilon}{\lambda}|\mathbf{Q}| \geq 0 \qquad (3.52)$$

we again find the estimate (1.43).

Skyrme's suggestion (1961) on the stability of hedgehog configurations (see Section 1.5) is in fact guaranteed by the estimate (1.43), which means that in any homotopy class the energy of field configurations has a lower bound. If this lower bound is attainable on a solution of field equations, then such a solution is stable in the Lyapunov sense. There are great difficulties in dealing with these problems by the standard methods of stability analysis. The existence of estimates for the energy functional through the use of topological invariants incredibly simplifies the problem. In the previous section we demonstrated by means of the generalized Hobart-Derrick theorem that stable solitons might exist only under some restrictions imposed on the initial perturbations of the fields. It is apparent that in the case of topological solitons, due to the topological classification of fields already discussed (see Chapter 2), those restrictions are natural. As a consequence, for field models with topological charges the results listed in the previous section might be applied. In view of this we summarize our previous consideration in the following:

Proposition 3.1. The availability of an attainable lower bound of energy, estimated from below by a positive definite function of a topological charge $\mathbf{Q}$, is sufficient for the stability of solitons in topologically nontrivial models.

Remark 3.1. One may find this proposition to contradict the statement of Theorem 3.2. The topological charge $\mathbf{Q}$ by definition does not change its value under continuous deformations so that its fixation must not impose any restrictions on initial perturbations. But this is the case only with respect to infinitesimally small perturbations, continuously deformed to zero, which do not take the solution out of the given homotopy class. On the other hand, by finite perturbations the value of the topological charge can be changed and those perturbations are not prohibited in Theorem 3.2. Remember that in this theorem we study the stucture of the second variation, which in accordance with Eq. (3.2) is a homogeneous functional. Thus fixing the topological charge substantially determines the sign-definiteness of the second variation for the Lyapunov functional. The aformentioned contradiction is illusory.

This kind of stability gained the special name **Topological Stability**. When an estimate of the type (1.4.3) is exact it was shown by Bogomol'nyi (1976) that it is possible to reduce the Euler-Lagrange equation for a given energy functional to a first order (self-duality) equation. The method used resembles that of reduction of a quadratic form to a perfect square, discussed in Section 3.1. To see the reason why this kind of trick does not work for Skyrme's functional we can use Eq. (3.52), where the perfect square has been already found out. From this expression it is obvious that the lower bound would be saturated if the first term in the middle part of Eq. (3.52) were to vanish identically. Then we have the corresponding first-order equations

$$\mathbf{L}_i = \frac{\lambda\epsilon}{2\sqrt{2}}\epsilon_{ijk}\,[\mathbf{L}_j, \mathbf{L}_k], \tag{3.53}$$

which are analogous to the self-duality ones for instantons. If there were solutions of Eq. (3.53) which at the same time were solutions of Eq. (1.44) with boundary conditions (1.31) then the problem of finding solutions in the Skyrme model could be drastically simplified. Unfortunately, this is not the case. Taking into account the integrability conditions (1.36), it is readily seen that Eq. (3.53) implies $\boldsymbol{L}_i = (\lambda\epsilon/2\sqrt{2})[\text{curl}\,\boldsymbol{L}]_i$, whence $\boldsymbol{L}_i \equiv 0$. From this we conclude that the absolute minimum of the Skyrme functional (if any) exceeds the lower bound given by Eq. (1.43) or (3.52). Keeping in mind this result, in what follows we use the common term "lower bound" in this sense.

Thus to prove the Skyrmion stability we have to use different methods to show that the lower bound, the existence of which is supported by the estimate (1.43), can be attainable. As will be demonstrated in Part II by means of direct methods of the calculus of variations, the attainability of the lower bound can be rigorously proved in the first homotopy class or for configurations with unit topological charge. For configurations from higher homotopy classes the available answer is not so definite at this time.

Part II

Methods for the Study of Skyrmions

4. The Principle of Symmetric Criticality

"My work always tried to unite the true with the beautiful, but when I had to choose one over the other, I usually chose the beautiful."

Hermann Weyl

4.1 Some Auxiliary Information

The theory of solitons along with the further development of the beautiful concept of complete integrability for (1+1)-models brought into focus the problem of analytical investigations in higher dimensional field theories. In spite of frequent efforts to extend the complete integrability or a similar concept to the realistic (3+1)-dimensional case, reported results are still far from successful. On the other hand the appearance of more powerful computers led to visible progress in the lattice approach in field theory. The latter caused a widespread belief that it is possible to solve almost any problem even in such complicated theories as QCD or the Standard Model, when using a 'clever enough' computerized lattice complex. There are, however, already some indications that the reality is not so straightforward. Definitely now it is not yet the time when solutions of physical problems may be reduced to an electronic engineering problem of the type of how to construct a needed lattice. Computer simulations can provide us with a reliable answer only in the case when a likelihood description (a "scenario") of physical phenomena is available. From that point, attempts to extract nucleon observables from direct lattice calculations in the QCD are similar to 'predictions of a match score on the basis of the rules of the game only.' As an example in support of this opinion, we have a relative failure to get the correct value for the gluonic contribution in the "proton spin" puzzle directly from the lattice QCD calculations (Jaffe and Manohar 1990; Mandula 1990). From the general logic it is doubtful that any extreme position could ever be acceptable and both analytical and lattice approaches must supplement each other. Having this in mind, beginning with this chapter we try to give an answer as to what in particular one can expect from already available analytical methods in the study of multi-dimensional solitons.

Being interested first of all in the Skyrme-type models we must keep in mind that in general, analytical study of such field theories is a complicated task for purely mathematical reasons. Actually, we have to deal with such discomfortable functionals as (1.39) or (1.42). They have as arguments the fields with a non-compact domain on one hand and are nonconvex with respect to derivatives of fields, on the other. For such kinds of functionals the standard direct methods of the calculus of variations are not applicable, and an appropriate mathematical machinery is at an initial stage of development. The further enumeration of mathematical difficulties, arising in that case, can be found in (Kapitanskii and Ladyzhenskaya 1983). Here we would rather concentrate on what it is possible to do in spite of those difficulties.

There are known attempts to modify in different ways the original Skyrme model in order to eliminate some troubles. For example in (Manton and Ruback 1986; Manton 1987) the model was considered based on the Skyrme Lagrangian, but with the fields living in a compact domain (the $\mathbf{S}^3$-manifold). It turns out that this modification definitely could not be regarded as a model of the baryon, but rather as a model of condensed nuclear matter[1]. We discuss this modification a bit further in Section 11.3.3. There are also methods which endeavor to extend the initial variational problem so that the resulting functional becomes a convex one (Ekeland and Temam 1976; Ioffe and Tikhomirov 1979). Whether those methods can be applied to the Skyrme functional is to date an open question. Thus in both approaches we see attempts to improve the variational problem itself. On the contrary, in the approach suggested by P. Lions (1984, 1985) one may find an attempt to develop direct variational methods in such a way that they would be applicable to the functionals under question. This approach is based on the so-called concentration-compactness principle. In application to the Skyrme model those methods were used in (Esteban 1986) to prove the existence of Skyrmions, under some rather restrictive conditions. Here we are not going to discuss those promising developments any more, and refer the interested readers to the cited literature.

In the traditional Lagrangian formalism approach in search for extremals or critical points for the given functional we must write the Euler-Lagrange equation and then by one method or another seek for admissible solutions. The Skyrme model Eqs. (1.44) are nonlinear partial differential equations and to date there is no available procedure to handle them directly. Thus we must look for eventual simplifications. One of the first ideas that comes to mind is to look for symmetries of the functionals under study. This idea proved to be a good one long ago. In particular, Weyl applied this symmetric approach in the derivation of the Schwarzschild solution of the Einstein gravitational field equations (Weyl 1951). In brief the assertion used was, that **extremals of symmetric functionals (if any) are realizable in the class of invariant functions.** Difficulties in analytical study of multi-dimensional solitons led to further development and better understanding of this idea. It was emphasized by Coleman in Appendix 4 to his lecture notes (Coleman 1977), that the problem of searching for multi-dimensional solitons can be drastically simplified if a variational principle $\delta S = 0$ is invariant under some symmetry group G. The obvious advantage is the contraction of the range of search for extremals to the class of fields, which are invariant under transformations from G. In Section 4.2 we give a more large-scale exposition of those statements, known in the

[1] It is remarkable that this modification from a general point of view was first suggested in (Dubrovin, Novikov, and Fomenko 1979).

literature as **The Principle of Symmetric Criticality** or **the Coleman-Palais Theorem**. For that purpose we use the fundamental article (Palais 1979) and our previous publications (Makhankov, Rybakov and Sanyuk 1989, 1992). To proceed we find it reasonable to provide the reader with some information on the symmetry properties of the Skyrme model energy functionals.

4.2 The Symmetry Group of the Skyrme Energy Functional

Let us start with the apparent symmetries of functionals in the Skyrme model. In accord with Section 2.4.1, terms in the Lagrangian (1.39) are both relativistic and chiral invariant by construction. It means that the most general symmetry group of the Skyrme model action might be the direct product of the Poincaré group (geometrical symmetries) and the $SU(2)_L \otimes SU(2)_R$ chiral group (internal symmetries). This group, unfortunately, is too wide and proves to be useless in the search for invariant field configurations. Anticipating once again the conclusions of the next section, let us restrict ourselves to the search for the maximal compact subgroup of the aforementioned general group. Still another limitation is that throughout this chapter we are only dealing with continuous symmetries. The further restrictions on admissible symmetry transformations come from the following observation:

Proposition 4.1. Since the Hamiltonian **H** is quadratic in the time derivatives of fields, Inf **H** is realizable in the class of static fields.

The proof of this proposition is trivial and comes, for example, from the direct minimization of the Hamiltonian (1.42) with respect to time derivatives of fields. The reader may take the checking of this statement as a simple exercise. For the sake of completeness the calculations of the proof are given in Appendix C.1. Thus in searching for Inf **H** we can restrict ourselves to static field configurations:

$$\partial_0 \phi_\rho = 0, \quad \text{or} \quad \partial_0 n^a = \partial_0 \Theta = 0, \quad \text{or} \quad \mathbf{L}_0^a = 0. \tag{4.1}$$

Taking into account conditions (4.1) and introducing the polar variables (β, γ) for a parametrization of the unit vector $\underline{\mathbf{n}}$ from Eq. (1.29):

$$n^1 + \mathrm{i} n^2 = \sin\beta(\boldsymbol{x}) \exp\{\mathrm{i}\gamma(\boldsymbol{x})\}; \quad n^3 = \cos\beta(\boldsymbol{x}), \tag{4.2}$$

one may rewrite the functional **H** from (1.42) in the following form:

$$\begin{aligned} \mathbf{H} = \int d^3x \Big\{ & \frac{1}{2\lambda^2} [(\nabla\Theta)^2 + \sin^2\Theta((\nabla\beta)^2 + \sin^2\beta(\nabla\gamma)^2)] \\ & + \epsilon^2 \sin^2\Theta\{(\nabla\Theta \times \nabla\beta)^2 + \sin^2\beta(\nabla\Theta \times \nabla\gamma)^2 \\ & + \sin^2\Theta \sin^2\beta(\nabla\beta \times \nabla\gamma)^2\}\Big\}. \end{aligned} \tag{4.3}$$

The properties of this functional[2] will be the main subject of interest throughout Part II. The corresponding expression for the topological charge $\mathbf{Q}$ from (1.37) would be

$$\mathbf{Q} = -\frac{1}{2\pi^2}\int d^3x \sin^2\Theta \sin\beta\{\nabla\Theta\cdot(\nabla\beta\times\nabla\gamma)\}. \tag{4.4}$$

To proceed further with the symmetry analysis we might consider as candidates for a compact invariance group of the energy functional (4.3) the direct product of the group of spatial rotations $SO(3)_S$ with the chiral group, namely:

$$\bar{G} = SO(3)_S \otimes SU(2)_L \otimes SU(2)_R, \tag{4.5}$$

with action on chiral fields (1.27) or (1.29) defined by

$$\mathbf{U}(\boldsymbol{x}) \to V\cdot\mathbf{U}(\hat{R}\cdot\boldsymbol{x})\cdot W^{-1}, \tag{4.6}$$

where $\hat{R}\in SO(3)_S$, V and W are arbitrary $SU(2)$-matrices. With allowance for the boundary conditions (1.31) the vacuum state $\mathbf{U} = \mathbb{1}$ is invariant under internal transformations in (4.6) if and only if $V = W$. Thus the internal part of the symmetry group (4.5) should be further restricted to the diagonal subgroup

$$G' = \mathrm{diag}[SU(2)_L \otimes SU(2)_R] \simeq SU(2)_I \simeq SO(3)_I\otimes \mathbb{Z}_2 \tag{4.7}$$

where the latter notation is used for the rotation group in the isospin space. Still further restrictions on the remaining group of symmetry $\tilde{G} = SO(3)_S\otimes SO(3)_I$ follow from the requirement that invariant fields must belong to a nontrivial homotopy class. Then **the change of the field under a spatial rotation must be compensated by an isospace transformation** (Romanov, Frolov and Schwarz 1978). The maximal symmetry group that admits nontrivial invariant fields is related in this case to an automorphism $\eta : SO(3)_S \to SO(3)_I$. This makes it possible to identify rotations in $\mathbb{R}^3$ and those in the isospace and instead of $\tilde{G}$ to consider

$$G = \mathrm{diag}[SO(3)_I \otimes SO(3)_S]. \tag{4.8}$$

The mutual orientation of the basis in the space $\mathbb{R}^3$ and the basis of Pauli matrices τ_i in isospace is determined up to a 3-rotation $\hat{R}$, which specifies the automorphism η.

To be sure that all essential continuous symmetries of the functional (4.3) are taken into account (Kundu, Rybakov and Sanyuk 1979) we must prove the following:

[2] Functionals with a similar structure are commonly used in the nematic liquid crystals theory, in ferromagnets models, nonlinear elasticity etc. This gives us hope that methods exposed here can also be found effective in application to the abovementioned nonlinear models.

Proposition 4.2. The group (4.8) is the maximal compact group of transformations admitted by the functional (4.3) for which the invariant fields belong to a nontrivial homotopy class or satisfy the condition $\mathbf{Q} \neq 0$.

Proof. Let us write the general definition of fields $\phi^o(\boldsymbol{x})$, invariant (more precisely, equivariant[3]) against transformations from a group G:

$$\phi^0(\boldsymbol{x}) = \phi^0_g(\boldsymbol{x}) = T_g \phi^0\,(g^{-1} \cdot \boldsymbol{x}); \quad g \in G, \tag{4.9}$$

where T_g is an operator in a representation space of group $G \ni g$. Denoting by $\hat{\mathbf{J}}_i$ and $\hat{\mathbf{I}}_i$ generators of spatial and isospatial rotations, respectively, we can rewrite Eq. (4.9) in a more precise form:

$$[\,\hat{\mathbf{J}}_i \;+\; \hat{\mathbf{I}}_i\,]\underline{\phi}^0 \;=\; 0; \quad i \;=\; 1,2,3. \tag{4.10}$$

If now, we try to at least slightly expand the group (4.8), then the invariant field $\underline{\phi}^0$ must also satisfy the equation

$$[\,\lambda \cdot \hat{\mathbf{J}}_k \;+\; \hat{\mathbf{I}}_k\,]\underline{\phi}^0 \;=\; 0; \quad \lambda \;=\; \text{const} \neq 1, \quad k - \text{fixed}, \tag{4.10a}$$

Substracting (4.10a) from (4.10), we come up to the condition

$$\underline{\hat{\mathbf{J}}}_k\, \phi^a \;=\; 0, \tag{4.10b}$$

which is just the condition on vector field ϕ^a to have poloidal structure[4]. Now if we put $k = 3$ in (4.10b), then for fields in the parametrization (4.2) it readily implies $\nabla\gamma = 0$, and thus from the expression for topological charge (4.4) we have $\mathbf{Q} = 0$. It is apparent that any other extension leads to the same result. □

Remark 4.1. Two symmetry groups are of special interest in light of our discussion:

$$\begin{aligned} G_1 &= \text{diag}[SO(3)_I \otimes SO(3)_S], \\ G_2 &= \text{diag}[SO(2)_I \otimes SO(2)_S]. \end{aligned} \tag{4.11}$$

We claim that groups G_1 and G_2 are the only maximal compact continuous groups which possess the equivariant fields with nontrivial topological charges in realistic (3+1)-dimensional models. The fields invariant against transformations from G_1 are spherically-symmetric fields of the hedgehog type; those invariant under the group G_2 form axisymmetric configurations.

---

[3] Given a mapping ϕ from a manifold X to a manifold Φ one says that it is **equivariant** with respect to a group G, the latter is supposed to act on X and on Φ as well, if the relation

$$T_g \phi(x) = \phi(g \cdot x); \quad x \in X, \quad \phi \in \Phi$$

holds. It is clear that for $T_g = \mathbb{1}$ the field ϕ would be an invariant one.

[4] Vector fields are called **poloidal** when their field lines lay completely in meridional planes (remember the electric field lines of a dipole).

In what follows we deal largely with the symmetry group G_1 and corresponding hedgehog configurations. The group G_2 together with axisymmetric field configurations will be discussed in Chapter 7 in relation to multi-baryon field configurations in the Skyrme model and existence of corresponding invariant configurations would be proved in Appendix D.2.

4.3 The Coleman-Palais Theorem

Frequently there are implicit appeals to the principle of symmetric criticality in mathematical and physical papers, as to a guiding principle in solving different problems. For a long time it was a sort of folklore, which was used without any rigorous formalization[5]. Dealing for example with a rotationally invariant Lagrangian, it was conceived as quite natural to look for rotationally symmetric extremals. In order to exploit this widespread conviction we start from mostly intuitive (and sometimes not quite valid in complete generality) assertions and then step by step specify formulations together with their range of validity. The reader with good mathematical skills might omit without any loss most of this section and go ahead to the final mathematically correct formulation of the principle due to (Palais 1979).

As the departure point we take Coleman's practical recipe of searching for ansätze in nonlinear problems (Coleman 1977). This recipe was suggested on the basis of the following simple observation. Let G_0 be a subgroup of the symmetry group G of the functional $\Phi[\phi]$; then we can divide any field $\phi(\boldsymbol{x})$ into invariant and noninvariant parts with respect to transformations from G_0:

$$\phi(\boldsymbol{x}) = \phi_{\text{inv}}(\boldsymbol{x}) + \phi_{\text{noninv}}(\boldsymbol{x}). \tag{4.12}$$

After the substitution of this decomposition into Φ and simple manipulations, it becomes apparent that $\delta\Phi$ would not contain linear terms with respect to $\phi_{\text{noninv}}(\boldsymbol{x})$. Because of that the functional $\Phi_0 = \Phi[\phi_{\text{inv}}]$, computed only for the invariant field configurations, would be automatically stationary with respect to noninvariant field variations. To illustrate it let us consider a simple

Example 4.1. Consider a functional $\Phi[\phi] = \Phi[-\phi]$, which is manifestly invariant under reflections $\phi \to -\phi$, e.g.

$$\Phi[\phi] = \int_{-\infty}^{\infty} dx\, F(\phi^2, \cos\phi, \ldots),$$

where F is an even function of ϕ or ϕ enters F via even functions or their combinations. The decomposition (4.12) for this case looks like the following

[5] Some early origins of this can be found in (Waterhouse, 1983).

$$\phi(x) = \phi_{(-)} + \phi_{(+)},$$

where

$$\phi_{(-)} = \tfrac{1}{2}\left[\phi(x) - \phi(-x)\right]; \quad \phi_{(+)} = \tfrac{1}{2}\left[\phi(x) + \phi(-x)\right].$$

The corresponding linear part of the increment of the functional would disappear by virtue of the evident property

$$\int_{-\infty}^{\infty} dx \{\phi_{(-)}(x) f\left(\phi_{(+)}(x)\right)\} = 0.$$

Thus **the Coleman recipe** can be formulated as: *In search of symmetric static solutions for field equations it is sufficient to reduce the initial variational problem $\delta\Phi[\phi] = 0$ to that for invariant fields only: $\delta\Phi[\phi_{inv}] = 0$. Solutions of this reduced problem would automatically satisfy the initial problem.*

Those suggestions were reformulated by Faddeev into **the Coleman principle** where it was asserted (Faddeev 1976): *If the functional $\Phi[\phi]$ is invariant against transformations from a continuous group G, defined by Eq. (4.9), then the extremum of the functional $\Phi[\phi_{inv}]$, realizable on invariant fields ϕ_{inv} would be the extremum with respect to noninvariant perturbations as well.* Both statements were given unproved in the cited papers with demonstrations of their validity by a number of examples. Following that manner and the paper (Palais 1979) we give here a well-known example from the course of mathematical physics, which at the same time can be related to the study of properties of the simplest chiral invariant Lagrangian (1.35) or (2.40).

Example 4.2. *Symmetric Critical Points of the Dirichlet Functional*
Let us consider the following problem: to find all harmonic functions with the domain $\Omega = \{\boldsymbol{x} \in \mathbb{R}^n | r_1 \leq \boldsymbol{x} \leq r_2\}$ that are radially symmetric, i.e. invariant under the rotation group $SO(n)$. According to Dirichlet's principle *a harmonic function* in Ω having boundary values c_i on $|\boldsymbol{x}| = r_i$ is a critical point of the Dirichlet functional

$$\Phi[\phi] = \int_{\Omega} d^n x \, |\nabla\phi(\boldsymbol{x})|^2. \tag{4.13}$$

Here we consider the Banach space $\mathbf{C}^2(\Omega)$ of all twice differentiable real-valued functions on Ω. A linear representation of $SO(n)$ in $\mathbf{C}^2(\Omega)$ is given by $g\phi(\boldsymbol{x}) = \phi(g^{-1}\boldsymbol{x})$, $g \in SO(n)$ and the orbits of $SO(n)$ on Ω are just the spheres concentric with the origin. Taking a submanifold $M \subset \mathbf{C}^2(\Omega)$ of all $\phi(\boldsymbol{x})$, which assume arbitrary but fixed values c_1 and c_2 on boundary spheres $|\boldsymbol{x}| = r_1$ and $|\boldsymbol{x}| = r_2$, respectively, it is convenient to consider the functional (4.13) as a map $\Phi : M \to \mathbb{R}$. If ϕ is a symmetric $SO(n)$-invariant function on M, then ϕ must take constant values on the orbits of $SO(n)$. On the other hand due to the structure of orbits of $SO(n)$ on Ω the function

$\phi(\boldsymbol{x}) \in \mathbf{C}^2(\Omega)$ is *a stationary point* if and only if it has the form $\tilde{\phi}(|\boldsymbol{x}|)$ for some $\tilde{\phi}(\boldsymbol{x}) \in \mathbf{C}^2([r_1, r_2])$. The set of stationary points of M under the action of $SO(n)$ might be defined as

$$\Sigma = \{\tilde{\phi}(\boldsymbol{x}) \in \mathbf{C}^2([r_1, r_2]) | \, \tilde{\phi}(r_i) = c_i\}. \tag{4.14}$$

To that end one may say that by fixing (4.14) we have chosen *the ansatz.* Now we are in a position to compute the variation of Φ in the vicinity of the set picked out by (4.14):

$$\delta\Phi = -2 \int d^n x \, \Delta \, \phi(r) \delta\phi(x). \tag{4.15}$$

The variation of ϕ one can take as a decomposition in terms of spherical harmonics Y_l^m, which are known to form a basis of the Laplace operator, as

$$\delta\phi(x) = \sum_{lm} a_{lm}(r) Y_l^m(\boldsymbol{n}), \tag{4.16}$$

with coefficients $a_{lm}(r)$ being functions of radial variable, and where $\boldsymbol{n} = \boldsymbol{n}(\vartheta, \alpha)$ stands for angular variables.

Substituting (4.16) into (4.15) and taking into account the orthonormality property for spherical harmonics one finds

$$\delta\Phi = -\frac{2}{\sqrt{4\pi}} \int d^n x \, \Delta \, \phi(r) a_{00}(r),$$

since

$$\oint Y_l^m(\boldsymbol{n}) \, d\Omega = 0; \quad l \neq 0,$$

where $d\Omega$ is an angular part of the volume element. Therefore it takes place the separation of angular variables $\boldsymbol{r} = r\boldsymbol{n}$ and one can write

$$\Phi[\phi]\Big|_{\Sigma} = \mathrm{Vol}(\mathbf{S}^n) \int_{r_1}^{r_2} dr \, r^{n-1} \cdot \left(\tilde{\phi}'(r)\right)^2, \quad \tilde{\phi}'(r) \equiv \frac{d\tilde{\phi}(r)}{dr}. \tag{4.17}$$

Instead of partial differential Euler-Lagrange equations for the Dirichlet functional (4.13) in general we will get the ordinary differential equation, satisfied by the extremal $\tilde{\phi}$ of the functional (4.17):

$$\left(\tilde{\phi}'(r) \cdot r^{n-1}\right)' = 0 \quad \text{or} \quad r\tilde{\phi}'' + (n-1)\tilde{\phi}' = 0, \tag{4.18}$$

with the easily obtained general solution

$$\tilde{\phi}(r) = a + br^{2-n} \text{ for } n \neq 2, \quad \tilde{\phi}(r) = a + b\log(r) \text{ for } n = 2. \tag{4.19}$$

Remember that in general the equation (4.18) is only the necessary condition for $SO(n)$-invariant function $\tilde{\phi}(r)$ to realize an extremum. The invariance group $SO(n)$ of the functional (4.13) is a compact group for which the principle of symmetric criticality is valid (Palais 1979). Therefore Eq. (4.16) proves to be at the same time the sufficient condition for the extremum of the functional (4.13). Thus one can conclude that if the principle is valid we have the powerful tool of the so-called G-reduction of the multi-dimensional variational problems.

The validity of the principle in different situations was considered with mathematical rigour by R. Palais (1979). He proved some theorems which validate the use of the principle under fairly general circumstances which one encounters in theoretical and mathematical physics problems. We are not going to adapt those rather abstract theorems for the physical audience, knowing that for most of our readers it will be enough to understand in what situations they can use this tool[6]. To formulate this answer let us use the notations of Example 4.2, so that by M we mean a smooth ($\mathbf{C}^\infty$) manifold on which a group G acts by diffeomorphisms (a smooth G-manifold). Let $\Phi : M \to \mathbb{R}$ be a smooth G-invariant functional on M with the property that $\Phi = \text{const}$ on the orbits of G. **The Principle of Symmetric Criticality is valid if:**

1. M is a Riemannian manifold on which G acts isometrically, meaning that for each $g \in G$ the corresponding action on each tangent space TM_x (with an inner product $\langle , \rangle_x$) $Dg_x : TM_x \to TM_{gx}$ preserves inner product.
2. M is a real or complex Hilbert space on which G acts by orthogonal or unitary transformations (a special case of 1.)
3. G is a compact group
4. G is a connected semi-simple Lie group and M is a finite-dimensional real analytic manifold on which G acts analytically, or, as a special case, for a finite dimensional representation of a semi-simple Lie group

Here may be the very place to give (without a proof) the promised formulation of **The Coleman-Palais theorem** in a general abstract form (Palais 1979):

Theorem 4.1. Let $\Phi[\phi]$ be a G-invariant functional, which means that $\Phi[g\phi] = \Phi[\phi]$. Denote by $F_0 = \{\phi_0\}$ the set of G-invariant fields and let $\tilde{F}_0$ be the dual to F_0 subset of invariant vector fields $\Upsilon \equiv (\delta\Phi/\delta\phi)[\phi_0]$. If $\phi_0 \in F_0$ is an extremal of $\Phi[\phi]$ on F_0, then it is the true extremal with respect to noninvariant perturbations under condition that $\tilde{F}_0 \cap F_0^* = \emptyset$, where F_0^* is the annihilator of F_0.

Some explanatory remarks on terms used in this formulation, together with an illustrative example are found in Appendix C. Here we prefer to give a more practical statement, which at the same time provides us with a sort of strategy in searching for critical points of invariant functionals (Rybakov 1988; Makhankov, Rybakov and Sanyuk 1989, 1992).

[6] Definitions of mathematical terms which would be used here can be found partly in Appendix B and in the literature cited therein.

Theorem 4.2. Assume that a G-invariant field ϕ_0 realizes $\min \Phi[\phi]$ in the class of invariant fields. **If the functional $\Phi[\phi]$ is convex** with respect to derivatives taken at the point ϕ_0, then the field ϕ_0 realizes **the true minimum** of $\Phi[\phi]$ with respect to noninvariant perturbations as well.

Proof. Let us take the second variation of Φ at the point ϕ_0 and present it in the form of the scalar product in the Hilbert space $L_2(\mathbb{R}^3)$ (see Section 3.1 for definitions):

$$\delta^2\Phi = (\xi, \hat{K}(\boldsymbol{x})\xi), \quad \text{where} \quad \xi = \delta\phi = \phi - \phi_0. \tag{4.20}$$

Here $\hat{K}$ is the Jacobi operator. As we have assumed a convexity of Φ with respect to $\partial_i\phi_0$, $\hat{K}$ is an elliptic operator and from the G-invariance of Φ it follows that

$$\hat{K}(\boldsymbol{x}) = T_g\,\hat{K}(g^{-1}\boldsymbol{x})\,T_g^{-1}, \quad g \in G. \tag{4.21}$$

Then it is possible to obtain the expression for $\hat{K}$ in terms of the Casimir operators of G, denoted by $\hat{\mathbf{C}}_\alpha$. Taking for instance G_1 from Eq. (4.11) as the symmetry group of the functional $\Phi[\phi]$ with generators $\hat{\Gamma}_i = \hat{\mathbf{J}}_i + \hat{\mathbf{I}}_i$, we can present the Casimir operators in the form $\hat{\mathbf{C}}_\alpha = \hat{\Gamma}_i^2$.

Now we have to exploit some properties of elliptic operators. In particular we use the monotonic dependence of $\hat{K}$ on $\hat{\mathbf{C}}_\alpha$, which implies that eigenvalues, say λ, of the operator $\hat{K}$ increase with increasing eigenvalues μ_k of $\hat{\mathbf{C}}_\alpha$. In other words it can be expressed as

$$\frac{d\lambda}{d\mu_k} > 0 \quad \text{when} \quad \mu_k > 0. \tag{4.22}$$

From the other side, for an invariant eigenfunction ψ_0 of $\hat{K}$ we will have

$$\hat{\mathbf{C}}_\alpha\psi_0 = 0, \quad \text{as} \quad \hat{\Gamma}_i\psi_0 = 0.$$

Meantime for a noninvariant eigenfunction ψ_k the eigenvalue equation is

$$\hat{\mathbf{C}}_\alpha\psi_k = \mu_k\psi_k.$$

Note that due to ellipticity of operator $\hat{K}$ it is sufficient to consider only the positive branch of spectrum $\mu_k > 0$. With the assumption the spectrum of $\hat{K}$ is positive on invariant functions, i.e. $\lambda_0 > 0$, then from Eq. (4.20) it follows that $\lambda > \lambda_0 > 0$.

□

The strategy based on this theorem consists of two stages:

1. To study the convexity of the given functional $\Phi[\phi]$, using *the method of phase space extension* (see Chapter 5) or other direct minimization methods
2. To study the existence of minimum in invariant class of fields $\{\phi_0\}$, using direct variational methods (see Chapter 6).

As we shall see later, at this time to fulfill this plan entirely in the Skyrme model is possible only in the first homotopy class for Skyrmion configurations with $|\mathbf{Q}| = 1$. For states from higher homotopy classes with $|\mathbf{Q}| > 1$ this plan is not yet completed in its full generality (see Chapter 7).

4.4 The Structure of Invariant Fields (Ansätze)

To get an explicit form of invariant fields (ansätze) in the Skyrme model with respect to transformations from the group (4.8) we must rewrite invariance conditions (4.9) or (4.10) in a concrete representation for generators of spatial and isospatial rotations:

$$-\mathrm{i}(\boldsymbol{x} \times \nabla)_i \mathbf{U} + [\frac{\tau_i}{2}, \mathbf{U}] = 0. \tag{4.23}$$

Substituting the chiral field $\mathbf{U}$, in the parametrization (1.27), into (4.23) we are able to rewrite the latter in components as

$$-\mathrm{i}\,\epsilon^{ijk} x_j \partial_k \phi_0 + \tau_a(\epsilon^{ijk} x_j \partial_k \phi^a - \epsilon^{ila}\phi_l) = 0. \tag{4.24}$$

or as the equivalent system of two equations

$$\epsilon^{ijk} x_j \partial_k \phi_0 = 0, \tag{4.25}$$

$$\epsilon^{ijk} x_j \partial_k \phi^a - \epsilon^{ila}\phi_l = 0. \tag{4.26}$$

Saturating Eq. (4.26) with ϕ_a we obtain the equality

$$\phi_a \partial_k \phi^a \epsilon^{ijk} x_j = 0, \tag{4.27}$$

which implies

$$\phi_a \partial_k \phi^a = x_k \zeta(\boldsymbol{x}) \quad \text{or} \quad \phi^a = e^a f(r), \quad r = |\boldsymbol{x}|, \tag{4.28}$$

where e^a is a unit vector. Then we saturate Eq. (4.26) with x_i and obtain

$$\epsilon^{ila}\phi_l \cdot x_i = 0, \tag{4.29}$$

from which we deduce in accordance with Eq. (4.28) that

$$\phi_a = x_a \eta(r) = e_a f(r). \tag{4.30}$$

From Eq. (4.30) it follows that $e^a = x^a/r$, i.e.

$$\phi^a = \frac{x^a}{r} \cdot f(r). \tag{4.31}$$

In the same way from (4.25) we deduce the structure of the zeroth component: $\phi_0 = \varphi(r)$.

Taking into account the condition (1.9) we come to the general solution of the Eq. (4.23) which has the form

$$\phi_o = \cos\Theta(r)\,; \quad \phi_i = \frac{x_i}{r}\cdot\sin\Theta(r)\,; \quad r = |\boldsymbol{x}|; \quad i = 1,2,3, \tag{4.32}$$

which is the well-known Skyrme's "hedgehog" ansatz (1.57). The corresponding "hedgehog" chiral fields are

$$\mathbf{U}(\boldsymbol{x}) = \exp\left\{\mathrm{i}\,\frac{x^a\tau^a}{r}\Theta(r)\right\} \tag{4.33}$$

and by definition (1.32) the components of the G_1-invariant left chiral currents could be written

$$\mathbf{L}_i^a = (\delta_i^a - \frac{x^a x_i}{r^2})f_1(r) + \frac{x^a x_i}{r^2}f_2(r) + \epsilon^{aik}\frac{x_k}{r}f_3(r). \tag{4.34}$$

Here $f_i(r)$ are spherically symmetric functions.

In subsequent chapters we shall see that on the fields (4.32) – (4.34) the absolute minima of the functional (4.3) is realized for the unit value of topological charge $|\mathbf{Q}| = 1$. Here we continue to give some other examples of the Coleman-Palais theorem applications in deducing explicit structures of invariant fields.

The same machinery can be used to obtain the explicit structure of G_1-invariant gauge fields in the gauge Skyrme model (Kundu, Rybakov and Sanyuk 1979). The invariant gauge fields may be presented in a way analogous to Eq. (4.34) decomposition:

$$\boldsymbol{A} = \frac{i}{2}\left\{A_1(r)\underline{\tau} + [A_1(r) - A_2(r)]\underline{\boldsymbol{n}}(\underline{\tau}\cdot\underline{\boldsymbol{n}}) + A_3(r)(\underline{\tau}\times\underline{\boldsymbol{n}})\right\}, \tag{4.35}$$

where $n^a = x^a/r$ and $A_i(r)$ are spherically symmetric functions.

Recall that when extending the internal symmetry group of the Skyrme model (for example to $SU(3)$), we have even more possibilities for constructing invariant fields. Let $\mathbf{U} \in SU(3)$; then the condition (4.23) of the G_1-invariance can be written in the form

$$-i(\boldsymbol{x}\times\nabla)_i\mathbf{U} \;+\; [\boldsymbol{\Lambda}_i\,,\,\mathbf{U}] \;=\; 0, \tag{4.36}$$

where $\boldsymbol{\Lambda}_i$ are generators of a representation of a $SO(3)_I$-subgroup of $SU(3)$. There are two nonequivalent sets of these generators:

$$\boldsymbol{\Lambda}_1 = \left\{\frac{\lambda_1}{2}, \frac{\lambda_2}{2}, \frac{\lambda_3}{2}\right\}, \qquad \boldsymbol{\Lambda}_2 = \left\{\lambda_2, \lambda_5, \lambda_7\right\}, \tag{4.37}$$

where λ_a are Gell-Mann's matrices. For the first set of generators in (4.37) as the solution of Eq. (4.36) we have invariant fields of the form

$$\mathbf{U}_1(\boldsymbol{x}) = \exp\big(\mathrm{i}\Gamma\varphi(r)\big)\exp\left\{\mathrm{i}\,\frac{x^a\Lambda_1^a}{r}\Theta(r)\right\}, \tag{4.38}$$

where $\sqrt{3}\,\Gamma = \mathrm{diag}\,[1,1,-2]$. The solution (4.38) is expressed through two spherically symmetric functions $\varphi(r)$, $\Theta(r)$ and can be regarded as a direct generalization of the "hedgehog" ansatz (4.33). In fact, the topological charge for these configurations $\mathbf{Q} = n \in \mathbb{Z}$, if the boundary conditions $\Theta(0) = n\pi, \Theta(\infty) = 0$ are imposed.

For the second set of generators in (4.36) the solution of the invariance Eq. (4.36) is more complicated and has been found by two groups (Romanov, Frolov and Schwartz 1978; Balachandran et al. 1984). It takes the form

$$\mathbf{U}_2(\boldsymbol{x}) = \mathrm{e}^{\mathrm{i}\psi(r)} + \mathrm{i}\Lambda\, \mathrm{e}^{-\frac{1}{2}\psi(r)} \cdot \sin\chi(r) + \Lambda^2 \left(\mathrm{e}^{-\frac{1}{2}\psi(r)} \cdot \cos\chi(r) - \mathrm{e}^{\mathrm{i}\psi(r)}\right), \quad (4.39)$$

where $\Lambda \equiv \left(\boldsymbol{\Lambda}_2 \cdot \boldsymbol{x}\right) \cdot r^{-1}$, $\psi(r)$ and $\chi(r)$ are radial functions. The topological charge for the configurations (4.39) is equal to

$$\mathbf{Q} = \frac{2}{\pi} \cdot [\chi(0) - \chi(\infty)] = 2n, \quad n \in \mathbb{Z}, \quad (4.40)$$

where $\chi(r)$ satisfies the same boundary conditions as Θ. Thus the minimal charge for configurations (4.39) is equal to two. For this reason the corresponding states have been called **dibaryons**. It was emphasized by Balachandran et al. (1984), that this quasistable dibaryon state is in many respects analogous to the H-meson state found by Jaffe in the quark model. More information on dibaryon properties and also on a hierarchy of spherically symmetric ansätze can be found in the lecture notes (Balachandran 1986).

To conclude, let us once more draw the reader's attention to the fact that though the principle of symmetric criticality is a useful tool in searching for ansätze, it is not able to answer some actual questions:

1. Whether the obtained minimum in the class of invariant fields is the absolute one.
2. Whether the minimum of the functional exists (in all given formulations of principles the existence of minimum was just assumed).

5. Absolute Minimum of the Energy Functional

"... Nature is complicated and nonlinear equations are complicated, and therefore Nature should be simulated by means of nonlinear equations..."

John Wheeler

We start with a remark on the importance of the absolute minimum state in a particular aspect to validate the interpretation of Skyrmions as baryons. Our claim is that if the Skyrme model admits the existence of a ground state with the least mass for isovector fields from the first homotopy class, that fact can be regarded as a valuable support to the Skyrme hypothesis. Remember that the proton, which is the longest-living hadron, has the least mass among all particles with a nontrivial baryonic charge. In what follows we demonstrate that in fact the Skyrmion is the absolute minimum energy configuration in the first homotopy class. To obtain this result first of all we use one of the direct minimization methods – the method of extending the original phase space of the model. Afterwards we apply the spherical rearrangement method to derive the structure of fields with the lowest energy (see Section 5.3).

5.1 Method of Extending the Phase Space

The idea of this method consists in introducing a set of auxiliary variables (*extended variables*) in such a way that we can minimize the functional under study, say (4.3), as an ordinary composite function of several variables. To explain this main idea let us remind ourselves of the procedure of minimization of a composite function $F(x) = f\big(x, y(x)\big)$, where in its turn $y = y(x)$ is a smooth given function.

Example 5.1. *Gelfand's "Valley Method"*
There are two approaches in minimizing the function $F(x) = f\big(x, y(x)\big)$:

a. To minimize $F(x)$ as a function of one variable "x"
b. To minimize $f(x, y)$ as a function of two variables "x" and "y"

In the latter approach we "expand" the number of variables, taking "y" as an independent one. Then the set of variables $\{x, y\}$ is called *the extended variables*. To relate this example to the situation discussed in the subsequent

Theorem 5.1, let us consider the profile of the function $f(x,y)$ as a surface in three-dimensional space (x,y,f), displayed in Fig. 5.1 . The surface should contain a "valley" as otherwise the function $f(x,y)$ would not have a minimum with respect to extended variable "y".

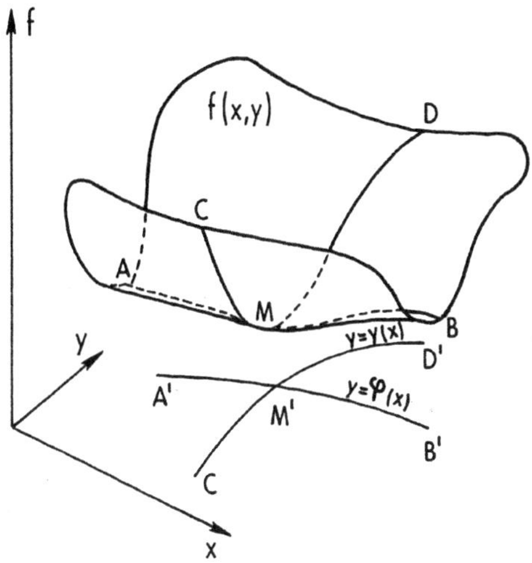

Fig. 5.1. The profile of the function $f(x,y)$ in the space of extended variables. The curve AMB corresponds to "the bottom of the valley"

To obtain the minimum of $f(x,y)$ we have to pass along the curve $y = y(x)$ and find the minimal point on the corresponding profile CMD. But we can also follow the other strategy: come down to "the bottom of the valley" – the profile AMB, and find the minimum point passing along "the stream line" $y = \varphi(x)$. The function $y = \varphi(x)$ can be found as the solution of the equation

$$\frac{\partial f(x,y)}{\partial y} = 0. \tag{5.1}$$

We demand the uniqueness of this solution, which corresponds to minimization of $f(x,y)$ over the extended variable. There are two possible situations:

1. The curves $y = y(x)$ and $y = \varphi(x)$ (or their images on the surface AMB and CMD) cross before we attain the minimum point M along the profile AMB and we have min $F(x)$ located on "the slope of the valley."
2. The curves $y = y(x)$ and $y = \varphi(x)$ cross at the minimal point M. (We, naturally, are not interested in situations when those curves do not cross at all.)

To exclude (but not to improve) the undesirable situation 1, we have first to find the minimal point on "the stream line" AMB and then to check if it is a minimum of the function $F(x) = f\big(x, y(x)\big)$.

Generalizing this method for the Skyrme functional (4.3), which for convenience we write once again

$$\mathbf{H} = \int d^3x \left\{ \frac{1}{2\lambda^2} \Big[(\nabla\Theta)^2 + \sin^2\Theta\big((\nabla\beta)^2 + \sin^2\beta(\nabla\gamma)^2\big) \Big] \right.$$
$$+ \epsilon^2 \sin^2\Theta \big\{ (\nabla\Theta \times \nabla\beta)^2 + \sin^2\beta(\nabla\Theta \times \nabla\gamma)^2$$
$$\left. + \sin^2\Theta \sin^2\beta(\nabla\beta \times \nabla\gamma)^2 \big\} \right\}, \tag{5.2}$$

as generalized coordinates (analogous to "x" in the above example) one may take the set of "polar" variables on $\mathbb{S}^3$, namely: $\{\omega\} = \{\Theta(\boldsymbol{x}), \beta(\boldsymbol{x}), \gamma(\boldsymbol{x})\}$ (Rybakov and Sanyuk 1981, 1991, 1992; Rybakov 1982). Then a set of **extended variables** would be $\{\sigma\} = (\{\omega\}, \{\nabla\omega\})$, where the role of "$y$" we attribute to angles between gradients $\nabla\Theta, \nabla\beta, \nabla\gamma$ - $\{\upsilon, \kappa, \chi\}$. Evidently those angles, as presented on Fig. 5.2, might be expressed as some functions of initial phase variables $\{\sigma\}$, making use of formulas of spherical trigonometry (Rybakov and Sanyuk 1981). We will need the corresponding expressions in Chapter 7, where they are written in explicit form.

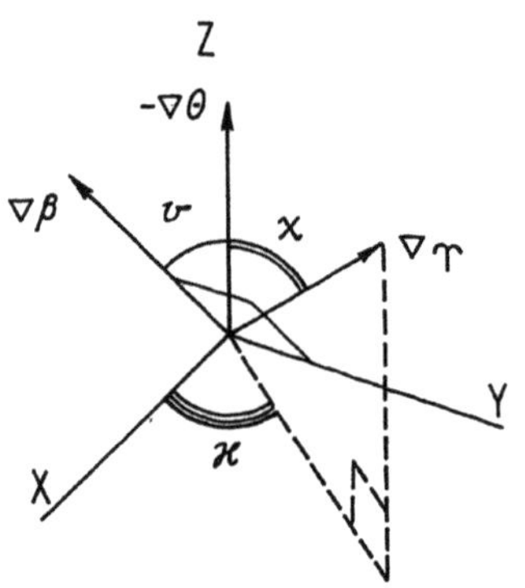

Fig. 5.2. The angles $\{\upsilon, \kappa, \chi\}$ between gradients $\nabla\Theta, \nabla\beta, \nabla\gamma$ as extended variables for the Skyrme model functional (4.3)

In accord with Example 5.1 we treat the supplementary variables as independent. Then, denoting by $\mathcal{H}$ the integrand of the functional (5.2) (the analog of the function $f(x, y)$), it is apparent that the following inequality should hold:

$$\min_{\{\omega\}} \int d^3x\, \mathcal{H}(\{\sigma\}) \geq \min_{\{\sigma\}} \int d^3x \min_{\{\upsilon,\kappa,\chi\}} \mathcal{H}(\{\sigma\}). \tag{5.3}$$

On the basis of the inequality (5.3) and the experience we have got in Example 5.1 we can give without a proof the following (Rybakov and Sanyuk 1981):

Theorem 5.1. If there exists **a unique minimum** in the right-hand side of the inequality (5.3), i.e. a unique minimum of the functional with integrand $\mathcal{H}(\{\sigma\})$ previously minimized over extended variables $\{\upsilon, \kappa, \chi\}$, which coincides with a minimum of the initial functional $\int d^3x\,\mathcal{H}\left(\{\sigma\}\right)$, then **this minimum will be the absolute one.**

The integrand $\mathcal{H}(\{\sigma\})$ does not depend on derivatives of angle variables $\{\upsilon, \kappa, \chi\}$. Thus, fulfilling our promise, we can minimize it with respect to supplementary variables as an ordinary function of several variables and obtain the following minimum conditions:

$$\frac{\partial \mathcal{H}}{\partial \upsilon} = 0, \qquad \frac{\partial \mathcal{H}}{\partial \kappa} = 0, \qquad \frac{\partial \mathcal{H}}{\partial \chi} = 0. \tag{5.4}$$

The only solution of Eqs. (5.4) consistent with the boundary condition $\Theta(\infty) = 0$ is

$$\upsilon = \kappa = \chi = \frac{\pi}{2} \tag{5.5}$$

or, equivalently,

$$(\nabla\Theta \cdot \nabla\beta) = (\nabla\Theta \cdot \nabla\gamma) = (\nabla\gamma \cdot \nabla\beta) = 0. \tag{5.6}$$

To verify the validity of solution (5.5) or (5.6) without lengthy and tedious calculations one can introduce notations:

$$\boldsymbol{X} = \nabla\Theta, \quad \boldsymbol{Y} = \sin\Theta\,\nabla\beta, \quad \boldsymbol{Z} = \sin\Theta\sin\beta\,\nabla\gamma, \quad \epsilon' = \epsilon\,\mathrm{sgn}\mathbf{Q}. \tag{5.7}$$

Taking into account the conclusions of Chapter 3 we consider instead of (5.2) the Lyapunov functional for the Skyrme model $\mathbf{H} - \mu|\mathbf{Q}|$ where $\mu = 12\,\pi^2\epsilon/\lambda\sqrt{2}$ is taken from the estimate (1.43). Using notations (5.7) we can represent the latter functional in the form

$$\begin{aligned} \mathbf{H} - \mu\,|\mathbf{Q}| = \int d^3x \Bigg\{ \left(\frac{1}{\sqrt{2}\lambda}\boldsymbol{X} + \epsilon'\,(\boldsymbol{Y}\times\boldsymbol{Z})\right)^2 \\ + \left(\frac{1}{\sqrt{2}\lambda}\boldsymbol{Y} + \epsilon'\,(\boldsymbol{Z}\times\boldsymbol{X})\right)^2 + \left(\frac{1}{\sqrt{2}\lambda}\boldsymbol{Z} + \epsilon'(\boldsymbol{X}\times\boldsymbol{Y})\right)^2 \Bigg\}. \end{aligned} \tag{5.8}$$

From the expression (5.8) it is readily seen that the minimum corresponds to the antiparallel orientation of the following pairs of vectors: $\boldsymbol{X}$ and $(\boldsymbol{Y} \times \boldsymbol{Z})$, $\boldsymbol{Y}$ and $(\boldsymbol{Z} \times \boldsymbol{X})$, $\boldsymbol{Z}$ and $(\boldsymbol{X} \times \boldsymbol{Y})$. This is equivalent to the condition (5.6), which can be used to rewrite (5.2) in a simpler form:

$$\begin{aligned} \mathbf{H} &= \int d^3x \Bigg\{ (\nabla\Theta)^2 \left[\frac{1}{2\lambda^2} + \epsilon^2 \sin^2\Theta((\nabla\beta)^2 + \sin^2\beta(\nabla\gamma)^2)\right] \\ &\quad + \frac{1}{2\lambda^2}\sin^2\Theta\left[(\nabla\beta)^2 + \sin^2\beta(\nabla\gamma)^2\right] + \epsilon^2\sin^4\Theta\sin^2\beta(\nabla\beta)^2(\nabla\gamma)^2 \Bigg\} \\ &\equiv \int d^3x \Big\{ (\nabla\Theta)^2\, F(|\nabla\beta|, |\nabla\gamma|, \Theta, \beta) + G(|\nabla\beta|, |\nabla\gamma|, \Theta, \beta) \Big\}. \end{aligned} \tag{5.9}$$

In terms of Example 5.1 we have just "come down to the bottom of the valley." Our next task is to move along "the stream" AMB in the search of the minimal point M. As a helpful tool we will use the shperical symmetrization or spherical rearrangement method (Rybakov and Sanyuk 1981, 1991, 1992; Rybakov 1982).

5.2 The Spherical Rearrangement Method

This method is commonly used in solving the extremum problems in geometrical theory of functions. The roots of this method originate with J.Steiner, whose name is usually associated with the parallel-axis theorem from textbooks in mechanics (see e.g. Pólya and Szegö 1951, Luttinger 1973). A special case of the general Steiner's symmetrization is **the spherical rearrangement method**, consisting of a transition from $f(\boldsymbol{x})$ to a spherically symmetric monotonous function $f_R(r)$, where as usual $r = |\boldsymbol{x}|$. The conditions, which are to be imposed on those transitions, read: for an arbitrary number $N \geq 0$ there must be satisfied the equality

$$\text{mes}\,\{\boldsymbol{x} : f(\boldsymbol{x}) \geq N\} = \text{mes}\,\{r : f_R(r) \geq N\}, \tag{5.10}$$

where the notation mes stands for the Lebesgue measure.

Definition 5.1. *If the initial function $f(\boldsymbol{x})$ is nonnegative and monotonically tends to zero at spatial infinity, then the corresponding spherical rearrangement $f_R(r)$ is called* **the decreasing rearrangement**. *In the opposite case, if $f(\boldsymbol{x})$ is a monotonically increasing function, the corresponding rearrangement $f_R(r)$ is called* **the increasing** *one, with the inequality in (5.10) being changed to the opposite one ($\leq$).*

The equimeasurability condition Eq. (5.10) can be conveniently rewritten through the step function

$$\Xi(u) = \begin{cases} 1, & u \geq 0 \\ 0, & u < 0 \end{cases}$$

in the following way

$$\int d^3x\, \Xi\big(f(\boldsymbol{x}) \quad N\big) = \int d^3x\, \Xi\big(f_R(r) \quad N\big). \tag{5.11}$$

Differentiating Eq. (5.11) with respect to N we have:

$$\int d^3x\, \delta\big(N - f(\boldsymbol{x})\big) = \int d^3x\, \delta\big(N - f_R(r)\big), \tag{5.12}$$

and after multiplication by an arbitrary function $F(N)$ and the subsequent integration, we obtain the following useful relation:

$$\int d^3x\, F(f(\boldsymbol{x})) = \int d^3x\, F(f_R(r)), \tag{5.13}$$

known by the name, *the Cavaliery principle.* In Appendix C.3 can be found an illustration of the spherical rearrangement method when proving the validity of the inequality

$$\int d^3x\, |\nabla f(\boldsymbol{x})|^2 \geq \int d^3x\, |\nabla_r f_R(r)|^2, \tag{5.14}$$

where ∇_r denotes the radial part of the operator ∇.

Now we are equipped to check the requirements of Theorem 5.1 in application to the functional (5.9), which as we remember describes "the bottom of the valley." According to the strategy, described at the end of Chapter 4, we need a mathematical tool to move along "the bottom" in the search for the minimum. Useful in this respect is the following theorem (Pólya and Szegö 1951; Coleman, Glaser and Martin 1978; Glaser, Martin, Grosse and Thirring 1976):

Theorem 5.2. Consider three scalar nonnegative functions $u(\boldsymbol{x}) \geq 0,\ v(\boldsymbol{x}) \geq 0,\ w(\boldsymbol{x}) \geq 0$. Assuming the spherical rearrangements $u_R(r)$ and $v_R(r)$ to be the monotonically decreasing, and on the contrary, $w_R(r)$ to be the monotonically increasing rearrangement (see Definition 5.1), then the following inequalities hold:

$$\int d^3x\, |\nabla u(\boldsymbol{x})|^2 \geq \int d^3x\, |\nabla_r u_R(r)|^2, \tag{5.15}$$

$$\int d^3x\, v(\boldsymbol{x}) w(\boldsymbol{x}) \geq \int d^3x\, v_R(r) w_R(r). \tag{5.16}$$

To apply Theorem 5.2 to the functional (5.9), let us introduce the following auxiliary functions

$$\begin{aligned} &u(\boldsymbol{x}) = \Theta(\boldsymbol{x}), \quad w_1(\boldsymbol{x}) = \cos^2\frac{\Theta(\boldsymbol{x})}{2}, \quad w_2(\boldsymbol{x}) = \cos^4\frac{\Theta(\boldsymbol{x})}{2}, \\ &v_1(\boldsymbol{x}) = \sin^2\frac{\Theta(\boldsymbol{x})}{2}\Big(\frac{2}{\lambda^2} + 4\epsilon^2(\nabla\Theta)^2\Big)\cdot\big[(\nabla\beta)^2 + \sin^2\beta(\nabla\gamma)^2\big], \\ &v_2(\boldsymbol{x}) = 16\,\epsilon^2\sin^2\frac{\Theta(\boldsymbol{x})}{2}\sin^2\beta(\nabla\beta)^2(\nabla\gamma)^2. \end{aligned} \tag{5.17}$$

Strictly speaking, to proceed we need to know the behavior of solutions $\Theta(\boldsymbol{x})$ of the equation of motion (1.44), when the chiral fields are parametrized through (1.29) and (4.2). But instead one can assume the required behaviour of these function with a subsequent verification of such an assumption. The appropriate analysis (see e.g. Section 6.1) confirms that on the interval $\Theta \in [0, \pi]$ the functions w_i introduced in Eq. (5.17) are monotonically increasing, and respectively the functions v_i are monotonically decreasing.

At the next step we present the functional (5.9) in terms of the auxiliary functions (5.17):

$$\mathbf{H} = \int d^3x \left\{ \frac{1}{2\lambda^2} |\nabla u(\boldsymbol{x})|^2 + v_1(\boldsymbol{x}) w_1(\boldsymbol{x}) + v_2(\boldsymbol{x}) w_2(\boldsymbol{x}) \right\}. \tag{5.18}$$

Theorem 5.2 tells us that under the performed spherical rearrangements we actually abate the value of the energy functional due to the inequality:

$$\mathbf{H} \geq \int d^3x \left\{ \frac{1}{2\lambda^2} |\nabla_r u_R(r)|^2 + v_{1R}(r) w_{1R}(r) + v_{2R}(r) w_{2R}(r) \right\}. \tag{5.19}$$

This inequality demonstrates that the spherical rearrangement is the right tool to move "along the bottom of the valley" in search of the minima. The validity of the inequality (5.19) is easily checked on the basis of the evident form of functions from Eq. (5.17) and also from the inequalities (5.15) and (5.16) as the requirements of Theorem 5.2 are satisfied when $\Theta \in [0, \pi]$.

Still again from the explicit form of the functions $v_1(\boldsymbol{x})$ and $v_2(\boldsymbol{x})$ (see Eq. (5.17)) it follows that **the necessary conditions for the performance of the required spherical rearrangement** $v_i(\boldsymbol{x}) \to v_{iR}(r)$ are:

$$(\nabla\beta)^2 = \varrho_1^2(r), \tag{5.20}$$

$$\sin^2\beta\,(\nabla\gamma)^2 = \varrho_2^2(r). \tag{5.21}$$

Summing up all previous considerations, we can state that if the conditions (5.6), (5.20) and (5.21) are compatible, then the functional in the right-hand side of the inequality (5.3) has **the unique minimum.** Thus, in accord with the statement of Theorem 5.1, to prove the existence of the absolute minima of energy in the Skyrme model it is sufficient to establish the consistency of the above listed conditions and then to check if the minimum thus obtained coincides with the minimum from the left hand side of the Eq. (5.3). This is what we are going to do in the next section.

5.3 Skyrmion as the Absolute Minimiser of the Energy

We write the gradients $\nabla\Theta, \nabla\beta$ and $\nabla\gamma$ in the spherical coordinates (r, ϑ, α) and substitute them into the orthogonality condition (5.6). As a result we at once realize that the "polar" variables β and γ are to depend only on angular variables ϑ and α, i.e.

$$\beta = \beta\,(\vartheta, \alpha), \qquad \gamma = \gamma\,(\vartheta, \alpha). \tag{5.22}$$

Substitution of (5.22) into Eq. (5.20) and Eq. (5.21) leads to the following system of equations:

$$\left(\frac{\partial\beta}{\partial\vartheta}\right)^2 + \frac{1}{\sin^2\vartheta}\left(\frac{\partial\beta}{\partial\alpha}\right)^2 = \varrho_1^2(r) r^2 = \text{const} \equiv C_0, \tag{5.23}$$

$$\sin^2\beta\left[\left(\frac{\partial\gamma}{\partial\vartheta}\right)^2+\frac{1}{\sin^2\vartheta}\left(\frac{\partial\gamma}{\partial\alpha}\right)^2\right]=\varrho_2^2(r)r^2=\text{const}\equiv C_1. \tag{5.24}$$

The Eq. (5.23) readily implies

$$\left(\frac{\partial\beta}{\partial\alpha}\right)=\text{const}\equiv C_2, \tag{5.25}$$

so that we have

$$\left(\frac{\partial\beta}{\partial\vartheta}\right)^2=C_0-\frac{C_2^2}{\sin^2\vartheta}. \tag{5.26}$$

It is apparent that we must set $C_2=0$, otherwise for $\sin\vartheta=0$ we will obtain a singularity in the last term of Eq. (5.26). Whence, we get

$$\beta=C_3\,\vartheta+C_4, \tag{5.27}$$

which meets all previous requirements.

From one of the conditions (5.6), namely from $(\nabla\gamma\cdot\nabla\beta)=0$ we have

$$\frac{\partial\beta}{\partial\vartheta}\cdot\frac{\partial\gamma}{\partial\vartheta}+\frac{1}{\sin^2\vartheta}\frac{\partial\beta}{\partial\alpha}\cdot\frac{\partial\gamma}{\partial\alpha}=0. \tag{5.28}$$

The second term in (5.28) is equal to zero due to the form (5.27) and thus from (5.28) we deduce that $\gamma=\gamma(\alpha)$. Taking all that into account we can rewrite Eq. (5.24) in the form:

$$\frac{\partial\gamma}{\partial\alpha}=\text{const}\equiv C_5, \tag{5.29}$$

$$\sin^2\beta=C_6\,\sin^2\vartheta. \tag{5.30}$$

As $\beta\in[0,\pi]$ and $\vartheta\in[0,\pi]$ it is clear from (5.30) that $C_6=1$ and $\beta=\vartheta$. On the other hand from (5.29) up to a constant rotation we deduce, that $\gamma=m\alpha$, where $m\in\mathbb{Z}$. To summarize, we can assert that the conditions (5.6), (5.20) and (5.21) are compatible on solutions of the form:

$$\Theta=\Theta(r),\qquad\beta=\vartheta,\qquad\gamma=m\alpha. \tag{5.31}$$

Substituting the obtained ansatz (5.31) into the functional (5.9) we reduce it to the form:

$$\mathcal{E}=\frac{4\pi\epsilon}{\lambda}\int_0^\infty dx\left\{\left(\frac{d\Theta}{dx}\right)^2\left(\frac{x^2}{2}+(m^2+1)\sin^2\Theta\right)\right.$$
$$\left.+\frac{m^2+1}{2}\sin^2\Theta+\frac{m^2\sin^4\Theta}{x^2}\right\}, \tag{5.32}$$

where we have put $r=\epsilon\lambda x$. The corresponding expression for the topological charge from Eq. (4.4) is

$$\mathbf{Q} = -\frac{1}{2\pi^2} \int\limits_{\Theta(0)}^{\Theta(\infty)} d\Theta \, \sin^2\Theta \int\limits_0^{\pi} d\vartheta \, \sin\vartheta \int\limits_0^{2\pi m} d\gamma. \tag{5.33}$$

Note, that the required spherical rearrangement for the function $\Theta(\boldsymbol{x})$ i.e. $\Theta(\boldsymbol{x}) \to \Theta_R(r)$ do exists if $\Theta(\boldsymbol{x})$ is a monotonically decreasing function. On the other hand the term $\sin^4\Theta/x^2$ in (5.32) is regular if and only if $\Theta_{\max} = n\pi$, whence $\mathbf{Q} = m \cdot n \in \mathbb{Z}$.

This means that the topological charge $\mathbf{Q}$ can take any integer value. But as far as we remember, the spherical rearrangements w_{1R} and w_{2R} would be increasing only under the condition $\Theta \in [0, \pi]$. Hence, to meet the requirements of Theorem 5.2 we have to restrict the admissible values for $\mathbf{Q}$ by setting $n = 1$, or equivalently, we have to choose the boundary condition on $\Theta(x)$ in the following form:

$$\Theta(0) = \pi, \qquad \Theta(\infty) = 0. \tag{5.34}$$

Only after this can we be sure that the minimum in the right hand side of the inequality (5.3) will be the unique one.

The next restriction comes from the requirements of Theorem 5.1, that the minimum of the left-hand side of the inequality must also be unique. It is straightforward to see that the only possibility to achieve this is to set $\mathbf{Q} = m = 1$. Assuming, for example, that $\mathbf{Q} = 2$, we learn at once that there is one configuration with $m = 2$ and with one maximum value of the function $\Theta(x)$, which is equal to π. But there is also another configuration with two maxima $\Theta(x) = \pi$ and $m = 1$ (the configuration with two infenitely separated Skyrmions). Thus we have the contradiction with Theorem 5.1 requirements.

One may obtain the same result, using the Coleman-Palais theorem from Chapter 4. Actually by means of calculations similar to those performed in Section 4.4 one can prove that min $\mathcal{E}$ on the field configuration (5.31) would be invariant if and only if $m = 1$. This verification can be a good exercise for the reader, who must write the invariance conditions (4.23), using the expression for generators in the spherical coordinates, and then to substitute the chiral field $\mathbf{U}$ from (1.29), using for the parametrization of the vector $\mathbf{n}$ "polar" variables in the form (5.31). After some algebra one inevitably comes to the conclusion that the invariance conditions are satisfied only for $m = 1$. To resume all the above considerations let us formulate the central result, following (Rybakov and Sanyuk 1981):

Theorem 5.3. The absolute minimum of the energy functional (5.2) in the first homotopy class is realized on the spherically-symmetric configurations:

$$\Theta = \Theta(r), \qquad n^a = \frac{x^a}{r}, \qquad r = |\boldsymbol{x}|, \tag{5.35}$$

with the boundary conditions (5.34) and with the unit topological charge $\mathbf{Q} = 1$. The configuration (5.35) in particular coincides with Skyrme's hedgehog ansatz (1.57).

6. The Existence of Skyrmions

"If you can't join 'em, beat 'em"

Julian Schwinger

Here we accomplish our study of the minimal energy configurations in the first homotopy class of the Skyrme model and also prove the existence of spherically-symmetric configurations in higher homotopy classes. In the next chapter we will see that the latter, contrary to the Skyrmion configuration, are not minimal energy configurations and hence should not be stable. Recall that up to this point we have gotten the positive answer only to the first of those questions listed at the end of Chapter 4. Whether the absolute minimum configurations (Skyrmions) do exist as solutions of the Euler-Lagrange equations is still an open question. This question can be formulated, in an equivalent form, as whether the lower bound of the energy functional is attainable, implying that Skyrmions are stable field configurations. We start with a frontal attack on the problem, trying to solve the Skyrmion field equation directly. In a moment we will learn that only numerical or approximate solutions to the Euler-Lagrange equation (6.2) are available. To get a more definite answer we need to use more powerful tools: direct methods of the calculus of variations.

6.1 The Field Equations for Skyrmions

The conclusions made in Section 5.3 tell us that the energy functional for the Skyrmion configurations can be obtained from the expression (5.32) by setting $m = 1$ so that

$$\mathcal{E} = \frac{4\pi\epsilon}{\lambda} \int_0^\infty dx \left\{ \left(\frac{d\Theta}{dx}\right)^2 \left(\frac{x^2}{2} + 2\sin^2\Theta\right) + \sin^2\Theta + \frac{\sin^4\Theta}{x^2} \right\}. \tag{6.1}$$

The corresponding Euler-Lagrange equation might be considered as the field equation for the Skyrmion configuration. It reads

$$\frac{d^2\Theta}{dx^2}(x^2 + 4\sin^2\Theta) = \sin 2\Theta\left[1 - 2\left(\frac{d\Theta}{dx}\right)^2 + \frac{2\sin^2\Theta}{x^2}\right] - 2x\frac{d\Theta}{dx}. \tag{6.2}$$

Though the Eq. (6.2) for the chiral angle $\Theta(x)$ is an ordinary differential equation and much simpler than the general Eq. (1.44), no analytical solution to this problem is available to date. Numerical simulations for solutions of Eq. (6.2) with the boundary conditions

$$\Theta(0) = \pi, \qquad \Theta(\infty) = 0, \tag{6.3}$$

as mentioned in Section 1.4, were pioneered by Skyrme himself (Skyrme 1962). Later on these calculations were repeated by, among others, (Nisichenko 1981[1]; Adkins, Nappi and Witten 1983, Jackson and Rho 1983). The results of numerical simulations are presented on Fig. 6.1.

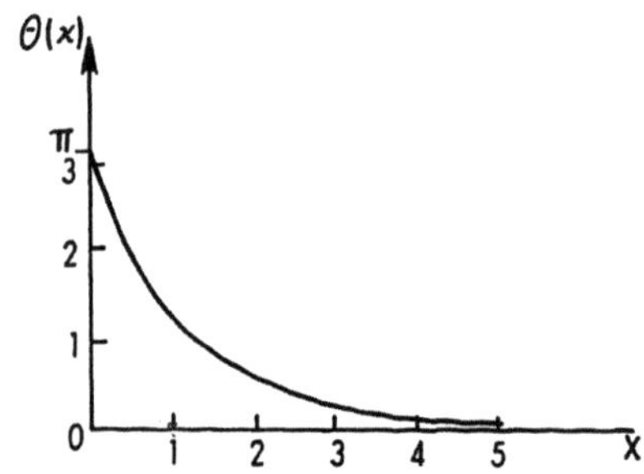

Fig. 6.1. The profile of the chiral angle $\Theta(x)$ as the numerical solution to the boundary problem (6.2) and (6.3)

The standard procedure to obtain numerical solutions to the above boundary problem is to use the shooting or relaxation methods. Hence we need the knowledge of the asymptotic behavior of $\Theta(x)$ near the origin and at spatial infinity. This behavior can be readily deduced from (6.2), rewritten for example, in the limit case for $x \to \infty$, when $\Theta(x) \to 0$ as:

$$x^2\Theta''(x) + 2x\Theta'(x) - 2\Theta(x) = 0, \quad \text{where} \quad \Theta' \equiv \frac{d\Theta}{dx}. \tag{6.4}$$

The solution of the asymptotic Eq. (6.4) is straightforward

$$\Theta(x \to \infty) = \eta x^{-2}, \tag{6.5}$$

where η is a constant. Note that this inverse quadratic rate of decreasing of the solutions at spatial infinity might be considered insufficient for our initial aim to describe structures localized in space. To improve this asymptotic behavior we can add to the initial Lagrangian (1.39) or to the functional (6.1) a mass term of the form $m_\pi x^2(1 - \cos\Theta)$. This addition provides the Yukawa-type asymptotic behavior for the chiral variable $\Theta(x)$ with the exponential fall as $x \to \infty$. In the same manner for $x \to 0$ we have $\pi - \Theta(x) \equiv \xi(x) \to 0$ which satisfies the asymptotic equation:

$$\xi\xi'' + \xi'^2 = \xi^2 x^{-2} \tag{6.6}$$

[1] In this thesis Skyrme's results were re-obtained as a special case of a more general problem of numerical simulations for the gauge Skyrme model.

with corresponding solution

$$\xi(x) = \zeta x, \quad \text{or} \quad \Theta(x \to 0) = \pi - \zeta x, \quad \text{where} \quad \zeta = \text{const.}$$

An interesting dependence of the energy spectrum on the value of the topological charge $\mathbf{Q} = N$ has been obtained in numerical studies (Nisichenko 1981) of solutions to Eq. (6.2) with the boundary conditions $\Theta(0) = N\pi$, $\Theta(\infty) = 0$, which holds with the high accuracy $\approx 1\%$ when $N < 10$:

$$\mathcal{E}_N \approx \frac{N(N+1)}{2} \cdot \mathcal{E}_1, \tag{6.7}$$

where $\mathcal{E}_N$ is the energy value for spherically-symmetric configurations from the Nth homotopy class, and $\mathcal{E}_1$ is the Skyrmion energy with $\mathbf{Q} = 1$. This rotator-like energy spectrum (6.7) has been studied in great detail in the paper (Bogomol'nyi and Fateev 1983), where it was demonstrated that for $N \to \infty$ the formula (6.7) should be changed to

$$\mathcal{E}_N = \left\{a_0\, N^2 + a_1\, N + a_2 + \mathcal{O}\left(\frac{1}{N}\right)\right\} \cdot \mathcal{E}_1. \tag{6.8}$$

For the coefficients a_i there were obtained the following values: $a_0 = 4.155$ and $a_1 = 3.626$.

In addition to the numerical studies there are some approximate solutions available to Eq. (6.2), which one can obtain by choosing a trial function with parameters defined through variational methods. To obtain a zeroth-approximation Θ_0 to solutions of (6.2) we can use, for instance, a trial function

$$\cos\Theta_0(x) = 1 - 2\varpi(x), \quad \text{with} \quad \varpi(x) = 2e^{-x/\alpha} + e^{-2x/\alpha}. \tag{6.9}$$

Substituting (6.9) into the energy functional (6.1) and minimizing the latter with respect to the parameter of the trial function we come to the following zeroth-approximation:

$$\Theta_0(x) = 2\arccos\bigl[1 - \exp(-x/a)\bigr], \quad \text{with} \quad a = 0.94510062. \tag{6.10}$$

The comparison of the corresponding energy value $\mathcal{E}_0 = 8.4676$ with that obtained in numerical calculations $\mathcal{E}_{num} = 8.2067$ gives the accuracy $\delta \approx 3.2\%$. The next approximation one can get by setting

$$\Theta = \Theta_0 + \delta\Theta, \quad \text{where e.g.} \quad \delta\Theta = A\,\sin\Theta_0 + B\,\sin 2\Theta_0. \tag{6.11}$$

Through the same procedure one can obtain the values of parameters A, B. These approximate solutions might be useful for those problems where we need some (at least approximate) knowledge of the behavior of the chiral angle $\Theta(x)$ when numerical results are insufficient.

Nevertheless, with none of these methods can one be sure of the existence and stability of the Skyrmion configurations. On the contrary, numerical methods in application to nonlinear equations of the type (6.2) require justification themselves. Previous considerations lead to the conclusion that the basic approach usually used to solve variational problems is not effective in our case. It should be noted that these difficulties are inherent in the standard variational approach to boundary problems when we need to know solutions of differential equations with imposed boundary conditions not in a small neighbourhood of some point, but rather solutions in some fixed region, (see e.g. Chapter 8 in Gelfand and Fomin 1963). Another possible approach, known by the name of **the direct method in the calculus of variations** proves to be more effective for such problems. The basic idea is the following: as Eq. (6.2) is the Euler-Lagrange equation for the functional (6.1), instead of looking for solutions of Eq. (6.2) in inverse we can try to prove somehow that the functional (6.1) has an extremum for a sufficiently smooth admissible function $\Theta(x)$. Then this very fact proves that the differential Eq. (6.2) has a solution, satisfying the boundary conditions (6.3). In other words, as we cannot find any analytical solution to the Eq. (6.7), we can try to prove its existence. Before doing that let us give a brief account of the appropriate direct method in the calculus of variations.

6.2 The Direct Method in the Calculus of Variations

In this section we will use standard calculus notations. Let us consider a functional $\Phi(x)$, which is given on elements of a Banach space $X \ni x$, with the properties defined in what follows. The problem is to find the minimum of this functional. The central idea of the presented direct method comes from a generalization of **the Weierstrass theorem**, that *a continuous function $f(x)$ attains its minimum on a compact set.* We have to find (necessary and sufficient) conditions of applicability of the Weierstrass theorem in the case of a functional $\Phi(x)$.

Let us restrict ourselves to a situation, wherein the functional $\Phi(x)$ is bounded from below, i.e. when there exists such a constant A, that $\Phi(x) > A$. This signifies the existence of the lower bound of the functional

$$\mathrm{Inf}_{x \in X}\, \Phi(x) = \Phi_0. \tag{6.12}$$

Still another consequense is that there exists a minimizing sequence $\{x_n\}$ of elements $x_n \in X$, such that

$$\lim_{n \to \infty} \Phi(x_n) = \Phi_0. \tag{6.13}$$

Thus the first task can be formulated as: *to construct a minimizing sequence* $\{x_n\}$, i.e. to find some set $M \subset X$ with the property, that for $x_n \in M$ the corresponding sequence $\{x_n\}$ converges in M in some sense. In brief, we are to establish the existence of the limit point (function) $x_0 \in M$ for the sequence $\{x_n\}$. Finally, we have to find out if the following equality holds:

$$\Phi(x_0) = \Phi_0. \tag{6.14}$$

It is well known that the latter holds for certain in the case of continuous functions. But the continuity requirements show to be too restrictive when applied to functionals. It was Lebesgue, who noted that for the validity of Eq. (6.14) it is sufficient to demand for a weaker property, namely for *a semicontinuity of the functional* $\Phi(x)$ *at the point* x_0.

Recall that the continuity of $\Phi(x)$ might be equivalently expressed through the inequality

$$|\Phi(x_0 + h) - \Phi(x_0)| < \epsilon, \quad |h| < \delta(\epsilon). \tag{6.15}$$

In its turn the latter is equivalent to the two inequalities

$$-\epsilon < \Phi(x_0 + h) - \Phi(x_0) < \epsilon. \tag{6.16}$$

Definition 6.1. *A functional* $\Phi(x)$ *is called* **a lower semicontinuous** *at the point* x_0*, if only one of the two inequalities (6.16) holds, namely if*

$$\lim_{n\to\infty} \Phi(x_n) \geq \Phi(x_0). \tag{6.17}$$

Let us show that inequality (6.17) implies Eq. (6.14). Actually, from the definition of a minimizing sequence (6.13), we have as a consequence

$$\lim_{n\to\infty} \Phi(x_n) = \Phi_0 \leq \Phi(x_0), \tag{6.18}$$

where the last inequality follows from the fact that Φ_0 is the lower bound of $\Phi(x)$. Simultaneously, from Eq. (6.17) we deduce

$$\lim_{n\to\infty} \Phi(x_n) = \Phi_0 \geq \Phi(x_0). \tag{6.19}$$

The last two conclusions are compatible only if $\Phi(x_0) = \Phi_0$.

Thus, the direct method consists of the following stages:

1. The construction of a minimizing sequence $\{x_n\}$, $x_n \in X$
2. The proof of the convergence of the sequence $\{x_n\}$ to a limit point $x_0 \in M$
3. The proof of the semicontinuity of the functional $\Phi(x)$ at the limit point x_0.

An important special case is when a functional $\Phi(x)$ is defined on a Hilbert space, say L_2. Then we can choose as $M \subset L_2$ a ball $\| x \| \leq 1$ and in accord with the Tikhonov-Alaoglu theorem (see e.g. Akhieser and Glazman 1961-1963) there exists a sequence $\{x_n\}$, $x_n \in M$, which is weakly convergent to the vector $x_0 \in M$. Remember that a sequence is *weakly convergent* (with the notation $x_n \rightharpoonup x_0$) if for an arbitrary $h \in L_2$, $\lim_{n\to\infty} (h, x_n) = (h, x_0)$, where $(,)$ is the scalar product of elements in L_2. In this respect in searching for a minimum we have to check a more delicate property of the functional under study - **a weak semicontinuity**, instead of a semicontinuity of the functional $\Phi(x)$. In other words we have to be sure that the inequality (6.17) holds for weakly convergent sequences. The following theorem (Vainberg 1964) proves to be useful in this case.

Theorem 6.1. The positive scalar product (x, y) for vectors $x, y \in L_2$ is lower weakly semicontinuous.

Proof. Let us take two positive weakly convergent sequences of elements from L_2: $x_n \rightharpoonup x_0$, $y_n \rightharpoonup y_0$. We must show that

$$\lim_{n \to \infty} (x_n, y_n) \geq (x_0, y_0). \tag{6.20}$$

Note that we always have the right to regard the two sequences $\{x_n\}$, and $\{y_n\}$ to be monotonically increasing, as we may always redefine them in the following way:

$$x_n^* = \inf_m \{x_m | m \geq n\}, \qquad y_n^* = \inf_m \{y_m | m \geq n\}. \tag{6.21}$$

At the next step let us use the identity

$$(x_n, y_n) = (x_0, y_0) + (x_n - x_0, y_n - y_0) \\ + (x_n - x_0, y_0) + (x_0, y_n - y_0) \tag{6.22}$$

and note that as we assumed the weak convergence $x_n \rightharpoonup x_0$, $y_n \rightharpoonup y_0$ it follows that

$$(x_n - x_0, y_0) \to 0, \qquad (x_0, y_n - y_0) \to 0.$$

That is why

$$\lim_{n \to \infty} (x_n, y_n) = \lim_{n \to \infty} (x_n - x_0, y_n - y_0) + (x_0, y_0). \tag{6.23}$$

From (6.23) and the monotonic character of sequences $x_n \leq x_0$, $y_n \leq y_0$ we have the desired inequality (6.20). □

Corollary 6.1. In applications of Theorem 6.1 it is useful to represent functionals as a sum of positive scalar products of vectors from L_2.

6.3 An Outline of the Proof of the Skyrmion Existence

The idea of the proof is not a new one in physical applications (see e.g Romanov, Frolov and Schwarz 1978). Our exposition is largely based on previous papers (Rybakov 1979; Sanyuk 1979; Rybakov and Sanyuk 1981, 1991, 1992). Consider the energy functional for Skyrmions (6.1) and for convenience present it in the form

$$\mathcal{E} = \frac{4\pi\epsilon}{\lambda} \Phi[\Theta(x)], \tag{6.24}$$

where we have singled out the factor which is dimensionless and which is the only piece of (6.1) essential for further study:

$$\Phi[\Theta(x)] = \int_0^\infty dx \left\{ \left(\frac{d\Theta}{dx}\right)^2 \left(\frac{x^2}{2} + 2\sin^2\Theta\right) + \sin^2\Theta + \frac{\sin^4\Theta}{x^2} \right\}. \tag{6.25}$$

Remark 6.1. Remember that the Skyrmion configuration itself belongs to the first homotopy class with topological charge $\mathbf{Q} = 1$. The proof of the existence of Skyrmions as solutions of Eq. (6.2) can be extended to a more general situation, i.e. for spherically-symmetric configurations with an arbitrary value of topological charge $\mathbf{Q} = n$. We do not claim here that the minimum of the energy in each class is realized on these configurations. In fact this is not true already for the second homotopy class (see Chapter 7). Hence, the result we will obtain below for higher homotopy classes means that there exist spherically-symmetric configurations, but for $\mathbf{Q} > 1$ these configurations should be unstable.

After this remark we consider chiral variables $\Theta(x)$ satisfying the boundary conditions

$$\Theta(0) = n\pi, \qquad \Theta(\infty) = 0. \tag{6.26}$$

Here by the direct variational method we have to prove that $\operatorname{Inf}\Phi[\Theta]$ is attainable. According to Theorem 6.1 and the subsequent corollary we need to present (6.25) as a sum of scalar products. At the first stage it is convenient to consider our variational problem to be defined on a finite interval $[a, b] \in X$, with the following correspondence

$$\Phi[\Theta_n] = \lim_{a\to 0}\lim_{b\to\infty} \int_a^b dx\, F[\Theta_n, \Theta'_n] = \lim_{a\to 0}\lim_{b\to\infty} \Phi_{ab}; \quad ' \equiv \partial_x, \tag{6.27}$$

where $\{\Theta_n\}$ and $\{\Theta'_n\}$ are sequences of chiral variables and their derivatives, satisfying the appropriate boundary conditions

$$\Theta(a) = n\pi, \qquad \Theta(b) = 0. \tag{6.28}$$

The functional Φ_{ab} can be presented, for example, as

$$\Phi_{ab} = \sum_{i=1}^{4} \| h_i \|^2, \tag{6.29}$$

where the following notations are introduced

$$h_1 = \frac{x}{\sqrt{2}}\Theta_n', \qquad h_2 = \sqrt{2}\sin\Theta_n \cdot \Theta_n',$$
$$h_3 = \sin\Theta_n, \qquad h_4 = \frac{\sin^2\Theta_n}{x}. \tag{6.30}$$

Secondly we are to prove that these sequences (6.30) are convergent and that their limit points belong to the same subspace as the sequences themselves. A right useful way to do that is to derive the so-called *a priori estimates* on the function $\Theta(x)$ and its derivatives[2]. These estimates are called a priori because one can obtain them without knowledge of an exact solution of Eq. (6.2) with the boundary conditions (6.26). For example it is possible to see that

$$|\Theta'(0)| < C = \text{const} < \infty. \tag{6.31}$$

Here we omit these pure mathematical considerations in proof of (6.31) and refer the reader to (Rybakov, 1991). To establish some limitations on the behavior of $\Theta(x)$ and $\Theta'(x)$ at spatial infinity we note that the functional $\Phi[\Theta]$ is bounded from above. This fact is readily seen, for when constructing a minimizing sequence $\{\Theta_n\}$ one can take e.g. $\Theta_1(x) = n\pi\exp(-x)$, which gives $\Phi[\Theta_1] < \infty$. Hence, the following inequality holds

$$\int\limits_0^\infty dx\, x^2 \cdot \Theta'^2(x) \le C_0^2 < \infty. \tag{6.32}$$

Whence, for $b > 0$ we have according to the Schwarz inequality

$$|\Theta(b)| \le \int\limits_b^\infty dx |\Theta'(x)| \le \left(\int\limits_b^\infty \frac{dx}{x^2}\right)^{1/2} \left(\int\limits_b^\infty dx\, x^2 \cdot \Theta'^2\right)^{1/2} < C_0 b^{-1/2}, \tag{6.33}$$

and from the arbitrariness of the point b for $x \ge b \ge 0$ it follows that

$$|\Theta(x)| < C_0 \cdot x^{-1/2}. \tag{6.34}$$

The obtained inequalities (6.31) and (6.34) can be further specified if one takes into account that Eq. (6.2) can be rewritten in the form

$$\begin{aligned} &\left[\frac{x}{2}\Theta'^2\left(x^2 + 4\sin^2\Theta\right) - x\sin^2\Theta\left(1 + \frac{\sin^2\Theta}{x^2}\right)\right]' \\ &\qquad = -\frac{1}{2}x^2\Theta'^2 + \sin^2\Theta\left(2\Theta'^2 + \frac{\sin^2\Theta}{x^2} - 1\right). \end{aligned} \tag{6.35}$$

Integrating Eq. (6.35) in the region $a \ge x \ge b$ we get

$$\begin{aligned} &\left[\frac{x}{2}\Theta'^2\left(x^2 + 4\sin^2\Theta\right) - x\sin^2\Theta\left(1 + \frac{\sin^2\Theta}{x^2}\right)\right]_a^b \\ &\qquad = \int\limits_a^b dx \left\{-\frac{1}{2}x^2\Theta'^2 + \sin^2\Theta\left(2\Theta'^2 + \frac{\sin^2\Theta}{x^2} - 1\right)\right\}. \end{aligned} \tag{6.36}$$

---

[2] In practice the sequence of steps is quite the opposite. First of all one must to find out what a priori estimates are available for a given problem and only after that to choose the form of vectors in (6.30), being sure that on the basis of possible estimates the convergence of chosen sequences can be proved.

Then, setting $a \rightarrow 0, \quad b \gg 1$ in (6.36) and using the inequalities already obtained in (6.31) and (6.34), we derive

$$|\Theta'(b)| = \mathcal{O}(b^{-3/2}). \tag{6.37}$$

Still a further specification of the behavior of $\Theta(x)$ at spatial infinity one may obtain by making use of the integral form of Eq. (6.2)

$$\Theta(x) = \int_0^\infty ds\, G(x,s) \cdot (s^2 + 4\sin^2\Theta) \times \left\{ 8\sin^2\Theta\left(\frac{\Theta}{s}\right)' - 2\sin 2\Theta\left(\Theta'^2 - \frac{\sin^2\Theta}{s^2}\right) + \sin 2\Theta - 2\Theta \right\}, \tag{6.38}$$

where we have used the Green function for the linearized problem

$$G(x,s) = -\frac{s^2}{3} \times \begin{cases} s/x^2, & \text{for } x > s, \\ x/s^2, & \text{for } x < s. \end{cases} \tag{6.39}$$

Applying the estimate (6.37) from (6.38) with (6.39) we derive that $|\Theta(b)| = \mathcal{O}(b^{-3/2})$ and after the second performance finally arrive at the estimates:

$$|\Theta(b)| = \mathcal{O}(b^{-2}), \qquad |\Theta'(b)| = \mathcal{O}(b^{-3}). \tag{6.40}$$

Now we are equipped enough to prove by means of prepared estimates (6.31), (6.34) and (6.40) that a minimizing sequence $\{\Theta_n\}$ of smooth functions, which satisfy boundary conditions (6.28), would be bounded in the Sobolev space $W_2^1(a,b)$ with the norm

$$\| \Theta \| = \left(\int_a^b dx\, x^2 \cdot \Theta'^2 + \Theta^2(1) \right)^{1/2}, \quad a < 1 < b. \tag{6.41}$$

In accord with the Tikhonov-Alaoglu theorem from the boundod sequence $\{\Theta_n\}$ it is always possible to choose such a weakly convergent subsequence, so that $w - \lim \Theta_n = \Theta$ in $W_2^1(a,b)$, where by "w – lim" we denote a limit in the weak sense. The next useful fact (Sobolev 1963) is the compactness of embedding of $W_2^1(a,b)$ into the space of continuous functions $\mathbf{C}(a,b)$: $(W_2^1(a,b) \hookrightarrow \mathbf{C}(a,b))$. This fact tells us that $\{\Theta_n\}$ is strongly convergent to an element Θ^* in $\mathbf{C}(a,b)$.

To prove the lower weak semicontinuity of the functional Φ_{ab} we have to examine convergence of vectors h_i, introduced in (6.30). It is easy to check that all $h_i(\Theta_n)$ are vectors in Hilbert space $L_2(a,b)$. Functions h_3 and h_4 are continuous with respect to Θ, this means they converge to their limits.Furthermore, the sequence $\{\Theta'_n\}$ is weakly convergent to Θ'^* in $W_2^1(a,b)$, whence $h_1(\Theta'_n)$ is weakly convergent in $L_2(a,b)$ to the limit vector $h_1(\Theta'^*)$. At last by definition in (6.30) $h_2(\Theta_n)$ is the product of two sequences: $\{\sqrt{2}\,\Theta'_n\}$, which is weakly convergent in $L_2(a,b)$, and $\{\sin\Theta_n\}$, which in its turn is strongly convergent in $\mathbf{C}(a,b)$. The result of this product would be weakly convergent in $L_2(a,b)$. Thus we find that all vectors h_i, in terms of which we presented our functional in (6.29) as a sum of scalar products, are weakly convergent to their limits. Theorem 6.1 then proves that Φ_{ab} **is a lower weakly semicontinuous functional.**

Finally, we let a and b tend to zero and infinity, respectively, ($a \to 0$, $b \to \infty$) and notice that the functional Φ_{ab} is a continuous function when regarding it with respect to a and b, since by virtue of the estimates (6.31) and (6.40) it holds, that

$$\int_0^a dx\mathcal{H} = \mathcal{O}(a^3); \quad \int_b^\infty dx\mathcal{H} = \mathcal{O}(b^{-3}). \tag{6.42}$$

Therefore one is allowed to perform this limit transition, and as a consequence in view of (6.27) Inf $\Phi[\Theta]$ **is weakly attainable** in the space $W_2^1(0,\infty)$ for $\mathbf{Q} = n \in \mathbb{Z}$.

It remains only to check the regularity of the weak solution $\Theta_0(r)$, and thereby to make this result essential for physical applications (Rybakov and Sanyuk, 1981, 1991). The required regularity follows directly from a one more representation of Eq. (6.2):

$$\Theta'' = F(x,\Theta,\Theta'), \tag{6.43}$$

where

$$F = \left(x^2 + 4\sin^2\Theta\right)^{-1}\left\{\sin 2\Theta\left[1 - 2\left(\frac{d\Theta}{dx}\right)^2 + \frac{2\sin^2\Theta}{x^2}\right] - 2x\frac{d\Theta}{dx}\right\}.$$

Notice that for some $a > 0$, $x\Theta', \sin\Theta \in L_2(a,\infty)$, hence $F \in L_1(a,\infty)$ and Eq. (6.43) is equivalent to

$$\Theta'(x) = -\int_x^\infty dx\, F(x,\Theta,\Theta'), \quad x \geq a > 0. \tag{6.44}$$

The latter equation implyes the boundness of $\Theta'(x)$ for $x \geq a$, and thus the function F is bounded as well. It means that Θ' and F are smooth functions for $x \geq a$. At last, after sufficient repetition of this very logic, one deduces that Θ', F and Θ'' are infinitely smooth functions for $x \geq a$, or in other words $\Theta(x) \in \mathbf{C}^\infty(a,\infty)$. Letting $a \to 0$ and taking into account Eq. (6.31) one comes to the conclusion that the limiting function $\Theta_0(x)$ belongs to the space of real-valued analytic functions $\mathbf{C}^\infty(0,\infty)$. This ends the proof of existence of regular solutions (Skyrmions).

7. Multi-Baryon and Rotating Skyrmion States

"What is the difference between science and other ways of talk? In science you do not give up from the outset the hope that your concepts might in the end be a little clearer than in the beginning."

Niels Bohr

After the good helping of mathematics in previous chapters it seems necessary to have a short break for physics. Beyond any doubt the problems of bound states and interactions among Skyrmions are of great importance as long as one considers possible applications of the Skyrme model in particle and nuclear physics. For this reason Skyrme's previous results were reexamined at once after the renewal of interest in 1983. In a moment we shall see that the above questions are closely related to the problem of the existence of stable soliton structures in higher homotopy classes with $\mathbf{Q} \geq 2$. Remember that Skyrme himself obtained a negative answer to the problem of bound states (see Section 1.5) and this appears to be his only result in the field which was discovered to be erroneous under further investigations.

7.1 The Problems of Bound States and Interaction Among Skyrmions

We already mentioned in Chapter 5 the existence of stable bound states, called *dibaryons*, which was disclosed in the $SU(3)$ generalized Skyrme model by A.P. Balachandran et al. (1984). Then Braaten and Carson numerically found the bound states in a modified product-ansatz approximation, computing the static energy as a function of the relative isospin orientation and the half-separation of two Skyrmions (Braaten and Carson 1986a). Their results differed from Skyrme's, cited in Section 1.5, as the static energy of a two-Skyrmion state was found to be $\mathcal{E}^{(2)} = 2\,\mathcal{E}^{(1)} - 24\,\text{MeV}$, when the half-separation of Skyrmions is $(\boldsymbol{r}_1 - \boldsymbol{r}_2)/2 = 1.1\,\text{fm}$. In support of these results it was confirmed by further analysis in (Braaten and Carson 1988, 1989), that the $\mathbf{Q} = 2$ soliton could be identified with the deuteron, as the corresponding quantum state has correct quantum numbers and gives reasonable values for such observables as mean charge radius, magnetic and quadrupole moments. The values obtained for $\mathbf{Q} = 2$ solitons agreed with experimental data for the deuteron to within 30%. Later on the crude variational ansatz used in

(Braaten and Carson 1986a) was improved in series of papers (Verbaarschot et al. 1987; Kopeliovich and Stern 1987; Schramm, Dothan and Biedenharn 1988), where it was justified that the variational solution is indeed a $\mathbf{Q} = 2$ soliton with the lowest energy among configurations from the second homotopy class. The abovementioned results did much to rescue the nuclear physics reputation of the Skyrme model, as without allowance for bound states no one baryon model can deserve attention from the pragmatic point of view.

In order to make the above cited numerical results more visual, here we give a topological description of the system of two interacting Skyrmions, i.e. the $\mathbf{Q} = 2$ configurations, in the adiabatic approximation. The latter means that we are going to minimize the energy of the configuration after fixing the geometrical centers of the Skyrmions at some points, say, $\boldsymbol{r}_1$ and $\boldsymbol{r}_2$. The positions of the centers can be found from the condition

$$\Theta(\boldsymbol{r}_i) = \pi, \quad \text{or} \quad \mathrm{U}(\boldsymbol{r}_i) = \mathbb{1}. \tag{7.1}$$

On the other hand, Eq. (7.1) reads: the field configuration $\mathbf{U}$ in the neighborhoods of points $\boldsymbol{r}_1$ and $\boldsymbol{r}_2$ coincides with the field of a single Skyrmion Eq. (1.29) or Eq. (4.33), which we denote here by $\mathbf{U}_0$. That is why the surfaces $\Theta(\boldsymbol{r}_i) =$ const $\approx \pi$ are close to spherical surfaces (see Fig. 7.1).

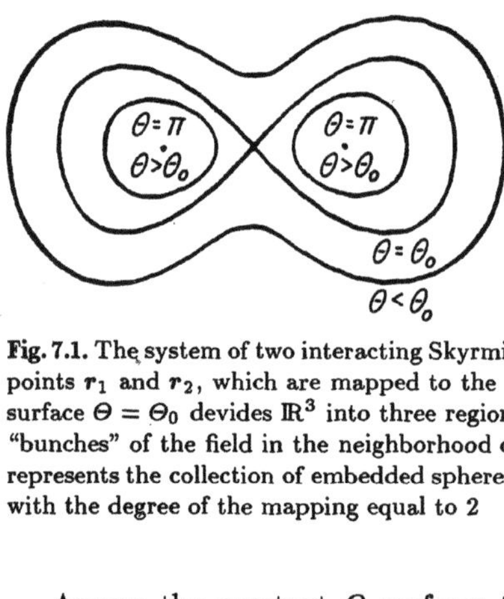

Fig. 7.1. The system of two interacting Skyrmions in $\mathbb{R}^3$, with centers defined as two singular points $\boldsymbol{r}_1$ and $\boldsymbol{r}_2$, which are mapped to the south pole of $\mathbb{S}^3 \approx SU(2)$, where $\Theta = \pi$. The surface $\Theta = \Theta_0$ devides $\mathbb{R}^3$ into three regions. Two internal ones ($\Theta > \Theta_0$) correspond to "bunches" of the field in the neighborhood of the centers, and the outer region ($\Theta < \Theta_0$) represents the collection of embedded spheres $\mathbb{S}^2$ that are mapped to the surface $\Theta =$ const with the degree of the mapping equal to 2

Among the constant Θ surfaces there exists a distinguished one corresponding to a certain value $\Theta = \Theta_0$ and homeomorphic to a connected sum of two spheres $\mathbb{S}^2$. This surface divides $\mathbb{R}^3$ into three regions: two internal ones ($\Theta > \Theta_0$), with points $\boldsymbol{r}_1$ and $\boldsymbol{r}_2$, and the outer region ($\Theta < \Theta_0$). And each surface $\Theta =$ const is mapped into some region $\Omega \subset SU(2)$, which is homeomorphic to the sphere $\mathbb{S}^2$ ($\mathbb{S}^2$-"circles" on the surface of the $\mathbb{S}^3$ sphere). This fact comes from the observation that after fixing Θ in the expression for the chiral field from Eq. (1.29)

$$\mathrm{U}(\boldsymbol{r}, t) = \exp\{i(\underline{\mathbf{n}} \cdot \underline{\tau})\Theta(\boldsymbol{r}, t)\} \tag{7.2}$$

the unit vector $\boldsymbol{n} \in \mathbf{S}^2$ remains free and thus performs the aforementioned mapping. From Fig. 7.1 it is clear that all connected parts of the constant Θ surfaces with $\Theta > \Theta_0$ are mapped into $\mathbf{S}^2$ with the degree 1. On the other hand all the constant Θ surfaces with $\Theta < \Theta_0$ are mapped into the sphere $\mathbf{S}^2$ with degree 2. For this reason the topology of the field configurations in the region $\Theta < \Theta_0$ coincides with the topology of the G_2-invariant configurations with $\mathbf{Q} = 2$, mentioned in Remark 4.1 and to be examined more closely in the next subsection. As the centers of the Skyrmions approach each other, Θ_0 increases and in the limit becomes equal to π. The field configuration obtained in this way is identical to an axisymmetric one with $\mathbf{Q} = 2$, which, however, does not correspond to the closest possible approach of the Skyrmions, as far as we have the toroidal distribution for the energy density of this state. The further reconciliation would mean a draining of the "mesonic fluid" (remember Skyrme's initial image from Section 1.1) into the central region, and this process would lead to the formation of a region with $\Theta > \pi$. As far as the latter region grows we will finally get the G_1-invariant field configuration with $\mathbf{Q} = 2$. But this would be the situation where Skyrme's result is valid, namely: the energy of the $\mathbf{Q} = 2$ hedgehog configuration $\mathcal{E}^{(2)} \approx 3\,\mathcal{E}^{(1)}$, where $\mathcal{E}^{(1)}$ is a single Skyrmion energy. Thus there would be a strong repulsion between the Skyrmions (with the energy of an order of magnitude $\mathcal{E}^{(1)}$).

The standard description of the Skyrmions' interactions is based on *"the product ansatz" approximation* Eq. (1.58), which in a more precise form might be written as

$$\mathbf{U}_{12} = \mathbf{U}_0^A(\boldsymbol{r} - \boldsymbol{r}_1) \cdot \mathbf{U}_0^B(\boldsymbol{r} - \boldsymbol{r}_2). \tag{7.3}$$

Here $\mathbf{U}_0^A = A \cdot \mathbf{U}_0 \cdot A^\dagger$, $\mathbf{U}_0^B = B \cdot \mathbf{U}_0 \cdot B^\dagger$; where $A, B \in SU(2)$ and $A^\dagger$ denotes the Hermitian conjugate of the element A. If the matrices A and B are different, then Eq. (7.3) describes Skyrmions with relative rotation in isospace. In particular, from Fig. 7.1 it follows that the maximal attraction between Skyrmions arises when the relative rotation is through angle π about an axis perpendicular to the line of separation of two Skyrmions. Indeed, the spherical constant Θ surfaces in the limit $\Theta \to \Theta_0$ must touch each other at points that are mapped to the same $SU(2)$ element, so that precisely the rotation of the spheres through angle π is obtained. Thus by means of relatively simple topological considerations (Rybakov 1984; Rybakov and Sutantio 1985) we arrive at exactly the same conclusions achieved in (Braaten and Carson 1986a) through numerical investigations.

Remark 7.1. From the above considerations it is clear that the configuration (7.3) faithfully reproduces the interactions between Skyrmions at large distances and gives only a qualitatively correct description at small distances. At the same time, one cannot extract from that kind of description, for example, a formation of a G_2-invariant configurations at the intermediate distances. These conclusions are in agreement with the fact that the Skyrme model description is the simplest low-energy approximation to QCD, when one takes into account only the leading order terms in the $1/N_c$-expansion. Because of this one cannot expect to get a complete picture of hadron interactions in the framework of this model. Nevertheless, the main qualitative features of the interactions may be correctly reproduced, as it has been demonstrated.

7.1.1 The Invariant Fields in Higher Homotopy Classes

Now it is high time to perform the promised symmetric analysis of admissible invariant field configurations in higher homotopy classes with $\mathbf{Q} \geq 2$. In this analysis it proves to be convenient to distinguish *the classes of isotopic mappings* corresponding to "single-particle" states (Kozhevnikov, Rybakov and Fomin 1988).

Definition 7.1. *Two fields (or mappings) ϕ_1 and ϕ_2: $\mathbb{R}^3 \rightarrow SU(2)$ are called* **isotopic** *if there exist homeomorphisms $\eta_1 : \mathbb{R}^3 \rightarrow \mathbb{R}^3$ and $\eta_2 : SU(2) \rightarrow SU(2)$ such that $\phi_2 = \eta_2 \cdot \phi_1 \cdot \eta_1^{-1}$*

In order to construct invariant representatives of isotopic classes, we identify a certain point in $\mathbb{R}^3$ with the center of a particle and associate with it a decomposition of $\mathbb{R}^3$ into a collection of embedded surfaces homeomorphic to $\mathbb{S}^2$. Suppose that among these surfaces there are singular ones that are mapped to a single point on $SU(2)$ and hence effectively divide $\mathbb{R}^3$ into n spherical layers, each of which is entirely mapped onto SU(2). The $\mathbb{S}^2$ surfaces in the layers go over into their images (also homeomorphic to two dimensional spheres) with degrees m_i. Thus the class of isotopic mappings is determined by the set of integers $\{n, m_i\}$, $i = 1, 2, \ldots, n$, and is contained in the homotopy class with $\mathbf{Q} = \sum_{i=1}^{n} m_i$. The convenience of this tool for a more delicate study of the structure of the minimizing fields will be evident from what follows. Now for each class of isotopic mappings we find a compact symmetry group that admits invariant fields in the framework of this class, then by virtue of the Coleman-Palais theorem (see Section 4.3) the expressions obtained for invariant fields will be configurations that satisfy the previous equation of motion (1.44).

In Section 4.2, when deducing the maximal symmetry group, we identified spatial and internal rotations by means of an automorphism

$$\eta : SO(3)_S \rightarrow SO(3)_I .$$

The easily verified statement is that for the hedgehog ansatz Eq. (4.32) or Eq. (4.33) the orbits of the group $SO(3)_S$ are mapped to the orbits of the group $SO(3)_I$ with degree $m = 1$, and the topological charge of the hedgehog configurations (4.33) is equal in general to the number of layers: $\mathbf{Q} = n$. Another obvious claim is that the case $m \geq 2$ is not allowed by the $SO(3)$ symmetry. Hence, to describe configurations with $m \geq 2$ in the framework of invariant fields it is necessary to restrict the group $SO(3)$ to $SO(2)$. In the same manner as we have proved Proposition 4.2 it is possible to verify that

the maximal compact group, which possesses invariant fields with the degrees of mappings $m \geq 2$ is

$$G_2 = \mathrm{diag}[SO(2)_I \otimes SO(2)_S]. \tag{7.4}$$

The reader might feel it necessary to verify this as a small exercise.

To obtain the structure of field configurations invariant against the transformations from the group (7.4), we choose the group of rotations about the Z axis as $SO(2)_S$, and we relate $SO(2)_I$ to the generator[1] σ_3. Homomorphisms of these groups are determined by the relation $\varphi_I = m\varphi_S$, $m \in \mathbb{Z}$, where by φ_S and φ_I we denote the angles of the spatial and internal rotations, respectively. By direct calculations analogous to those performed in Section 4.4, one can obtain the following general solutions of Eqs. (4.9) for the group G_2, which is the axisymmetric ansatz:

$$\Theta = \Theta(r, \vartheta), \quad \beta = \beta(r, \vartheta), \quad \gamma = m\alpha, \tag{7.5}$$

where (r, ϑ, α) are spherical coordinates in $\mathbb{R}^3$, $m \in \mathbb{Z}$. This is another good exercise for the thoughtful reader (see also Appendix C.2). Here we would rather arrive at the same result in a different way. Note that the G_m-invariant fields (where m specifies the above homomorphism) determine a mapping of an $SO(2)_S$ orbit to an $SO(2)_I$ orbit, which is described by functions $\Theta(\rho, z)$, and $\beta(\rho, z)$ of the cylinder coordinates ρ and z. For the orbits themselves the condition of G_m invariance leads to the relation $\gamma = m\varphi + \xi$, where ξ is some angle that is constant for a given orbit. Whence the complete expression for G_m-invariant fields in the cylinder coordinates is

$$\Theta = \Theta(\rho, z), \quad \beta = \beta(\rho, z), \quad \gamma = m\varphi + \xi(\rho, z). \tag{7.6}$$

The boundary conditions on the function β for the configurations of the considered type have the form

$$\beta(0, z > 0) = 0, \qquad \beta(0, z < 0) = \pi. \tag{7.7}$$

Summarizing, we must bring to the reader's attention, that besides the axisymmetric invariant field configurations (7.5) and (7.6) in each homotopy class with $\mathbf{Q} \geq 2$ we still have the hedgehog spherically symmetric configurations with $m = 1$ and $\mathbf{Q} = n \geq 2$. Formulas (7.5) and (7.6) demonstrate that in the case of axial symmetry several realizations of the charge $\mathbf{Q} \geq 2$ configurations are allowed. On the basis of Definition 7.1 it can easily be checked that within the first homotopy class these two types of invariant fields - the hedgehog and axisymmetric configurations - are isotopic (identical).

[1] Thus we fix axes of rotations in the space and isospace, but further results do not depend on a particular choice.

7.2 Minima of the Energy Functional in Higher Homotopy Classes

Let us show that it is possible to get the form of the ansatz (7.5) by direct minimization of the energy functional (5.2). Our exposition of this result will largely be based on the papers (Rybakov 1988; Kozhevnikov, Rybakov and Fomin 1988; Makhankov, Rybakov and Sanyuk 1989, 1992). In agreement with our previous considerations in Chapters 3 and 5 we start by seeking for the minimum of the Lyapunov functional for the Skyrme model

$$\begin{aligned}\mathbf{H}-\mu|\mathbf{Q}| = \int d^3x \Big\{ & \frac{1}{2\lambda^2}\Big[(\nabla\Theta)^2+\sin^2\Theta\big((\nabla\beta)^2+\sin^2\beta\,(\nabla\gamma)^2\big)\Big] \\ & +\epsilon^2\sin^2\Theta\big\{(\nabla\Theta\times\nabla\beta)^2+\sin^2\beta\,(\nabla\Theta\times\nabla\gamma)^2 \\ & +\sin^2\Theta\sin^2\beta\,(\nabla\beta\times\nabla\gamma)^2\big\} \\ & +\frac{\mu}{2\pi^2}\sin^2\Theta\sin^2\beta\,\nabla\Theta\cdot(\nabla\beta\times\nabla\gamma)\Big\}\,, \end{aligned} \tag{7.8}$$

where as before $\mu = 12\,\pi^2\epsilon/\lambda\sqrt{2}$ is taken from the estimate (1.43). We exploit once again *the method of extending the phase space* from Section 5.1, which consists of regarding the field functions and their derivatives as independent and then proving that the minimum obtained in this way is indeed a critical point of the original functional.

We fix the value of the topological charge $\mathbf{Q} = N$ and take once again the angles $\{\upsilon,\kappa,\chi\}$ between gradients $\nabla\Theta, \nabla\beta, \nabla\gamma$ as extended variables. Those angles are presented in Fig. 5.2, and can be expressed as functions of gradients, making use of formulas of spherical trigonometry. Then the part of the integrand in Eq. (7.8) which depends on the angles υ, κ and χ can be written as

$$\begin{aligned} F = & -C\left(\cos\upsilon\cos\chi+\sin\upsilon\sin\chi\cos\kappa\right)^2 \\ & +A\,\sin^2\upsilon+B\,\sin^2\chi-2D\,\sin\upsilon\sin\chi\sin\kappa, \end{aligned} \tag{7.9}$$

where we denote

$$\begin{aligned} & A=\epsilon^2\sin^2\Theta(\nabla\Theta)^2(\nabla\beta)^2, \quad B=\epsilon^2\sin^2\Theta\sin^2\beta(\nabla\Theta)^2(\nabla\gamma)^2, \\ & C=\epsilon^2\sin^4\Theta\sin^2\beta(\nabla\beta)^2(\nabla\gamma)^2,\ D=\frac{\mu}{4\pi^2}\sin^2\Theta\sin\beta|\nabla\Theta||\nabla\beta||\nabla\gamma|. \end{aligned} \tag{7.10}$$

Let us also note that from the definition of the topological charge $\mathbf{Q} = n \neq 0$ as the degree of mapping $\mathbb{R}^3 \to \mathbb{S}^3$ there follows an important property of angular variables on the sphere $\mathbb{S}^3$. Namely, in terms of the spherical coordinates (r,ϑ,α) we find that Θ and β are to be periodic functions of α, and γ must satisfy the following boundary condition:

$$\gamma(\alpha+2\pi)=\gamma(\alpha)+2m\pi, \quad m\in\mathbb{Z}. \tag{7.11}$$

The condition (7.11) signifies that the sphere $\mathbf{S}^2$ has been covered at least m times. Therefore $\gamma = \gamma_1 + \gamma_2$, where $\gamma_1 = m\alpha$ and γ_2 is a function periodic in α.

We first find the critical points of the function (7.9) with respect to the angles χ and κ. For this it is sufficient to set $D = 0$ in Eq. (7.9), since the critical points of the function F and those of the initial functional (5.2) must be identical. Minimizing the rest of Eq. (7.9) we obtain the equations

$$\begin{aligned} &B \sin 2\chi - 2C \cos\delta\big(\sin\upsilon \cos\chi \cos\kappa - \cos\upsilon \sin\chi\big) = 0, \\ &\cos\delta \sin\upsilon \sin\chi \sin\kappa = 0, \end{aligned} \tag{7.12}$$

where we have used the notation

$$\cos\delta \equiv \cos\upsilon \cos\chi + \sin\upsilon \sin\chi \cos\kappa. \tag{7.13}$$

Since $\mathbf{Q} \neq 0$ the expression for the topological charge $\mathbf{Q}$ in terms of angles $\{\upsilon, \kappa, \chi\}$ from Eqs.(7.8) and (7.9) implies $\sin\upsilon \sin\chi \sin\kappa \neq 0$, and hence, the unique required solution of the Eq. (7.12) is

$$\cos\chi = \cos\kappa = 0, \tag{7.14}$$

or equivalently

$$\big(\nabla\Theta \cdot \nabla\gamma\big) = \big(\nabla\beta \cdot \nabla\gamma\big) = 0. \tag{7.15}$$

With allowance for (7.15) the functional (7.8) takes the form

$$\begin{aligned} \mathbf{H} - \mu|\mathbf{Q}| = \int d^3x \, \Big\{ &\frac{1}{2\lambda^2}\Big[(\nabla\Theta)^2 + \sin^2\Theta((\nabla\beta)^2)\Big] \\ &+ (\nabla\gamma_1 + \nabla\gamma_2)^2 \sin^2\Theta \sin^2\beta \Big[\frac{1}{2\lambda^2} + \epsilon^2\big((\nabla\Theta)^2 + \sin^2\Theta(\nabla\beta)^2\big)\Big] \\ &+ \epsilon^2 \sin^2\Theta(\nabla\Theta \times \nabla\beta)^2 + \frac{\mu}{2\pi^2} \sin^2\Theta \sin^2\beta(\nabla\Theta \times \nabla\beta) \cdot \nabla\gamma_1\big)\Big\} \end{aligned} \tag{7.16}$$

By an analysis analogous to that used in Section 5.1, we convince ourselves that the minimum of the functional (7.16) is attained if the vectors $\nabla\gamma_1$ and $\nabla\gamma_2$ are antiparallel, i.e., for

$$\gamma = \gamma(\alpha). \tag{7.17}$$

Finally it follows from (7.15) and (7.17) that

$$\frac{\partial\Theta}{\partial\alpha} = \frac{\partial\beta}{\partial\alpha} = 0. \tag{7.18}$$

With the help of Eq. (7.17) and Eq. (7.18) the Euler-Lagrange equation for $\gamma(\alpha)$ reduces to

$$\frac{\partial^2\gamma}{\partial\alpha^2} = 0, \tag{7.19}$$

whence $\gamma = m\alpha$. Thus we find the following structure of the required solution in terms of spherical coordinates (r, ϑ, α):

$$\Theta = \Theta(r, \vartheta); \quad \beta = \beta(r, \vartheta); \quad \gamma = m\alpha,$$

which coincides with Eq. (7.5) completely and with Eq. (7.6) for $\xi = 0$. The topological charge for these configurations takes the value

$$\mathbf{Q} = mn, \quad \text{where} \quad n = \max\left(\frac{\Theta}{\pi}\right). \tag{7.20}$$

It remains to verify that the function (7.9) does indeed have a minimum at the point (7.14). For this we find its second variation with respect to κ and χ:

$$\begin{aligned} \tfrac{1}{2}\delta^2 F = (\delta\chi)^2 &(D \sin v - B - C \cos^2 v) \\ &+ (\delta\kappa)^2 (D \sin v - C \sin^2 v) - \delta\chi\, \delta\kappa\, C \sin v. \end{aligned} \tag{7.21}$$

This quadratic form will be positive definite (see Section 3.1) if the following inequality holds

$$2D \sin v > B + C + \left[B^2 + C^2 + 2BC \cos 2v\right]^{1/2}. \tag{7.22}$$

The condition (7.22) must be verified locally by substituting in it the required soliton solution. The proof of existence of axisymmetric solitonic configurations (7.5) would be given in Appendix D.2.

It might be interesting to see the correspondence between the obtained configuration (7.5) and the hedgehog or Skyrmion configuration (5.31) or (5.35), where

$$\Theta = \Theta(r); \quad \beta = \vartheta; \quad \gamma = \alpha, \tag{7.23}$$

with the values of the topological charge

$$\mathbf{Q} = n = \max\left(\frac{\Theta}{\pi}\right) = \frac{\Theta(0)}{\pi}. \tag{7.24}$$

For this we note, taking into account Eq. (7.14), that the function (7.9) takes the form

$$F(v) = A \sin^2 v + B - 2D \sin v, \tag{7.25}$$

and therefore its critical point, which coincides with the critical point of the Skyrme functional, is

$$v = \frac{\pi}{2}. \tag{7.26}$$

The same value for v was obtained in Eq. (5.5). At the same time the second variation of F is

$$\delta^2 F = 2(\delta v)^2 (D - A). \tag{7.27a}$$

and the condition of its positive definiteness reduces to the inequality

$$D > A. \tag{7.27b}$$

Neither the condition (7.22) nor the condition (7.27b) can be verified up to the moment analytically, as even for the hedgehog configuration we do not have any exact solution (see the discussion in Section 6.1). Thus we can verify the conditions for the second variation to be positive definite only numerically, on the basis of an appropriate choice of trial functions. Numerical data on the profile of function $\Theta(r)$ show that the inequality (7.27) holds only for Skyrmions with $|\mathbf{Q}| = 1$. Hence, this result agrees with the statement of Theorem 5.3. It should be emphasized that though we do not know an exact form of an analytical solution to Eq. (6.2) we definitely know that the required solutions do exist and the Skyrmion configuration is the true minimum of the energy functional in the first homotopy class (see Section 6.3).

The situation in higher homotopy classes with $\mathbf{Q} \geq 2$ is essentially less definite. First of all we have here a more complicated situation, as we do not have a unique class of extremals and it is necessary to compare the energy values of hedgehog (7.23) and axisymmetric (7.5) configurations. To perform this comparison of the energies let us remember that, with the increasing topological charge N, the energy of the spherically-symmetric configuration (7.23) increases in proportion to $N(N+1)$ (see formulas (6.7) and (6.8)). At the same time the energy in each homotopy class is bounded from below in terms of the topological charge only linearly, in accord with the estimate (1.43). Hence, the absolute minimum of the energy for higher homotopy classes seems not to be realized in the class of spherically-symmetric functions with $m = 1$, and we have to consider axially symmetric fields with $m \geq 2$.

Direct minimization of the functional (7.16) on the axisymmetric configurations (7.6) or (7.5) leads to a two-dimensional vector problem for the functions Θ and β. Along the lines of the proof displayed in Section 6.3, it is possible to prove the existence of regular axisymmetric configurations in higher homotopy classes as well. The sketch of this proof is given in Appendix D.2. To get more detailed information for the energy values one still has to turn to the method of trial functions. Numerical calculations in (Kopeliovich and Stern 1987; Verbaarschot 1987) show that the relation $\mathcal{E}_{1,m} < m\,\mathcal{E}_{1,1}$ holds with good accuracy up to $m = 5$. Here the notation $\mathcal{E}_{n,m}$ stands foi the energy of configurations, specified by integers n and m as defined in the previous section, e.g. $\mathcal{E}_{1,1}$ corresponds to the Skyrmion configuration. In particular, the value $\mathcal{E}_{1,2} \simeq 1.92\,\mathcal{E}_{1,1}$ was found for the second homotopy class. This is the only available confirmation of the fact that **the energy minimum in higher homotopy classes, or equivalently, for multi-baryon states is realized on the axisymmetric configurations.**

The interpretation of these results is that passing from the first to higher homotopy classes causes the minimal-energy configurations lose their symmetry. In the second homotopy class the spherically symmetric configurations are no longer extremals, but rather saddle points of the energy functional. Already there are available supercomputer calculations (Braaten, Carson and Townsend 1990) of the minimal-energy configurations for $\mathbf{Q} = 3, 4, 5, 6$. For the convenience of the reader we display in Fig. 7.2 the structures obtained, presenting surfaces of constant baryon number density in higher homotopy classes.

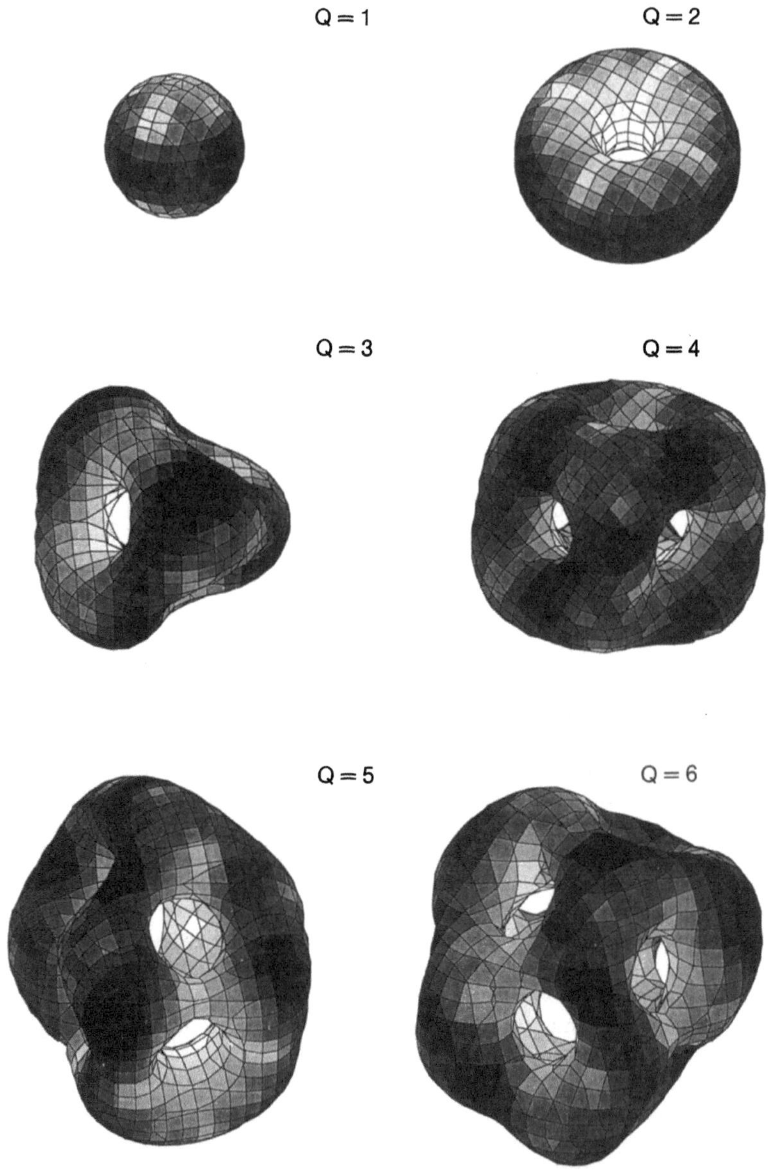

Fig. 7.2. The surfaces of constant baryon number density in homotopy classes with $Q = 3, 4, 5, 6$, obtained by calculations on a Cray-2 supercomputer, as reported in (Braaten, Carson and Townsend 1990)

From Fig. 7.2 it is obvious that for $\mathbf{Q} = 3,4,5,6$, thus obtained minimal-energy structures are not axisymmetric any longer, and we observe the phenomena of the further loss of symmetry. Already in the third homotopy class the minimal-energy configurations do not have any continuous symmetry, but instead a discrete (in particular, the tetrahedral) symmetry. Some dicsussion of these results may be found in (Sanyuk 1992). Without going into further detail here, we should say that the structure of the minimal-energy configurations is actually an interesting and an important problem and merits special study, both numerically and, (if one is too clever or lucky) analytically.

It should also be noted that the aforementioned calculations were performed in order to attest to the validity of the approach proposed in (Braaten and Carson 1986b, 1986c), which might be considered as an alternative approach to conventional nuclear physics. The central idea of Braaten-Carson's approach is that nuclei might arise as quantum states of solitons in higher homotopy classes with the topological charge $\mathbf{Q}$, which now may be interpreted as the atomic number. It is too early to discuss the validity of this promising proposal, as only initial steps have been made in that direction. But some remarks of general character are in order. First of all it is possible to recognize that once again Skyrme's "Mesonic Fluid" idea is developed at a new level for the study of nuclear structure. From that point of view the appearance of toroidally shaped solitons would not seem to be an accident, but rather support for Skyrme's initial image of the nucleon as a vortex in the "Mesonic Fluid." It is also amazing that the structures in Fig. 7.2 are somewhat similar to drawings in Fig. 1, representing Kelvin's various "Vortex Atoms" configurations. A possible way to understand the last similarity is to regard all those distinct physical situations as universal manifestations of Helmholtz's "Wirbelbewegung." One may say that **even ancient ideas do not die, if they are valid!**

7.3 The Rotating Skyrmion

An adequate description of the rotating excitations of Skyrmions is another important problem which certainly deserves a great deal of attention and also leads to axisymmetric field configurations. To extract the static properties of baryons in the framework of the soliton approach in general, we use the Bogolubov method of collective coordinates for the quantum description of the soliton's excitations. We treat this method in detail in the next chapter and here only remind the reader that in the pioneering paper of Adkins, Nappi and Witten (1983) the quantum description is based on *the rigid-body rotation ansatz*:

$$\mathbf{U}(\boldsymbol{x},t) = A(t) \cdot \mathbf{U}_0(\boldsymbol{x}) \cdot A^{\dagger}(t), \tag{7.28}$$

where $A(t) \in SU(2)$ and $\mathbf{U}_0(\boldsymbol{x})$ is the hedgehog ansatz (4.32). It should be noted that the substitution (7.28) has some essential shortcomings, which were reported in (Bander and Hayot 1984; Braaten 1984; Kudryavtsev and Martemyanov 1985). The main objections are that Eq. (7.28) does not allow the obvious deformations of the Skyrmion under rotations to be taken into account and the function $U(\boldsymbol{x},t)$ defined by (7.28) cannot be a solution of the equation of motion (1.44). Because the collective coordinates method is a semi-classical method of quantization, the latter incompatibility must be considered as a serious difficulty to the validity of the exploited quantization scheme. On the other hand, the rigid-body rotation of the Skyrmion in the isotopic space contradicts the "Mesonic Fluid" model, which, as we remember, was the background idea for the Skyrmion. Therefore the configuration (7.28) seems to be meaningful only for slow rotations, the property to be supposed within the frame of semi-classical approximation. As was noticed in (Potapov and Rybakov 1989) and will be explained in detail below, the above supposition can be well founded only for the nucleon state, but it fails for the case of Δ-resonance as the angular velocity approaches the critical value $\omega_c = m_\pi$ that corresponds to the radiative regime.

In a development of these notes let us consider rotations of the Skyrmion which will be compatible with the equation of motion (1.44) and at the same time will allow deformations of the static configuration. As a particular realization of such a motion we can choose a stationary rotation about Z axis, given by a substitution:

$$\frac{\partial}{\partial t} = -\omega\frac{\partial}{\partial \alpha} \quad \text{or} \quad \alpha \to \alpha - \omega t. \tag{7.29}$$

When dealing with large deformations of the rotating Skyrmion it proves appropriate to describe its motion by the axially symmetric distributions:

$$\phi_0 = \cos\Theta(r,\vartheta), \qquad \boldsymbol{\phi} = \sin\Theta(r,\vartheta)\boldsymbol{n}, \qquad \boldsymbol{\phi} = (\phi_1,\phi_2,\phi_3), \tag{7.30}$$

where the unit vector $\boldsymbol{n}$ is determined by the polar coordinates

$$\beta = \beta(r,\vartheta), \qquad \gamma = \alpha - \omega t. \tag{7.31}$$

Effectively this leads to the following transformation of the Skyrme Lagrangian from Eq. (1.39):

$$\begin{aligned}\mathcal{L}_{\text{rot}} = &-\left\{\frac{1}{2\lambda^2} + \epsilon^2\sin^2\Theta\sin^2\beta\left(\frac{1}{r^2\sin^2\vartheta} - \omega^2\right)\right\}\\ &\times\left[(\partial_r\Theta)^2 + \frac{1}{r^2}(\partial_\vartheta\Theta)^2 + \sin^2\Theta\left((\partial_r\beta)^2 + \frac{1}{r^2}(\partial_\vartheta\beta)^2\right)\right]\\ &-\frac{1}{2\lambda^2}\sin^2\Theta\sin^2\beta\left(\frac{1}{r^2\sin^2\vartheta} - \omega^2\right) - \frac{m_\pi^2}{\lambda^2}(1-\cos\Theta)\\ &-\sin^2\Theta\frac{\epsilon^2}{r^2}\left(\partial_r\Theta\partial_\vartheta\Theta - \partial_r\beta\partial_\vartheta\beta\right)^2.\end{aligned} \tag{7.32}$$

Here we added the pion mass term, proportional to $(1-\cos\Theta)$. The necessity of this addition should be clear from what follows.

From the structure of the Lagrangian (7.32) it is straightforward to find after rewriting it in terms of the dimensionless variables

$$r \to r\epsilon\lambda, \quad \omega \to \frac{\omega}{\epsilon\lambda}, \quad m_\pi \to \frac{m}{\epsilon\lambda},$$

that we get a typical combination $1-\omega^2 r^2 \sin^2\vartheta$, which contains ω^2. This means that for $r \ll \omega^{-1}$ we can perform a decomposition in terms of ω^2, taking the latter as a small parameter and the spherically symmetric Skyrmion configuration $\{\Theta_0, \vartheta\}$ as a first approximation. Then one can present the functions Θ and β in the form

$$\Theta(r,\vartheta) \simeq \Theta_0 + \omega^2\,\Theta_1(r,\vartheta), \qquad \beta(r,\vartheta) = \vartheta + \omega^2\,\beta_1(r,\vartheta). \tag{7.33}$$

The perturbations Θ_1 and β_1 in (7.33) might be taken as expansions in terms of the Legendre polynomials:

$$\Theta_1(r,\vartheta) = \sum_{n=0}^{\infty} a_n(r) P_{2n}(\vartheta), \qquad \beta_1(r,\vartheta) = \sum_{n=1}^{\infty} b_n(r) P^1_{2n}(\vartheta). \tag{7.34}$$

If we restrict ourselves to terms in the Lagrangian (7.32), which are quadratic with respect to a_n and b_n, then the only nontrivial amplitudes in decompositions (7.34) would be a_0, a_1, b_1 (check this!). In this approximation we obtain an effective separation of angular variables and can easily proceed with further calculations.

Finally, for $r \gg \omega^{-1}$ the corresponding equations of motion are nearly linear. Let us exchange the field variables for

$$X = \sin\Theta\cos\beta, \quad Y = \sin\Theta\sin\beta \tag{7.35}$$

and take into account terms in the Lagrangian (7.32) up to the fourth order in X and Y. Straightforward analysis shows that solutions in the form Eq. (7.34) will conform to the following asymptotic behavior of the variables in Eq. (7.35) introduced:

$$X = \cos\vartheta(a + b\cos^2\vartheta), \quad Y = \sin\vartheta(c + d\cos^2\vartheta), \tag{7.36}$$

where a, b, c, d are radial functions, such that for $r \to \infty$

$$X \sim \frac{1}{r}\exp(-mr), \quad Y \sim \frac{1}{r}\exp(-m'r), \quad \text{where} \quad m' = \sqrt{m^2 - \omega^2}. \tag{7.37}$$

By matching the fields (7.33) with (7.36) at some radius $r = r_0$ it is possible to obtain a close approximation to the true field configuration.

It should be emphasized that the Skyrmion must rotate rather slowly in order to meet the condition $\omega < m$, as otherwise the configuration will be unstable (it will radiate). Actually, for $\omega > m$ we have a typical radiative asymptotic form

$$Y \sim \frac{1}{m''r}\exp(\mathrm{i}\,m''r), \quad \text{where} \quad m'' = \sqrt{\omega^2 - m^2}. \tag{7.38}$$

As the fields with asymptotic behavior of the type (7.38) are not quadratically integrable, a soliton configuration with $\omega > m$ is unrealizable (the field configurations would have infinite energy). The Skyrmion is unstable in this regime as it will start to radiate pions. The substantial contribution to the rotational excitation energy of the Skyrmion is due to the large distance region. A crucial aspect of this contribution is readily seen in the situation when $\omega \to m$ for $m' \to 0$ and the Skyrmion configuration takes the least among available asymptotics for the component Y. It appears that for a nucleon state the inequality $\omega \ll m$ holds for sure, as far as the asymptotic behavior $\Theta \sim \exp(-mr)/r$ conforms to the one-pion-exchange potential. For higher excitations (Δ-resonance) we have $\omega \lesssim m$ and the radiation regime conditions play the role of an essential restrictive factor.

We will proceed further with the advocated approach in the next section, where we will implement the above ideas in a quantization scheme. This activity relies on the hope that a substitution of the axially-symmetric configurations described above instead of the Skyrmion configuration into Eq. (7.28) will lead to some improvements in matching the results of the Skyrme approach with experimental data.

8. Quantization of Skyrmions

"Whatever they're made out of, when we find out later, they'll behave like this"

Richard Feynman

This is the next and the inevitable step towards a practical implementation of the Skyrme approach. The quantization procedure, which was found acceptable to this problem in the pioneering paper (Adkins, Nappi and Witten 1983), was known long ago as **Bogolubov's method of collective coordinates.** Note, that this method was originally developed and applied in (Bogolubov 1950) to *the polaron problem*, which has some close parallels to the description of a nucleon state in the soliton approach.

For a purely pedagogical purpose let us recall this analogy between the physical situation in the low-energy sector of the QCD and the 'low-energy' solid state physics (D'yakonov and Mirlin 1988). The solid state crystalline structure is built from electrons and nuclei, which are fundamental objects and which are bounded through the Coulomb interaction. Nevertheless an adequate description of the low-temperature phenomena in solids was found to be in terms of electrons with an effective mass m^* (polarons) interacting with lattice vibrations (phonons). It is remarkable that the appearence of these effective degrees of freedom is a consequence of the spontaneous breaking of the translational invariance of the system (the crystal).

Although the QCD first principles are formulated in the language of quarks and gluons, it became apparent, that this formulation is inappropriate at the distances 1–2 fm, which are characteristic for the hadron world On the other hand at these distances such essentially nonperturbative QCD phenomena as the color confinement and the spontaneous breaking of chiral symmetry come into play[1]. It means that the quark and the gluonic degrees of freedom are not essential in the hadron sector and we have to look for their effective substitutes. In the chiral approach a nucleon is a localized state of three "constituent" quarks with the total effective mass M bounded in the self-consistent pion field. This is again very close to the polaron definition, which we regard as a localized state of an electron with an effective mass in the self-consistent field of lattice vibrations. The above analogy might be considered as a justification or at least gives us some foundations for the application of the collective coordinate method to the Skyrmion quantization.

[1] For a brief account of the present status of the QCD see Section 9.1.

8.1 Bogolubov's Method of Collective Coordinates

The central idea of this semiclassical method consists in the possibility to separate the quantum fluctuations around a classical solution into two classes: the modes generated by symmetries of the action or the Hamiltonian of the classical system (the symmetry modes) and the modes, which are orthogonal to the previous. The symmetry modes might be considered as exceptional by purely physical reasons. The main feature of these modes is, that an infinitely small amount of energy is sufficient to generate them, and hence, the large amplitude fluctuations from the classical configuration are characteristic to the symmetry modes. It means that they would provide the leading contributions to quantum excitations around the classical ground state, and hence, to leading order only symmetry modes can be treated quantum mechanically exactly. Being applied this approach leads to the problem of the so-called zero modes, which arise whenever the classical solution is non-invariant under symmetries of the classical Hamiltonian. One of the ways out is the application of the collective coordinate method. In advance we should stress that thus developed quantization scheme would be by no means complete, because we retain just a few modes out of a possible infinite number of modes. In spite of all physical reasoning, this scheme must be rather considered as the simplest approximation to the quantum description of field systems. Nevertheless, in quantizing field theories with extended objects we do not have too much possibilities to choose from and thus have to follow this approximate scheme with all its lacks and defects.

The collective coordinate method consists in a change of variables leading to the extraction of the variables, which appear to be parameters of the symmetry group for the Hamiltonian of the system. Usually Bogolubov's transformations are performed to detach these *collective coordinates*. The modern exposition of the Bogolubov method and its subsequent generalizations in the strong coupling theory might be found in (Khrustalev, Razumov and Taranov 1980; Rajaraman 1982, Chapter 8). In application to the Skyrme model we recommend as the most understandable the following papers (Balachandran et al. 1985, Balachandran, Marmo, Skagerstam and Stern 1991; Balachandran 1986a,b; Chemtob 1985; Wudka 1985; Jain and Wadia 1985). Remark that some of these papers deal with the $SU(3)$ Skyrme model with the additional Wess-Zumino term in the action, in accord with the modern achievements in the field. For simplicity, here we expose the pure quantization scheme for the hedgehog configuration in the original $SU(2)$ model and leave all the modern trends for the subsequent chapters.

The Skyrme model Hamiltonian (1.42) is manifestly invariant under global translations as well as under rotations in either space or isospace. At the same time, the hedgehog ansatz (4.33) is not invariant under spatial $SO(3)$ or either isospin $SU(2)$ rotations, when they are performed separately. From discussion in Chapter 4 we learn that the configuration (4.33) is invariant only under the combined spatial and isospin rotations. Let us present this fact in a visual form, using for isospin rotation a matrix $A \in SU(2)$, then

$$\begin{aligned}\mathbf{U}_A(\boldsymbol{x}) &= A\cdot\mathbf{U}_0(\boldsymbol{x})\cdot A^\dagger = \exp\left\{\mathrm{i}A\boldsymbol{\tau}A^\dagger\cdot\frac{\boldsymbol{x}}{r}\Theta(r)\right\}\\ &= \exp\left\{\mathrm{i}\boldsymbol{\tau}\cdot R(A)\frac{\boldsymbol{x}}{r}\Theta(r)\right\} = \mathbf{U}(R\boldsymbol{x}) = \mathbf{U}_R(\boldsymbol{x}),\end{aligned} \tag{8.1}$$

where $R(A)$ is the matrix from the adjoint representation of $SU(2)$, defined as

$$R_{ij}(A) = \tfrac{1}{2}\mathrm{Tr}(\tau_i A\,\tau_j\,A^\dagger). \tag{8.2}$$

To check the latter statement it is easy to show, that the matrices R are real, orthogonal, and have unit determinant, hence they belong to the group $SO(3)$ and also form the adjoint representation of $SU(2)$. In the context of this section we must interpret this fact as a confirmation that the hedgehog configurations (4.33) belong to the class of degenerate in the energy solutions, as, due to aforementioned invariance of the Hamiltonian, configurations characterized by different A (or $\hat{R}$) would have the same energy value. On the other hand, the relation (8.2) tells us, that these two different configurations are to be identified, and hence we should quantize the single set of collective coordinates (either $\{A\}$ or $\{R\}$). Note, that both configurations are physically allowed, as they satisfy the boundary conditions $\mathbf{U}\to\mathbb{1}$ at spatial infinity.

To perform the quantization, in general, means to break down the observed degeneracy of classical configurations. Quantization of the translational degrees of freedom is essentially trivial: the energy eigenstates are given by linear momentum eigenstates. For quantization of the vibrational excitations we refer the reader to (Parmentola 1984; Meissner and Zahed 1986; Konoplich, Kudryavtsev, Martemyanov and Rubin 1988). Here we restrict ourselves to the rotational fluctuations of the type

$$\mathbf{U}(\boldsymbol{x},t) = A(t)\cdot\mathbf{U}_0\cdot A^\dagger(t), \tag{8.3}$$

with $\mathbf{U}_0$ standing for the classical hedgehog configuration (4.33) and with the matrix $A(t)\in SU(2)$ as **the collective coordinate for the Skyrmions**. Now we substitute (8.3) into the Lagrangian of the Skyrme model[2]

$$L = \int d^3x\left\{-\frac{F_\pi^2}{16}\mathrm{Tr}\,\mathbf{L}_\mu{}^2 + \frac{1}{32e^2}\mathrm{Tr}\,[\mathbf{L}_\mu,\mathbf{L}_\nu]^2\right\}, \tag{8.4}$$

and integrate out the degrees of freedom other than the collective coordinates $A(t)$. This leads to the effective Lagrangian in terms of the collective coordinates:

$$L(A,\dot{A}) = -M_{\mathrm{cl}} + I(\mathbf{U}_0)\,\mathrm{Tr}(\dot{A}\dot{A}^\dagger), \tag{8.5}$$

where $\dot{A}$ means dA/dt, M_{cl} is the classical Skyrmion mass which in accordance with (6.1) for the Eq. (8.4) takes the form

[2] For further applications it is convenient to use here the third expression from Eq. (1.39) for the Lagrangian density.

$$M_{\rm cl} = \frac{\pi F_\pi}{e} \int\limits_0^\infty dx \left\{ \left(\frac{d\Theta}{dx}\right)^2 \left(\frac{x^2}{2} + 4\sin^2\Theta\right) + \sin^2\Theta + \frac{2\sin^4\Theta}{x^2} \right\}. \quad (8.6)$$

The language of the rotational motion is quite natural here and one can regard

$$A^\dagger \dot{A} = \frac{\tau_i}{2} \mathrm{Tr}\left(\tau_i \cdot A^\dagger \cdot \dot{A}\right) \quad (8.7)$$

as an "angular velocity" and I as the "moment of inertia" for the rotating Skyrmion

$$I(\mathbf{U}_0) = \frac{2\pi}{3e^3 F_\pi} \int\limits_0^\infty dx\, x^2 \sin^2\Theta(x) \left\{ 1 + 4\left[\left(\frac{d\Theta}{dx}\right)^2 + \frac{\sin^2\Theta}{x^2} \right] \right\}. \quad (8.8)$$

Numerical calculation of integrals in (8.6) and (8.8) provides with the following value for the mass $M_{\rm cl}$ and the moment of inertia I (Adkins, Nappi and Witten 1983):

$$M_{\rm cl}(\mathbf{U}_0) \simeq 36.5 \frac{F_\pi}{e}; \quad I(\mathbf{U}_0) \simeq 106.7 \frac{1}{F_\pi e^3}. \quad (8.9)$$

Now we are ready to pass to the quantization procedure immediately, i.e. to introduce a set of canonical variables and to write the Hamiltonian in terms of these variables. The quantum states of the particle, corresponding to the classical Skyrmion solution, will arrise as the eigenfunctions of the Hamiltonian. As it was demonstrated by Balachandran (1986b) the quantization of the effective Lagrangian (8.5) is quite ambiguous, as it can be quantized in three different ways:

a. As for purely bosonic states,
b. As for purely fermionic states,
c. As for both bosonic and fermionic states.

As far as terms "bosonic" and "fermionic" are refered to the angular momenta associated with the states and we are not going to consider here issues involving statistics of these states, it is more appropriate to understand them as "tensorial" and "spinorial", respectively. The exposition in the next section is largely borrowed from the lecture notes (Balachandran 1986b; Makhankov, Rybakov and Sanyuk 1989) and the review (Meissner and Zahed 1986).

8.2 Canonical Quantization of Skyrmions

Since the collective variables $A(t) \in SU(2)$, they can be locally parametrized in two different ways:

$$A = a_0[\xi] + \mathrm{i}\,\boldsymbol{\tau} \cdot \boldsymbol{a}[\xi] = \begin{pmatrix} a_0 + \mathrm{i}a_3 & a_2 + \mathrm{i}a_1 \\ -a_2 + \mathrm{i}a_1 & a_0 - \mathrm{i}a_3 \end{pmatrix}. \quad (8.10)$$

Either we consider the set of parameters $(a_0, \boldsymbol{a})$ with the constraint $a_0^2 + \boldsymbol{a}^2 = 1$ and perform the quantization of a constrained dynamical system, or we introduce three independent variables $\{\xi^i\}$, $i = 1, 2, 3$, on $\mathbf{S}^3$ e.g.

$$a_0 = \cos\xi_1, \quad a_1 = \sin\xi_1 \sin\xi_2,$$

$$a_2 = \sin\xi_1 \sin\xi_2 \cos\xi_3, \quad a_3 = \sin\xi_1 \sin\xi_2 \sin\xi_3.$$

In terms of variables ξ^i we can treat our dynamical system as an unconstrained and apply the usual canonical quantization scheme. The results obtained in both quantization procedures for the spectrum, wave functions and algebra of operators completely agree. Thus it is enough only to expose here one of them, for example, the second one. For the first approach one may consult with (Adkins, Nappi and Witten 1983) and the lecture notes (Adkins 1987).

Using the fact that $\det A = 1$ we can also introduce nonsingular matrices N_{ij} and K_{ij} (see Appendix D.3) by the relations

$$A^{-1}[\xi(t)] \cdot \frac{\partial A[\xi(t)]}{\partial \xi^i} = \frac{\mathrm{i}}{2} N_{ij} \tau_j, \tag{8.11}$$

and

$$A[\xi(t)] \cdot \frac{\partial A^{-1}[\xi(t)]}{\partial \xi^i} = \frac{\mathrm{i}}{2} K_{ij} \tau_j. \tag{8.12}$$

Now A in the Lagrangian (8.5) is $A[\xi(t)]$ and $\dot{A}$ is $(\partial A / \partial \xi^i)\dot{\xi}^i$. In terms of ξ^i the Lagrangian density reads

$$\mathcal{L}_{Sk}(\xi, \dot{\xi}) = -M_{\mathrm{cl}} - \tfrac{1}{2} I(\mathbf{U}_0)\, \dot{\xi}^i N_{ij} N_{kj} \dot{\xi}^k, \tag{8.13}$$

while the canonical momentum conjugate to ξ^i is

$$P_i = \frac{\partial \mathcal{L}}{\partial \dot{\xi}^i} = -a(\mathbf{U}_0)) N_{ij} N_{kj} \dot{\xi}^k. \tag{8.14}$$

By definition, ξ_i and P_i satisfy the canonical Poisson brackets,

$$\{\xi_i, \xi_j\} = \{P_i, P_j\} = 0, \quad \{\xi_i, P_j\} = \delta_{ij}. \tag{8.15}$$

According to the standard Dirac's quantization prescription for an unconstrained system we pass to the quantum theory by exchange of the Poisson brackets (8.15) for commutators

$$[\hat{\xi}_i, \hat{\xi}_j] = [\hat{P}_i, \hat{P}_j] = 0, \quad [\hat{\xi}_i, \hat{P}_j] = \mathrm{i}\,\delta_{ij}. \tag{8.16}$$

Here $\hat{\xi}_i$ and $\hat{P}_i$ are operators in the Hilbert space $\mathcal{H}$ of square integrable functions on $SU(2)$ with the inner product

$$(\psi \cdot \phi) = \int_{SU(2)} d\mu(A)\, \psi^*(A)\, \phi(A), \tag{8.17}$$

where $d\mu(A)$ is the invariant Haar measure on $SU(2)$. Quantization as itself consists in finding an irreducible representation of the appropriate algebra of observables, defined on this Hilbert space $\mathcal{H}$. Since we quantize the rotational excitations of the classical configuration, it is natural to pass to the "angular" observables, which could be, for instance, angular momenta and generators of $SU(2)$ flavor (isospin) rotations, or the spin and isospin operators. The two latter follow from the usual Noether construction:

$$\hat{\mathbf{J}}_i = -\mathrm{i}\,\tfrac{1}{4}I(\mathbf{U}_0)\,\mathrm{Tr}\,(\tau_i \dot{A}^{-1}A) = N_{ij}^{-1}P_j, \tag{8.18}$$

$$\hat{\mathbf{I}}_i = -\mathrm{i}\,\tfrac{1}{4}I(\mathbf{U}_0)\,\mathrm{Tr}\,(\tau_i \dot{A}A^{-1}) = K_{ij}^{-1}P_j. \tag{8.19}$$

They fulfill the standard angular momentum algebra on $\mathbf{S}^3$ (Zahed and Brown 1986), i.e.

$$\left[\hat{\mathbf{J}}_i,\, \hat{\mathbf{J}}_j\right] = \mathrm{i}\,\epsilon_{ijk}\hat{\mathbf{J}}_k, \quad \left[\hat{\mathbf{I}}_i,\, \hat{\mathbf{I}}_j\right] = \mathrm{i}\,\epsilon_{ijk}\hat{\mathbf{I}}_k, \tag{8.20}$$

with $\mathbf{J}^2 \triangleq \mathbf{I}^2$ and with action on A as

$$\left[\hat{\mathbf{J}}_i,\, A\right] = A\cdot\frac{\tau_i}{2}, \quad \left[\hat{\mathbf{I}}_i,\, A\right] = -\frac{\tau_i}{2}\cdot A. \tag{8.21}$$

The latter relations mean that $\hat{\mathbf{J}}_i$ and $\hat{\mathbf{I}}_i$ induce $SU(2)_R$ and $SU(2)_L$ rotations respectively[3]. The Hamiltonian of the quantum system in terms of operators (8.18) and (8.19) takes the diagonalized form

$$\mathbf{H} = M_{\mathrm{cl}} + \frac{1}{2I(\mathbf{U}_0)}\cdot\hat{\mathbf{J}}^2 = M_{\mathrm{cl}} + \frac{1}{2\,I(\mathbf{U}_0)}\cdot\hat{\mathbf{I}}^2. \tag{8.22}$$

This coincides with the Hamiltonian of a spherical top in $SU(2)$ and one can expect that the spherical states $|I = J, I_3, J_3 >$ constitute a convenient basis for spanning the Hilbert space $\mathcal{H}$.

To proceed we need to find a representation of the angular momentum algebra (8.20) on $\mathcal{H}$. It is generally assumed that this representation must be irreducible. There are, however, two inequivalent irreducible representations of the algebra (8.20), which give rise to the ambiguity of the Skyrmion quantization. The form of the rotating Skyrmion configuration (8.3) is the same for both matricies A and $-A$. One may consider the latter matrix as the result of a 2π rotation of the Skyrmion configuration:

$$\psi(A\exp[\mathrm{i}2\pi\tau_3/2]) = \psi(-A). \tag{8.23}$$

Then for the quantum wave function either the condition

$$\psi(-A) = \psi(A), \tag{8.24}$$

[3] Note that if instead of generators of $SU(2)$ flavor rotations $\hat{\mathbf{J}}_i$ and $\hat{\mathbf{I}}_i$ ("left" $SU(2)$ generators) to consider corresponding angular momentum generators ("right" $SU(2)$ generators) the right hand side of commutation relations (8.20) will obtain the minus sign. For more details consult with (Balachandran, Marmo, Skagerstam and Stern 1991).

must hold for all Skyrmions, indicating that Skyrmions are bosons: $\psi \in \mathcal{H}^{(B)}$, or instead

$$\psi(-A) = -\psi(A), \tag{8.25}$$

Skyrmions are fermions in this case and $\psi \in \mathcal{H}^{(F)}$. Here $\mathcal{H}^{(B)}$ and $\mathcal{H}^{(F)}$ denote the Hilbert spaces of bosonic and fermionic states, respectively. From the standard course of quantum mechanics (Landau and Lifshits, 1977) we know that in the $|A>$ representation the wave function of the basis states are the usual spin wave functions, which can be expressed in terms of the group representation matrices as

$$\langle A|I = J, I_3, J_3\rangle = \left(\frac{2J+1}{2\pi^2}\right)^{1/2} (-1)^{J+J_3} D^J_{-J_3,I_3}(A), \tag{8.26}$$

where $D^J_{-J_3,I_3}(A)$ is an irreducible representation of $SU(2)$ so that

$$D^0(A) = I, \quad D^{1/2}(A) = A, \dots \tag{8.27}$$

The last formula in (8.27) reflects the fact that the fundamental ($J = 1/2$) spinor representation of $SU(2)$ is just $SU(2)$ itself. Using, for example the second parametrization for A in (8.26), after proper normalization one can find from (8.26) the following wave functions for the proton and the neutron

$$\begin{aligned} \langle A|p, 1/2\rangle &= \frac{-\mathrm{i}}{\pi}(a_1 + \mathrm{i}a_2); \\ \langle A|p, -1/2\rangle &= \frac{-1}{\pi}(a_0 - \mathrm{i}a_3); \\ \langle A|n, 1/2\rangle &= \frac{1}{\pi}(a_0 + \mathrm{i}a_3); \\ \langle A|n, -1/2\rangle &= \frac{\mathrm{i}}{\pi}(a_1 - \mathrm{i}a_2). \end{aligned} \tag{8.28}$$

The Δ wave functions are already cubic in the a's:

$$\begin{aligned} \langle A|\Delta^{++}, 3/2\rangle &= \frac{\sqrt{2}\mathrm{i}}{\pi}(a_1 + \mathrm{i}a_2)^3; \\ \langle A|\Delta^{+}, 1/2\rangle &= \frac{-\sqrt{2}\mathrm{i}}{\pi}(a_1 + \mathrm{i}a_2)(1 - 3(a_0^2 + a_3^2)). \end{aligned} \tag{8.29}$$

In general, higher representations are polynomials in A, and from (8.21) follows, that

$$[J_3, A_{11}^n] = [J_3, A_{11}] \cdot A_{11}^{n-1} + A_{11} \cdot [J_3, A_{11}^{n-1}] = \frac{1}{2}nA_{11}^n. \tag{8.30}$$

This means that even polynomials in $A : A^2, A^4 \dots$ correspond to tensorial wave functions with integer spin, while odd polynomials in $A : A^1, A^3, A^5, \dots$ correspond to spinorial wave functions with a half integer spin.

In Chapter 11 we will see that aforementioned ambiguity in the quantum description of Skyrmions can be removed by embedding the $SU(2)$ hedgehog in a larger flavor space ($SU(3)_f$) and using the gauge constraints provided by the Wess-Zumino term to select the relevant physical states.

To conclude, let us say some words how to extract predictions for physical observables in the Skyrme model. It is clear, that the main problem is to calculate matrix elements and expectation values. Assume that we are interested in the values of an observable expressed by an operator $\hat{F}(A)$. Then the matrix elements of this operator in $\mathcal{H}$ can be expressed as an integral over $SU(2)$:

$$\langle J, J_3, I_3|\hat{F}(A)|J, J_3, I_3\rangle = \left(\frac{2J+1}{2\pi^2}\right) \int_{SU(2)} dA\, D^{J}_{J_3,I_3}(A) \cdot D^{J'}_{J'_3,I'_3}(A) \cdot F(A). \tag{8.31}$$

Some further results for baryon masses, sizes, coupling constants, magnetic moments etc. as calculated in the $SU(3)$ generalized Skyrme model approach will be discussed in Section 11.2.3.

8.3 The "Non-Rigid" Quantization of Skyrmions

From the previous discussion one can deduce that the profile function $\Theta(r)$ is unaltered in the aforementioned quantization procedure and the rotation (8.3) is in much like the rigid-body rotation in molecules and nuclei. Some limitations and shortcomings of such an approach were already listed in Section 7.3 and discussed in (Blaizot and Ripka, 1988; Li, Liu and Zhang, 1987; Wambach, Wyld and Sommermann, 1987). Following the initial Skyrme's idea to consider Skyrmions as extended fluid objects, one might expect the substantial influence of centrifugal effects. Then instead of a spherical shape for the rigidly rotating hedgehog configuration it looks more appropriate to consider a pancake or a toroidal (donut) shape of Skyrmion under rotations. To implement a simplest variant of such a "non-rigid" quantization it is natural to quantize the axially symmetric ansatz (7.30) and (7.31), using already two sets of collective coordinates. Some other "non-rigid" quantization procedures were used in (Braaten and Ralston, 1984; Verbaarschot, Walhout, Wambach and Wyld, 1987; Braaten and Carson, 1988; Kopeliovich, 1988) in application to baryon-number-two solutions of the Skyrme model.

To apply the collective coordinates method to the configuration (7.30) and (7.31) we consider the functions

$$\Theta = \Theta(\mathbf{R}^{-1}\boldsymbol{r}), \quad \boldsymbol{n} = \mathbf{T}\boldsymbol{n}(\mathbf{R}^{-1}\boldsymbol{r}), \tag{8.32}$$

with $\mathbf{R}$ and $\mathbf{T}$ being the rotation matrices in coordinate and isotopic spaces respectively. It is convenient to use the Euler angles parametrization of these matrices (Edmonds, 1957):

$$\mathbf{R} = \mathbf{R}(\alpha, \varphi, \varphi'), \quad \mathbf{T} = \mathbf{T}(\mu, \nu, \nu'). \tag{8.33}$$

Performing the diferentiation with respect to time t, we get from (8.33) the so-called cranking expressions

$$\dot{\Theta} = -[(\boldsymbol{\omega} \times \boldsymbol{r}) \cdot \nabla]\Theta, \quad \dot{\boldsymbol{n}} = -[(\boldsymbol{\omega} \times \boldsymbol{r}) \cdot \nabla]\boldsymbol{n} + (\boldsymbol{\Omega} \times \boldsymbol{n}), \tag{8.34}$$

where we have introduced the angular velocities $\boldsymbol{\omega}$ and $\boldsymbol{\Omega}$ in coordinate and isotopic space. From Eqs. (8.34) it follows that the profile function $\Theta(\boldsymbol{r})$ is not unaltered any more and in this sense we call such a rotation as "non-rigid". The body-fixed projection of $\boldsymbol{\omega}$ have the standard form:

$$\begin{aligned} \omega_1 &= \dot{\alpha} \cos\varphi' + \dot{\varphi} \sin\alpha \sin\varphi', \\ \omega_2 &= -\dot{\alpha} \sin\varphi' + \dot{\varphi} \sin\alpha \cos\varphi', \\ \omega_3 &= \dot{\varphi}' + \dot{\varphi} \cos\alpha, \end{aligned} \tag{8.35}$$

where the third axis has been chosen as the symmetry axis of the configuration (7.30). The analogous formulae for the isotopic angular velocity $\boldsymbol{\Omega}$ follow from (8.35) by the evident change: $(\alpha, \varphi, \varphi') \to (\mu, \nu, \nu')$.

Let us write down the Lagrangian density (7.32) for the non-rigidly rotating Skyrmion, using the dimensionless coordinates $\boldsymbol{r} \to \epsilon\lambda\boldsymbol{r}$:

$$\begin{aligned} \mathcal{L} &= \frac{1}{2}(\partial_\mu\Theta)^2 + \frac{1}{2}\sin^2\Theta\left[(\partial_\mu\boldsymbol{n})^2 - (\partial_\mu\Theta\partial_\nu\boldsymbol{n} - \partial_\nu\Theta\partial_\mu\boldsymbol{n})^2\right] \\ &\quad - \frac{1}{2}\sin^4\Theta[\partial_\mu\boldsymbol{n}\partial_\nu\boldsymbol{n}]^2 - 2m^2\sin^2\frac{\Theta}{2}, \end{aligned} \tag{8.36}$$

with $m = \epsilon\lambda m_\pi$. The Lagrangian for the collective motion could be obtained from (8.36) in the way analogous to that in Section 8.1, but now we substitute for time derivatives of Θ and $\boldsymbol{n}$ the cranking expressions (8.34) and integrate over space variables. The result in terms of collective coordinates $\boldsymbol{\omega}$ and $\boldsymbol{\Omega}$ reads

$$\begin{aligned} L(\boldsymbol{\omega}, \boldsymbol{\Omega}) &= \frac{A}{2}(\omega_1^2 + \omega_2^2) + \frac{B}{2}(\Omega_1^2 + \Omega_2^2) + \frac{C}{2}(\omega_3 - \Omega_3)^2 \\ &\quad - D(\omega_1\Omega_1 + \omega_2\Omega_2) - N, \end{aligned} \tag{8.37}$$

with the following notations:

$$\begin{aligned} A = \frac{\epsilon\pi}{\lambda}\int_0^\infty r^2 dr \int_0^\pi \sin\vartheta d\vartheta &\left\{(\partial_\vartheta\Theta)^2\left[1 + 2\sin^2\Theta\left((\nabla\beta)^2 + \frac{\sin^2\beta}{r^2\sin^2\vartheta}\right)\right]\right. \\ &+ \sin^2\Theta\left[(\partial_\vartheta\beta)^2 + \mathrm{ctg}^2\vartheta\sin^2\beta\right]K - 4\sin^2\Theta\,(\nabla\Theta\cdot\nabla\beta)\,\partial_\vartheta\Theta\partial_\vartheta\beta \\ &\left.- 2\sin^4\Theta\left[(\nabla\beta)^2(\partial_\vartheta\beta)^2 + \frac{\sin^4\beta\,\mathrm{ctg}^2\vartheta}{r^2\sin^2\vartheta}\right]\right\}, \end{aligned}$$

$$\begin{aligned} B &= \frac{\epsilon\pi}{\lambda}\int_0^\infty r^2 dr \int_0^\pi \sin\vartheta d\vartheta \sin^2\Theta \\ &\quad \times \left\{K(1+\cos^2\beta) - 2\sin^2\Theta\left[\left((\nabla\beta)^2 + \frac{\sin^2 2\beta}{4r^2\sin^2\vartheta}\right)\right]\right\}, \end{aligned}$$

$$C = \frac{2\epsilon\pi}{\lambda} \int_0^\infty r^2 dr \int_0^\pi \sin\vartheta d\vartheta \, \sin^2\Theta \sin^2\beta \left(K - \frac{2\sin^2\Theta\sin^2\beta}{r^2\sin^2\vartheta} \right),$$

$$D = \frac{\epsilon\pi}{\lambda} \int_0^\infty r^2 dr \int_0^\pi \sin\vartheta d\vartheta \, \sin^2\Theta \Big\{ K(\partial_\vartheta\beta + \sin\beta\cos\beta \mathrm{ctg}\vartheta)$$
$$- 2(\nabla\Theta\cdot\nabla\beta)\,\partial_\vartheta\Theta - 2\sin^2\Theta \left[\partial_\vartheta\beta(\nabla\beta)^2 + \frac{\sin^2\beta\sin 2\beta \mathrm{ctg}\vartheta}{r^2\sin^2\vartheta} \right] \Big\},$$

$$N = \frac{2\epsilon\pi}{\lambda} \int_0^\infty r^2 dr \int_0^\pi \sin\vartheta d\vartheta \Big\{ \left[(\nabla\Theta)^2 + \sin^2\Theta(\nabla\beta)^2\right]$$
$$\times \left(\frac{1}{2} + \frac{\sin^2\Theta\sin^2\beta}{r^2\sin^2\vartheta} \right) + \sin^2\Theta \left((\nabla\Theta\times\nabla\beta)^2 + \frac{\sin^2\beta}{2r^2\sin^2\vartheta} \right)$$
$$+ m^2(1-\cos\Theta) \Big\},$$

$$K = 1 + 2(\nabla\Theta)^2 + 2\sin^2\Theta \left((\nabla\beta)^2 + \frac{\sin^2\beta}{r^2\sin^2\vartheta} \right).$$

Taking into account Euler's formulae (8.35), we write down the Hamiltonian:

$$2\epsilon^2\lambda^2 \mathbf{H} = \frac{1}{AB - D^2} \cdot \left[B(\mathbf{J}'^2_1 + \mathbf{J}'^2_2) + A(\mathbf{I}'^2_1 + \mathbf{I}'^2_2) \right.$$
$$\left. + 2D\cdot(\mathbf{J}'_1\mathbf{I}'_1 + \mathbf{J}'_2\mathbf{I}'_2) \right] + \frac{\mathbf{J}'^2_3}{C} + 2\epsilon^2\lambda^2 N, \tag{8.38}$$

where the body-fixed projections $\mathbf{J}'_i$ of the angular momentum $\boldsymbol{J}'$ read

$$\mathbf{J}'_1 = p_\vartheta \cos\chi + (p_\varphi - p_{\varphi'}\cos\vartheta)\frac{\sin\chi}{\sin\vartheta},$$
$$\mathbf{J}'_2 = -p_\vartheta \sin\chi + (p_\varphi - p_{\varphi'}\cos\vartheta)\frac{\cos\chi}{\sin\vartheta}, \tag{8.39}$$
$$\mathbf{J}'_3 = p_{\varphi'}.$$

Similar expressions hold for the isotopic spin $\mathbf{I}'$, the constraint $\mathbf{J}'_3 + \mathbf{I}'_3 = 0$ being implemented due to the axial symmetry of the ansatz (7.30) and (7.31). The quantization prescriptions (analogous to those used in previous section) transform the angular momenta $\mathbf{J}', \mathbf{I}'$ into the operators satisfying the commutation relations of the form $[\hat{\mathbf{J}}'_i, \hat{\mathbf{J}}'_j] = -\mathrm{i}\,\epsilon_{ijk}\hat{\mathbf{J}}'_k$. The energy eigenvalue E for the state with maximum spin projection $J = J'_3 = -I'_3 = J' = I'$ can be deduced from (8.39) and (8.40):

$$E = \frac{1}{2\epsilon^2\lambda^2} \left\{ \frac{A + B - 2D}{AB - D^2} \cdot J + \frac{1}{C}J^2 + 2\epsilon^2\lambda^2 N \right\}. \tag{8.40}$$

The corresponding eigenfunction reads

$$\psi = \left(\frac{2J+1}{4\pi^2}\right) 2^{-(2J+1)}(1+\cos\vartheta)^J(1+\cos\mu)^J \exp[\mathrm{i}J(\varphi' - \nu')]. \quad (8.41)$$

Choosing $J = 1/2$ or $J = 3/2$ we obtain respectively the nucleon state and that of Δ-resonance. The eigenvectors with lower values of $\mathbf{J}'_3$ (larger values of $\mathbf{I}'_3$) can be readily obtained by applying the lowering (uprising) operators to the vector (8.41). It should be added that the value of classical angular velocity ω, which enters into the rotating configuration (7.30), can be found from the equation $J = \epsilon\lambda C \cdot \omega$ fixing the spin of the Skyrmion. The "non-rigid" quantization of Skyrmions advocated here might be regarded as one of the possiblities to avoid the criticism on the "rigid-body" quantization. At the same time calculations of physical observables, along the lines partly described in Section 11, give reasonable values within 30 % accuracy to experimental data (Potapov and Rybakov, 1991). Since these numerical results do not look like the main goal in our exposition, and moreover present by themselves a most popular subject for improvements in almost each recent paper on the Skyrme model applications, it can be a good excuse not to proceed in this topic any further.

Part III

Hadron Physics Applications

9. The Skyrme Model and QCD

" ... scientific truth should be presented in different forms, and should be regarded as equally scientific, whether it appears in the robust form and the vivid coloring of a physical illustration, or in the tenuity and paleness of a symbolical expressions."

James Clerk Maxwell

This Part presents an attempt to bring the most evident applications of the Skyrme model to the attention of the reader, who is not exactly an expert in the field. This motivation gives us the right to concentrate our efforts mostly on background ideas and principle questions, neglecting some technical details of calculations, which can be found in the literature, if necessary. Therefore we will proceed whenever possible on rather elementary level. On the equal footing we are not going to collect all known facts and achievements in the field of the Skyrme model applications in hadron and condense matter physics obtained during the last decade. Such a task obviously requires an enormous increasement of the volume and the only way out we find in references to the original papers and review literature. We hope that the content of this Part provides sufficient background so that the unsatisfied reader is enabled to consult the original literature without undue difficulty.

9.1 Express Review of the QCD Present Status

First of all let us list the basic landmarks on the way to the fundamental theory of strong interactions – Quantum Chromodynamics (QCD). In Sections 1.2 and 2.4 we have seen already the physical reasons for and one of possible ways of a generalization of the isotopic symmetry in the hadron world. A result was the appearence of the chiral $SU(2) \otimes SU(2)$ symmetric theories of mesons and baryons in the 1960's (the Skyrme model among them) with the subsequent development into the *Fundamental or Effective Lagrangian Approach* (Gasiorovich and Geffen, 1969). Another idea to enlarge the isotopic $SU(2)$ invariance to the $SU(3)$ symmetry in order to take into account the evidence of strange particles was suggested by Gell Mann and Neeman (1964). This was the famous **quark hypothesis** and it is a well-known story how these new degrees of freedom $q = (u, d, s)$ produced the correct classification of hadrons in terms of representations of the $SU(3)$ group. The mesons appeared in this scheme to be $q\bar{q}$, while the baryons – 3-quarks composites qqq. To the extent that the quark mass differences are less than the dynamical scale Λ

(necessery to bind quarks into hadrons), then the spectra of hadrons with the invariant vacuum should be approximately $SU(3)$ symmetric. In the present day language this global symmetry is called the flavor $SU(3)_f$ symmetry and one can consider this scheme as the initial step toward the QCD formulation.

The next step was motivated by a tremendous success of the Gell Mann - Neeman scheme in the prediction and the subsequent discovery of the Ω^- baryon. This achievement brought at the same time an interesting puzzle. In accordance with the $SU(3)$ classification, the Ω^- state is supposed to be the lowest ($L = 0$) state in the decuplet of spin 3/2 baryons. But $J = 3/2$ state made of three identical s-quarks, with strangeness - 3, cannot be an $L = 0$ state, since it violates the Pauli principle. The solution of this 'Ω^- puzzle' was found by Greenberg (1964), Bogolubov, Struminskii and Tavkhelidze (1965), Han and Nambu (1965), who introduced an extra label on the quarks – **color** – with three values $\{1, 0, -1\}$. This new characteristic of quark states (in much similar to the electric charge in Electrodynamics) enables one to construct an Ω^- state as an $L = 0$ bound state of three s-quarks with different values of the color charge. The Pauli principle is unviolated any more and, what is an equally important, this improvement is equivalent to the introduction of an exact local symmetry – the $SU(3)_c$ color symmetry – into the quark world (Fritzsch, Gell Mann and Leutwyler, 1973).

Recall that in Quantum Electrodynamics we have in principle an analogous picture: the charged particles (e.g. the electrons and positrons) are described via fermionic fields ψ and $\bar{\psi}$ by the Lagrangian

$$\mathcal{L}_{\text{QED}} = \bar{\psi}(i\gamma^\mu D_\mu - m)\psi - \tfrac{1}{4} F_{\mu\nu} F^{\mu\nu}, \tag{9.1}$$

where D_μ is the covariant derivative, defined through the $U(1)$-gauge field A_μ,

$$D_\mu\psi = [\partial_\mu - ieA_\mu]\psi \tag{9.2}$$

and $F_{\mu\nu}$ is the field strength tensor

$$F_{\mu\nu} = \partial_\mu A_\nu - \partial_\nu A_\mu. \tag{9.3}$$

The Lagrangian (9.1) examplifies the simplest locally invariant theory – the Abelian $U(1)$-symmetric gauge theory. The situation with quarks is more complicated; as the symmetry group $SU(3)$ is a non Abelian one and thus it requires non Abelian gauge fields to describe interactions between quarks. Fortunately, the construction of a locally invariant theory possessing a non Abelian symmetry was already studied by Yang and Mills (1954). But still there were serious obstructions on a formulation of a gauge theory of quark interactions in the provoke analogy with QED. The first obstacle was the lack of any evidence for free quarks (contrary to electrons and positrons) in experiments. By this reason for some period quarks were considered as purely mathematical constructions without any physical meaning. The experimental observation of scaling behavior in deep inelastic scattering at SLAC changed this opinion drastically. The formerly known formfactors of the elastic scattering of leptons on a proton decrease as p^{-4}, where p is the momentum transfers, and this behaviour is typical for the scattering on an extended proton. In SLAC experiments for deep enelastic formfactors it was observed the p^2-independent phenomena (Bjorken scaling), which is characteristic for a scattering on a pointlike part of proton. This observation supported the Gell Mann – Neeman hypothesis on existence of pointlike constituents of nucleons and could be interpreted as a quark "visualization". On the other hand, the SLAC experiments clearly signalled that the strong interactions looked remarkably weak at short distances. Thus it was the experimental evidence of a free particle behaviour for the constituents of hadrons.

The further confirmation to the view on QCD as a theory of strong interactions was due to the discovery of **the asymptotic freedom** by Politzer (1973) and Gross and Wilczek (1973 a,b; 1974). They proved the remarkable feature of non Abelian gauge theories to behave approximately as a free theory at short distances. Thus the quarks at short distances can be described as free fermionic colored constituents of hadrons, which are bound via the exchange of the massless gauge bosons (the octet of gluons). Then the formulation of the QCD dynamics is straightforward through evident generalization of expressions (9.1) – (9.3). The quark fields q_a^α, where α is a flavor label, $\alpha = \{u, d, s, \ldots N_f\}$, and $a = 1, 2, 3$ is a color label, are in the fundamental representation of both $SU(N_f)$ and $SU(3)_c$ groups. The QCD Lagrangian density is formally similar to (9.1)

$$\mathcal{L}_{\rm QCD} = \sum_\alpha \bar{q}_a^\alpha (\mathrm{i}\gamma^\mu D_\mu - m_\alpha)\, q_a^\alpha - \tfrac{1}{4}\,\mathrm{Tr}(G_{\mu\nu}G^{\mu\nu}), \tag{9.4}$$

where instead of (9.2) for covariant derivatives we have

$$D_\mu q_a^\alpha = \left[\delta_{ab}\partial_\mu - \mathrm{i}g\left(\frac{\lambda_i}{2}\right)_{ab} A_{\mu i}\right] q_b^\alpha, \tag{9.5}$$

and respectively for the gluon field strength components

$$G_{\mu\nu}^i = \partial_\mu\, A_\nu^i - \partial_\nu\, A_\mu^i + g f^{ijk}\, A_\mu^j A_\nu^k. \tag{9.6}$$

Here λ_i, $\quad i = \overline{1,8}$ are the Gell Mann matrices, f_{ijk} are the structure constants for the $SU(3)_c$ group, g is the color coupling constant and gauge gluon fields $A_\mu = A_\mu^i\, \lambda_i$ are in the adjoint representation of $SU(3)_c$.

There are still some unclear questions. The most discouraging one is that the exchange via massless gluons leads to long-range color forces. To avoid such an undesirable behaviour as a counterpart to asymptotic freedom the idea of **confinement** was launched by Weinberg (1973) among others. In brief, it was suggested that weak interactions among quarks at short distances grew progressively stronger if we try to pull the color objects apart. That is why one can observe hadrons only as color singlet states (colorless objects) and although hadrons are bound states of quarks and gluons, they do not feel the color force. This beautiful hypothesis of confinement remains to be proven in a rigorous fashion to the present day. Another one difficulty, closely related with the confinement, is the absence of a consistent quantization procedure in QCD. The commonly used procedure, which is analogous to that working in QED is definitely illegal in QCD as in any approximation there is no completely free quarks or gluons, as it is required. The gluonic fields, contrary to electromagnetic ones, do carry the color charge by themselves and thus are self-interacting essentially nonlinear fields. A self-consistent procedure for quantization of such a nonlinear theory is a longstanding unsolved problem.

Summarizing one can say that the present day QCD formulation embodied at one stroke all of the bits of theoretical and experimental information one had gathered about strong interactions. Note that mostly this information concerned with the high-energy region and then extrapolated to other QCD regions without any justification (Gell Mann, 1985). QCD dynamics can roughly be devided into three main energy regimes or phases:

a. **Short distances** of the order of 0.1 fm or smaller, (i.e. at high energies or momentum transfers in the multi GeV range) where asymptotic freedom phenomena enables a rather good description in terms of quark and gluon degrees of freedom. The effective coupling constant α_s is a small enough expansion parameter in this region. This allows one to use perturbation theory together with the renormalization group analysis to calculate various quantities at high energies with results that are entirely consistent with experimental data. This is **the asymptotic freedom phase**,
b. **Intermediate distances** (between 0.1 fm and 1 fm) of the order of confinement scale ($\Lambda^{-1} \approx 1$ fm), present the most difficult dynamical problem. Here there is a mixing between quark-gluon and hadronic degrees of freedom. This is the phase of quark-gluon plasma or chromoplasma,
c. **Large distances** of the order 1 fm and bigger (i.e. at low energies and momentum transfers up to 1 GeV) where there are only hadronic degrees of freedom. Quarks and gluons are invisibly confined inside hadrons, and QCD is a nonperturbative theory. Usually this domain is called **the confinement phase**.

Note that aforegiven classification of QCD phases is in much similar to that in molecular physics (with quite the opposite dependance on the distance): Weakly interacting quarks and gluons at short length scales one can consider as an "ideal gas" of these constituents. The intermediate domain will correspond to a sort of "real or van der Waals gas" of quarks and gluons, where interactions and sizes of quarks are to be taken into account. Finally at large distances we will have quarks condensed into a mesonic fluid with vortices as nucleon images. Remember the initial Skyrme's motivations from Chapter 1, which sound rather reasonable from that point of view.

In general, because of mathematical complexity of QCD it has proved very difficult to extract quantitative predictions at any energies or for any phase in this theory. Even those predictions that can be made (mainly concerning the asymptotic freedom phase) are hard to check experimentally. Most of tests of QCD validity, that have been carried out to date, are semi-quantitative. The mathematical difficulties grow exponentially when one approaches the most interesting for practical purposes the low energy region of QCD. This region, which in principle contains all Nuclear Physics phenomena, in particular should provide us with the information on the masses, lifetimes, magnetic moments, sizes, scattering rates, and other properties of strongly interacting particles. At present this is not the case, as there is no a systematic method for treating the low energy QCD and for extracting the necessary information directly from (9.4). The understanding here is mostly qualitative and relies heavily on the use of such effective quark-gluon models as the potential model, bag models, hybrid bag models etc. The connection between the asymptotic freedom domain and the phase of confinement presents an almost insurmountable problem, despite of some progress of lattice calculations, and there seems to be not much hope in the nearest future. These attempts to get the whole picture of strong interactions in terms of quarks and gluons overshadowed for more than decade the visible achievements of the phenomenological (or effective) Lagrangian approach. The latter approach, developed on the basis of current algebra in 1960's, in spite of its relative simplicity enabled one to obtain some reasonable answers for hadron observables.

The renewal of the interest in further development of effective Lagrangians came indirectly from 't Hooft's discovery of the implicit expansion parameter for the low energy QCD. We give a detailed account to that topic in what follows with a special emphasis to benefits in a description of phenomena from the confinement phase of QCD. Our exposition is in much borrowed from (Witten 1979; 1980).

9.2 $1/N$-Expansion

There are some problems in physics which does not contain any free parameter and for this reason there is no natural perturbation expansion. The most familiar example of this type is the hydrogen atom problem with the Hamiltonian

$$H = \frac{p^2}{2m} - \frac{e^2}{r}. \tag{9.7}$$

For small e^2 one can treat the potential energy term $-e^2/r$ as a perturbation, but it turns out that such an expansion will not work. In the first place e^2 is not dimensionless and because of that cannot be used as an expansion parameter. Ephitets 'large' or 'small' with respect to e^2 does not make much sense, since the value of e^2 depends on the choice of units. In the second place after a rescaling $r \to r/me^2$; $p \to p \cdot me^2$ the expression (9.7) becomes

$$H = me^4 \left(\frac{p^2}{2} - \frac{1}{r}\right) = me^4 \, H', \tag{9.8}$$

where the 'coupling constant' e^2 appears only in an overall factor me^4. Therefore except for the overall scale of length and energy the physics of hydrogen atom is independent of e^2. As it follows from (9.8) the hydrogen atom dynamics can be as well described by the reduced Hamiltonian H' with e^2 scaled out or without any free parameter. In all such cases the standard perturbation theory is unapplicable and we have either to look for an exact solution (if any) or to seek a numerical solution. Dealing with nonlinear realistic theories, such as QCD, it is hard to expect that an exact solution can be found, and on the other hand even a computer treatment might be formidably difficult and require an ability of supercomputers of future generations.

To proceed in a problem without an obvious free parameter sometimes it is possible to find an implicit one looking at quantities which are usually regarded as given and fixed, such as for example the dimension of the physical space. Tricks of this kind were used long before in 1950's by Berlin and Kac (1952) and proved to work effectively in different problems of statistical physics. For an up to date comprehensive review on the subject one may consult (Chatterjee, 1990). To demonstrate this idea in an evident form let us continue with the hydrogen atom. The recipe to find an implicit expansion parameter for the hydrogen atom in particular (or for Atomic Physics, in general) is quite simple. Instead of studying the problem in three dimensions with the rotational $O(3)$-symmetry, one considers atom in N dimensions with $O(N)$-symmetry. Without lose of generality we can restrict ourselves to s-wave states, so that the wave function ψ would be a function of r only. The Schrödinger equation reads

$$\left[-\frac{1}{2m}\left(\frac{d^2}{dr^2}+\frac{(N-1)}{r}\frac{d}{dr}\right)-\frac{e^2}{r}\right]\psi(r)=E\psi(r). \tag{9.9}$$

The term with first derivative in (9.9) might be eliminated by means of the redefinition of the wave function $\psi \rightarrow r^{-(N-1)/2}\psi$. Thus redefined Eq. (9.9) becomes

$$\left[-\frac{1}{2m}\frac{d^2}{dr^2}+\frac{(N-1)^2}{8mr^2}-\frac{e^2}{r}\right]\psi(r)=E\psi(r). \tag{9.10}$$

Under the rescale of the radial coordinate $r=(N-1)^2R$ in (9.10) we obtain

$$\frac{1}{(N-1)^2}\left[-\frac{1}{2m(N-1)^2}\frac{d^2}{dR^2}+\frac{N-3}{8(N-1)mR^2}-\frac{e^2}{R}\right]\psi(R)=E\psi(R). \tag{9.11}$$

as the equation of motion for a particle with an effective mass $M_{\text{eff}}=m(N-1)^2$, moving in an effective potential

$$V_{\text{eff}}=\frac{\gamma}{8mR^2}-\frac{e^2}{R};\quad \gamma=\frac{N-3}{N-1}. \tag{9.12}$$

From Eq. (9.11) it is readily seen that the problem simplifies drastically for large N. The effective mass M_{eff} becomes very large and because of that it will not be a great mistake to assume that the particle simply sits in a bottom of the effective potential well. Thus for large N we are ensured to treat the problem semi-classically. The ground state energy is assumed to be the absolute minimum of V_{eff}, i.e.

$$E = -\frac{1}{2} me^4 \cdot \frac{4}{(N-1)^2},$$

so that for $N = 3$ we get the exact binding energy $E = -(1/2)me^4$, obtained otherwise from the proper quantum mechanical treatment of the problem.

After this illustrative example we can pass to our main subject of interest – Quantum Chromodynamics. The basic fact that makes QCD a difficult theory to understand is that in the massless limit, which corresponds to the manifested chiral symmetry of the low energy QCD (for details see below), this is a parameter-free theory. Because of dimensional transmutation, the quark-gluon coupling $g(\Lambda)$ is a function of the QCD cutoff Λ, and can be factorized out of the starting dynamics through a simple rescaling of the gauge fields in (9.4)-(9.6) or through a machinery known as the renormalization group (Marciano and Pagels, 1978; Ynduráin, 1983). So contrary to Quantum Electrodynamics and in common with Atomic Physics problems, QCD does not exhibit an explicit expansion parameter, depriving us of the conventional insights of perturbation theory. As in the illustrative example nothing depends on the coupling constant except the overall scale of energies, and therefore perturbation theory cannot answer such unsolved problems as explaining confinement and predicting the mass spectrum. Thus we have a typical situation where the $1/N$-expansion might be appropriate.

The first step in that direction has been done by G. 't Hooft (1974a). Looking for an implicit expansion parameter in QCD he found that if we were to increase the number of colors N_c in QCD, then $1/N_c$ could be used as an appropriate expansion parameter. It means that instead of $SU(3)_c$ symmetry of ordinary QCD we have to consider $SU(N)_c$ symmetric theory with the gluon gauge fields as $N \times N$ matrices. This generalization (in common with the hydrogen atom example) leads to a simplification of QCD and the basic reason for this effect is quite understandable. For large N_c, the gluon field $(A_\mu)^i_j$ has ${N_c}^2 - 1$ components, where we have taken into account that the matrices $(A_\mu)^i_j$ are elements of $su(N_c)$ algebra and thus have to be traceless. Therefore there are ${N_c}^2 - 1$ gluon particles which might appear as intermediate states in Feynman diagrams for any process. Even after a specification of the color quantum numbers of the initial and the final states we have a large number (at least equal to N_c) of possibilities for the quantum numbers of the intermediate states of gluons. This fact leads to the appearance of large combinatoric factors in Feynman diagrams when we sum up over all intermediate states.

To see how it works in more details, let us, following Witten (1980), consider the lowest order contribution to the gluon "vacuum polarization" with diagram on Fig. 9.1

The double-line notations as introduced by 't Hooft (1974a) (see also Witten, 1979) make calculations of the combinatoric factors rather visual. Let us give a brief account on the origin of double-line notations. The gluon field $(A_\mu)^i_j$ has one upper index like the quark field q^i and one lower index like the antiquark field $\bar{q}_j$. Then to simplify combinatoric calculations (and for this purpose only) one may think of the gluon as a quark-antiquark combination:

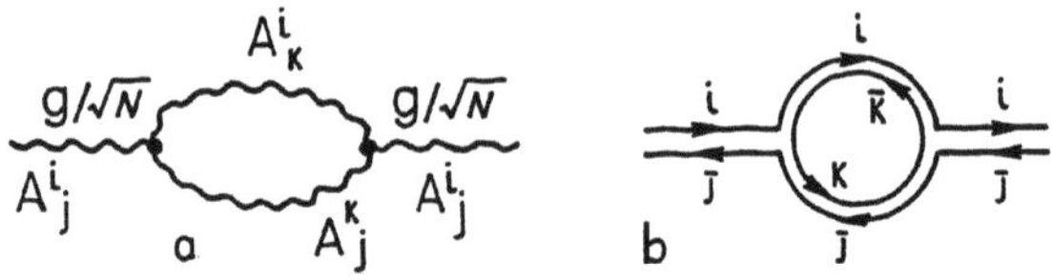

Fig. 9.1 a, b. The diagrams of the gluon contribution to the one-loop gluon vacuum polarization: (a) in standard and (b) in double-line notations

$(A_\mu)^i_j \asymp q^i \bar{q}_j$. In accordance with the standard representation of a quark or antiquark in a Feynman diagram as a single line with an arrow (the direction of the arrow distinguishing quark from antiquark) we should represent the gluon as a double-line, one line for the quark and one line for the antiquark (see Fig. 9.2).

Fig. 9.2. The double-line notations for quarks, antiquarks and gluons (for a visualization of combinatoric calculations only)

In the diagram in Fig. 9.1 we have two three-gluon vertices with the contribution $\mathrm{Tr}\, A_\mu A_\nu \partial_\mu A_\nu$, which in matrix notations can be written more explicitly as $A_\mu{}^i_j A_\nu{}^j_k \partial_\mu A_\nu{}^k_i$. Here the lower index of the first gluon field is contracted with the upper index of the second gluon field; the lower index of the second gluon field is contracted with the upper index of the third, and the lower index of the third field is contracted with the upper index of the first. In Fig. 9.1 b) in the double-line notations this is indicated schematically emphasizing the fact that the outgoing (antiquark) index of each gluon field is contracted with the incoming (quark) index of the next gluon field (in the clockwise direction). The color index lines at the edge of the diagram are contracted with those of the initial and final states. However, at the center of the figure, there is a closed color line (shown in the figure as having color k) that is contracted only with itself. The value of k remains unspecified even when the initial and final states are given. The sum over all possible k gives a combinatoric factor of N_c, associated with the diagram in Fig. 9.1.

If QCD is to have a smooth limit for large N_c, this factor of N_c must somehow be canceled. Otherwise the corrections to the gluon propagator would diverge for large N_c in proportion to N_c. All the other calculations would also give divergent results, and we could not construct a selfconsistent QCD theory for a large number of colors. To cancel the undesirable combinatoric factor of N_c we must remember that for each of two vertices where a gluon splits into two gluons or where two gluons recombine into one, there is a factor of the coupling constant g. One could control the contributions of diagrams for $N_c \to \infty$, if one chose the coupling constant to be $g/\sqrt{N_c}$, where g is to be fixed as $N_c \to \infty$. Then the factor of N_c cancels out of Fig. 9.1, because $N_c(g/\sqrt{N_c})^2 = g^2$, independent of N_c. So the choice of the coupling as $g/\sqrt{N_c}$ gives a smooth limit of this one-loop diagram.

There is a rather important observation due to 't Hooft (1974a), that in such an approach complicated diagrams with large number of vertices will vanish for large N_c, unless they have combinatoric factors large enough to cancel factors at each vertex. It turns out that only a certain class of diagrams with the largest combinatoric factors survives for large N_c, while all other diagrams vanish as $N_c \to \infty$. Consider, for example, two three-loops diagrams, as shown in Fig. 9.3.

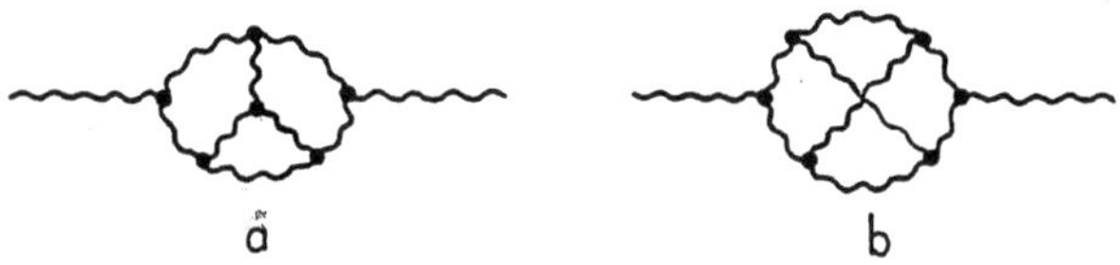

Fig. 9.3 a, b. The three-loops diagrams: (a) planar and (b) nonplanar

Both of them contain six interaction vertices and thus they have factors of $(g/\sqrt{N_c})^6$. However only the left diagram in Fig. 9.3 with three self-contracted loops survives as it has a combinatoric factor of N_c^3 from summing over the various intermediate states. Since $N_c^3(g/\sqrt{N_c})^6 = g^6$ is independent of N_c the left diagram has a smooth limit for large N_c. The right diagram in the same figure has only one large and tangled closed color loop leading to a single factor of N_c. This diagram is, therefore, of order $N_c(g/\sqrt{N_c})^6 = g^6/N_c^2$ and vanishes like $1/N_c^2$ as $N_c \to \infty$. The difference between these six-vertex diagrams is that the left one is **a planar diagram** and the right diagram is **a nonplanar**, as there is no way to draw the diagram 9.3b on the plane without line crossing (at points where there are no interacting vertices) in contrast to the diagram 9.3a. The most distinct form of this difference can be seen in the double-line notations presented in Fig. 9.4 for the same diagrams:

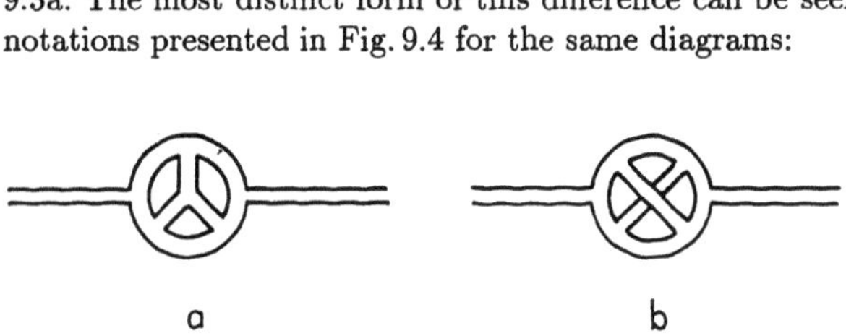

Fig. 9.4 a, b. The double-line notations for the three-loops diagrams from Fig. 9.3 (a) planar and (b) nonplanar

With a little practice, one can see that nonplanar Feynman diagrams always vanish at least like $1/N_c^2$ for large N_c. This observation can be formulated as the first "selection rule" in the large N_c limit:

1. Nonplanar diagrams are suppressed by factors of $1/N_c^2$. Another one "selection rule" comes from a consideration of the one-quark-loop contribution to the gluon propagator. This diagram is drawn in Fig. 9.5, both in the ordinary and double-line notations.

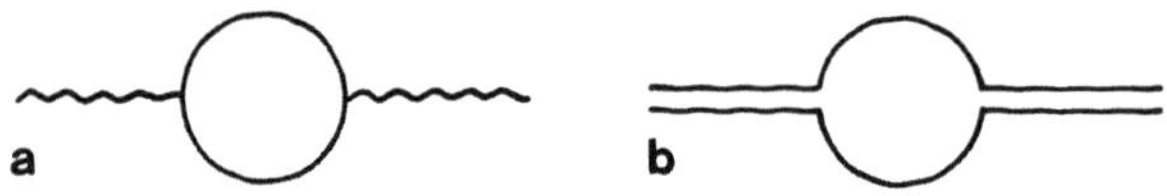

Fig. 9.5 a, b. The one-quark-loop contribution to the gluon self-energy, in the ordinary (a) and double-line notations (b)

This type of diagrams does not contain any closed color loop. By this cause the combinatoric factor would be $(g/\sqrt{N_c})^2 = g^2/N_c$ and the diagram contribution vanishes like $1/N_c$ for large N_c. In other words it comes out, that

2. Internal quark loops are suppressed by factors of $1/N_c$. In summary, both "selection rules" tell us that the leading diagrams for large N_c are the planar diagrams with a minimum number of quark loops (more precisely, with only a single quark loop which runs at the edge of the diagram). Here comes the first simplification of QCD in the large N_c limit, as we do not need any more to take into account contributions of all possible diagrams. Instead we can restrict ourselves to a more narrow (but still a rather vast) class of diagrams - to the planar diagrams. To predict values of physical observables in the direct manner we have to sum the planar diagrams, which is a very formidable task. An advantage of the $1/N_c$-expansion in QCD is, that even without being able to sum up the planar diagrams, we can get a considerable insight into the phenomenology of low energy QCD. The aforegiven "selection rules" are observed to be fairly well satisfied in Nature and therefore it is possible to obtain by means of the $1/N_c$-expansion an understanding of some qualitative properties of the strong interactions that are not otherwise well understood.

A further advancement one can obtain under the assumption that the confinement of quarks prolongs to hold for the large N_c QCD. It is equivalent to another assumption, that the planar diagrams sum up to give a confining theory, so that only colorless objects are observable. The detailed analysis performed by 't Hooft (1974a,b) and Witten (1979) resulted in an attractive picture for the low energy QCD with the following properties:

1. Mesons for $N_c \to \infty$ are free, stable and noninteracting. The number of meson states is infinite and the meson masses have smooth limits for large N_c.
2. Meson decay amplitudes are of order $1/\sqrt{N_c}$; meson-meson elastic scattering amplitudes are of order $1/N_c$ and are given by a sum of tree diagrams, involving the exchange, not of quarks and gluons, but of physically observed mesons. More generally, meson physics in the large N_c limit might

be described by the tree diagrams of an effective local Lagrangian, with local vertices and local meson fields. The meson masses and coupling constants in such a Lagrangian remain, unfortunately, unspecified.

Almost the same statements hold in the case of glueballs and therefore 't Hooft (1974a,b) claimed that **at the large N_c limit QCD would be equivalent to the theory of mesons and glueballs with an effective interaction of the order** $1/N_c$.

The next inevitable step in an understanding of the low energy QCD phenomenology belongs to Witten (1979), who extended the analogy between QCD for large N_c and weakly coupled local field theories to baryons as well as to mesons and glueballs. Witten used the fact, that weakly coupled nonlinear fields under some conditions might produce additional states, whose masses diverge, for a weak coupling α, like the inverse coupling $1/\alpha$. These states are known as solitons with the most familiar representative – the 't Hooft-Polyakov monopoles in the non Abelian gauge theory. The large N_c QCD analogues of the monopoles might be regarded as baryons with the masses of the order N_c, which can be written as $(1/N_c)^{-1}$, thus being inverse to the meson coupling constant $1/N_c$. The latter observation might be considered as a renewal at a new level of knowledge of Skyrme's old idea to describe baryons as soliton states of weakly interacting meson fields. It happened that the Skyrme model, which substantially predates the formulation of the main principles of QCD, appeared to be a simplest variant of an effective meson theory. Therefore the Skyrme model offers an important and a relatively simple test of validity for the $1/N_c$-expansion in the QCD framework.

Here arises the following natural question. What is the relation between the QCD fundamental Lagrangian (9.4) and an appropriate effective Lagrangian of mesons and glueballs in the confinement phase? As was mentioned in Section 9.1, the suprising success of QCD in the description of hadronic phenomena in the high-energy region has resulted in the belief in the universality of QCD. Then it looks like a natural step to derive the effective meson Lagrangian from Eq. (9.4). But there also exists another opinion that these two regions of hadronic phenomena represent different phases of state of hadronic matter and therefore instead of an abovementioned derivation one might rather to speak about a phase transition between these regions. All these still unsolved problems would be the main subject of the next section.

9.3 Effective Meson Theory from QCD

The derivation of effective actions (e.g. generalized Skyrme models) from the QCD fundamental Lagrangian (9.4) is a problem of much conceptual and practical relevance. Before going into discussion of some lines of attack on this problem, which is under an intensive study in the current literature, let us examine more carefully the statement of the problem as itself. Recall, that the QCD formulation (see Section 9.1) and its great success in the description of hadronic phenomena are mostly relevant to the high-energy region (perturbative regime). Therefore the universally adopted enlargement of the QCD domain of validity on the low energy region (nonperturbative regime) seems to be a rather courageous assumption. In all respects we prefer to understand the aforementioned derivation rather as a phase transition from a confining to a deconfining phase at scales of the order of $\Lambda^{-1} \simeq (300\,\mathrm{MeV})^{-1}$. In a manner this process is analogous to the derivation of the Landau-Ginzburg theory of superconductivity from the microscopic Bardeen-Cooper-Schrieffer theory.

9.3.1 The Low-energy QCD Attributes

Let us turn to commonly accepted basic elements of low energy QCD: chiral symmetry, the structure of the vacuum, the spontaneous breakdown of the global symmetry, and Goldstone pions as the relevant collective degrees of freedom[1]. Aside from the color, the QCD Lagrangian (9.4) involves three quark flavors (u, d, s), and in the low energy regime the QCD dynamics is dominated by the light flavor quarks (u, d). The masses of current *up* and *down* quarks at momentum transfer $p \approx 1\,\mathrm{GeV}$ are

$$m_u = (5.1 \pm 1.5)\,\mathrm{MeV}, \quad m_d = (8.9 \pm 2.6)\,\mathrm{MeV}, \tag{9.13}$$

i.e. almost negligible as compared to the QCD cutoff $\Lambda \simeq 300\,\mathrm{MeV}$ or to typical hadron masses of order 1 GeV. As for the strange quark with $m_s \approx 130\,\mathrm{MeV}$ the situation is not so straightforward. Nevertheless one can regard the zero quark mass limit as a reasonable (almost physical) approximation and in the absence of the Higgs fields[2] the QCD Lagrangian (9.4) with $m = 0$ will be invariant under separate unitary transformations on the flavor indices of the left handed and right handed quark fields, given as:

$$q = q_L + q_R; \quad q_L = \frac{(1+\gamma_5)}{2}\,q, \quad q_R = \frac{(1-\gamma_5)}{2}\,q. \tag{9.14}$$

In the group-theoretic language this global invariance group may be presented in the form

$$\begin{aligned} U(3)_L \otimes U(3)_R &= SU(3)_L \otimes SU(3)_R \otimes U(1)_L \otimes U(1)_R \\ &= SU(3)_L \otimes SU(3)_R \otimes U(1)_A \otimes U(1)_V, \end{aligned} \tag{9.15}$$

[1] Broadly speaking, the Vector Meson Dominance principle is included in this list of basic elements, but we are not going to exploit this idea in what follows. For interested reader we recomend the comprehansive review on the subject (Meissner, 1988).

[2] Remind, that in the standard $SU(3)_c \otimes SU(2) \otimes U(1)$ model of strong and weak interactions, the masses of the quarks are induced by the Higgs fields, which are responsible for the spontaneous breakdown of the weak group $SU(2) \otimes U(1)$ to the $U(1)$ group of electromagnetism. It is not quite accurate to state that quarks have no masses in the absence of Higgs fields if their bare masses are zero. Spontaneous breakdown of chiral symmetry will also contribute to the masses of the quarks. However, for the present exposition the significant point is that in the absence of Higgs fields the QCD Lagrangian (9.4) with $m = 0$ will have chiral symmetry.

where $U(1)_V$ and $U(1)_A$ are the vector and axial subgroups with generators which could be constructed from the generators of left and right handed transformations like (1.7) as the halfsum and halfdifference, respspectively (see Appendix A for details). At the quantum level the $U(1)_A$-symmetry is explicitly broken by the Adler-Bell-Jackiw-Bardeen anomaly (Adler, 1969; Bell and Jackiw, 1969; Bardeen, 1969), i.e.

$$\partial_\mu(\bar{q}\gamma^\mu\gamma_5 q) = \frac{g^2}{16\pi^2}\,\epsilon^{\mu\nu\lambda\rho}\mathrm{Tr}(G_{\mu\nu}G_{\lambda\rho}), \tag{9.16}$$

leaving us with the symmetry group

$$U(3)_L \otimes U(3)_R/U(1)_A = SU(3)_L \otimes SU(3)_R \otimes U(1)_V, \tag{9.17}$$

with transformations of the type

$$q \to \exp\{-\frac{\mathrm{i}}{2}\,\gamma_5\pi^k\lambda^k\}\,q, \tag{9.18}$$

which interchange states with different parities. Here π^k are chiral phases, λ^k are the $SU(3)$ generators. For the left and the right quark fields (9.14) the transformations (9.18) can be written in the form

$$q \to U^{-1/2}\cdot q_L + \mathrm{U}^{1/2}\cdot q_R; \quad \mathrm{U} = \exp\{\mathrm{i}\pi^k\lambda^k\}. \tag{9.19}$$

The spontaneous breaking of the global $SU(3)_L \otimes SU(3)_R$ symmetry means that the relative chiral phases of the left and right handed quarks became locally fixed so that phase functions $\pi^k(x)$ might be related with the octet of pseudo-Goldstone bosons $(\pi^0, \pi^\pm, \eta, K^0, K^\pm, \overline{K}^0)$. In the exact chiral limit these pseudoscalar mesons are massless and their actual masses (for example, $m_\pi \simeq 140\,\mathrm{MeV}$) are related to the non-zero but small current quark masses (9.13) through the PCAC relation (see Appendix A for details):

$$m_\pi^2 = -\frac{(m_u + m_d)}{f_\pi^2}\langle\bar{q}q\rangle, \quad f_\pi = \frac{F_\pi}{2}. \tag{9.20}$$

Here $\langle\bar{q}q\rangle$ denotes the quark pair condensate with an emperical value $\langle\bar{q}q\rangle = \langle\bar{u}u\rangle = \langle\bar{d}d\rangle \simeq -(250\,\mathrm{MeV})^3$. The existence of the nonvanishing quark condensate $\langle\bar{q}q\rangle$ (together with the gluon condensate $\langle G_{\mu\nu}G^{\mu\nu}\rangle$) in the physical vacuum of nonperturbative QCD implies that the chiral symmetry is spontaneously broken, since the scalar $\bar{q}q$ is evidently not invariant under the chiral transformations (9.18). On the other hand, the chiral condensate might be regarded as the order parameter for chiral symmetry breaking in the nonperturbative QCD vacuum which excludes free quark and gluon fields.

At spatial infinity $|x| \to \infty$ the meson fields $\pi^k(x) \to 0$ or equivalently from (9.19) $\mathrm{U}(x) \to \mathbb{1}$, where $\mathbb{1}$ is a 3×3 matrix. Under the transformations from the chiral group $G = SU(3)_L \otimes SU(3)_R$ the field $\mathrm{U}(x)$ is transformed as

$$\mathrm{U} \to W\cdot\mathrm{U}\cdot V^{-1}, \quad \text{where} \quad V \in SU(3)_R, \quad W \in SU(3)_L. \tag{9.21}$$

But the vacuum $\mathbf{U} = \mathbb{1}$ is invariant under (9.21) iff $V = W$, or in the other words under transformations from the diagonal subgroup

$$G_V = \mathrm{diag}[SU(3)_L \otimes SU(3)_R] \simeq SU(3)_I, \tag{9.22}$$

where $SU(3)_I$ is the three-flavor analog of the isospin group. From its definition, the field $\mathbf{U}(x)$ defines the orbit of the group G passing through the unit element $\mathbb{1}$ and thus takes the values on the homogeneous coset space $G/G_V \simeq SU(3)$. One might regard the above considerations as a justification of a one-to-one correspondence between the weak (phase) vacuum excitations, which parametrize the field $\mathbf{U}(x)$ through (9.19), and elements of the $SU(3)$ group manifold. In the adopted terminology, the fields $\mathbf{U}(x)$ taking values on a group manifold or on a homogeneous space are called **the principal chiral fields**. In accord with the previous argumentation one can find as appropriate to describe the low energy QCD dynamics in terms of chiral fields $\mathbf{U}(x)$. At the same time it proves that only mesonic degrees of freedom are essential in the nonperturbative region of QCD. In the case of three-flavored QCD the field manifold might be taken as $SU(3)$ group manifold, and for the two-flavored theory we have the $SU(2)$ manifold, homeomorphic to the sphere $\mathbb{S}^3$. This completes the available justification of the first Skyrme's conjecture (as listed in Section 1.5) from the QCD first principles.

On the same 'physical level of accuracy' it is possible to justify from the QCD point of view the other Skyrme's conjecture: the identification of the topological charge (1.37) with the baryon number. This result belongs to Balachandran, Nair, Rajeev and Stern (1982) and is based on some results of Goldstone and Wilczek (1981). Here it is exposed in accord with (Balachandran, Marmo, Skagerstam and Stern, 1991).

9.3.2 The Topological Charge as the Baryon Number

The spontaneous breakdown of chiral invariance can be expressed on the QCD language of quarks (gluons are not important in this context) as follows: the QCD ground state – the filled Dirac sea of quarks – corresponds to a nonzero value of the condensate $\langle \bar{q}q \rangle$. The pseudoscalar mesons are on the one hand the small oscillations of the chiral phase of this condensate, and on the other hand one identifies mesons with collective excitations of the Dirac sea of quarks. Being in its turn the collective excitation of mesonic fields, the topological soliton might be as well regarded as a notable fluctuation in the Dirac sea of quarks. Thus the problem is to find out how does this fluctuation affects reciprocally on the Dirac sea.

Let us start with the standard expression for the baryon current (the vector current) in terms of quarks for QCD with N_c colors

$$V_\mu = \frac{1}{N_c} \bar{q}\gamma_\mu q, \tag{9.23}$$

where the factor $1/N_c$ reflects the fact that each quark has baryon number $1/N_c$. The properties of the Dirac sea in the presence of the topological soliton can be studied in an approximate way by investigating the dynamics of second quantized quark fields in the presence of external classical chiral field $\mathbf{U}(x)$. Thus we need to calculate the expectation value

$$\langle V_\mu \rangle = B_\mu(U) \tag{9.24}$$

in the ground state when the quark field q is coupled to the given classical configuration U. This coupling can be described by the quark part of the Lagrangian density (9.4)

$$\mathcal{L}^{(q)}_{\mathrm{QCD}} = \mathrm{i}\,\bar{q}\gamma^\mu\partial_\mu q - m(\bar{q}_L U q_R + \mathrm{h.c.}), \tag{9.25}$$

with q_L and q_R defined in (9.14). Here U couples only to the flavor indices of quarks. The baryon current (9.24) obeys the conservation law $\partial_\mu B^\mu = 0$, which is valid independently of the dynamics of the U since V_μ is a conserved Noether current for (9.25). The general expression for such a B_μ, compatible with the conservation law, is

$$B_\mu(\mathbf{U}) = \beta J_\mu(\mathbf{U}) + \epsilon_{\mu\nu\lambda\rho}\partial^\nu\psi^{\lambda\rho}(\mathbf{U}), \tag{9.26}$$

where β is a constant, $\psi^{\lambda\rho}$ is any smooth function of $\mathbf{U}$ and $J_\mu(\mathbf{U})$ is the topological current, which an explicit form one needs to establish. As stated in (Balachandran, 1986) the expression (9.26) for B_μ is the most general for any nontrivial current on $\mathbb{S}^3$ which is conserved independently of the chiral field $\mathbf{U}$ dynamics.

Thus the baryon number of the Dirac vacuum

$$\int d^3x B_0 = \beta Q(\mathbf{U}) \tag{9.27}$$

is completely determined by the proportionality constant β, since the integral of the second term in (9.26) vanishes due to the smoothness of the function $\psi^{\lambda\rho}(\mathbf{U})$ on $\mathbb{S}^3$. To fix the value of β in (9.27), let us change variables in (9.25) as follows: $u_L = q_L$, $u_R = \mathbf{U}q_R$. The resulted expression is

$$\mathcal{L}^{(q)}_{\mathrm{QCD}} = \mathrm{i}\bar{u}_L\gamma^\mu\partial_\mu u_L + \mathrm{i}\bar{u}_R\gamma^\mu(\partial_\mu + \mathbf{U}\partial_\mu\mathbf{U}^\dagger)u_R - m(\bar{u}_L u_R + \mathrm{h.c.}). \tag{9.28}$$

It is reasonable to rewrite the second term in (9.28) as $\mathrm{i}\bar{u}_R\gamma^\mu(\partial_\mu - A_\mu)u_R$ with notations $A_\mu = -\mathbf{U}\partial_\mu\mathbf{U}^\dagger \equiv (-L_\mu)$ for a generic (externally prescribed) gauge potential for the group $SU(2)$. The introduced variables help us to visualize the "phenomenon" of the Dirac sea "polarization" with respect to the "right" and "left" Noether currents. The right hand currents obey the axial $U(1)$ anomaly

$$\begin{aligned}&\partial_\mu(\bar{u}_R\gamma^\mu u_R) - \mathrm{i}m(\bar{u}_L u_R - \bar{u}_R u_L)\\ &\quad = \frac{N_c}{8\pi^2}\epsilon^{\mu\nu\alpha\beta}\partial_\mu \mathrm{Tr}\left(A_\nu\partial_\alpha A_\beta + \frac{2}{3}A_\nu A_\alpha A_\beta\right)\end{aligned} \tag{9.29}$$

as for the left hand currents we have the standard divergence equation

$$\partial_\mu(\bar{u}_L\gamma^\mu u_L) - i\,m(\bar{u}_R u_L - \bar{u}_L u_R) = 0. \tag{9.30}$$

On adding (9.29) and (9.30), we find

$$\begin{aligned} \partial_\mu V^\mu &= \frac{1}{N_c}\partial_\mu(\bar{q}\gamma^\mu q) \\ &= \frac{1}{8\pi^2}\epsilon^{\mu\nu\alpha\beta}\partial_\mu \mathrm{Tr}\left(A_\nu\partial_\alpha A_\beta + \frac{2}{3}A_\nu A_\alpha A_\beta\right). \end{aligned} \tag{9.31}$$

We now take the expectation value of (9.31) in the fermion ground state with the result

$$\langle V^\mu\rangle = \frac{1}{8\pi^2}\epsilon^{\mu\nu\alpha\beta}\mathrm{Tr}\left(A_\nu\partial_\alpha A_\beta + \frac{2}{3}A_\nu A_\alpha A_\beta\right). \tag{9.32}$$

Taking into account the definitions for A_μ and for the topological current in the Skyrme model (1.38) we obtain

$$\langle V^\mu\rangle = -\frac{1}{24\pi^2}\epsilon^{\mu\nu\alpha\beta}\mathrm{Tr}(A_\nu A_\alpha A_\beta) = J_\mu. \tag{9.33}$$

This result might be regarded as a considerable support to Skyrme's conjecture. Notice, that the equality of baryonic and topological charges has been shown to all orders in the expansion in powers of $1/m$ by Balachandran, Nair, Rajeev and Stern (1982), and by Gipson (1984). D'yakonov and Petrov (1984) demonstrated the appearance of the baryonic charge of the Dirac vacuum due to the ultraviolet cutoff with the result in full agreement with (9.33).

9.3.3 Effective Chiral Lagrangians from QCD

Now we are prepared to return to the problem of direct derivation of the effective mesonic Lagrangian from QCD. For the sake of definiteness we stress once more, that we are to assume the universal character of QCD which is a hypothesis in a manner. But let us believe in the existence of a description of nuclear phenomena in which quarks and gluons are treated as primary objects. This assumption is equivalent in a sense to that of possibility to describe chemical or biological processes in terms of elementary particles. The plan of our exposition is the following: first of all we briefly describe the generally adopted scheme of derivation sparing the reader the technical details. Along this way we point out at appropriate places the scarcities and shortages of our knowledge and understanding of some nuclear physics phenomena, which limit up to now the progress in the field.

The starting point is the QCD generating functional Z_{QCD} associated to the vacuum amplitude in the absence of sources, i.e.

$$\begin{aligned} Z_{\mathrm{QCD}} &= \langle 0_+|0_-\rangle \\ &= \int d\mu[A]dq d\bar{q}\,\exp\left\{\frac{i}{\hbar}\int dx\,\mathcal{L}_{\mathrm{QCD}}(q,\bar{q},A)\right\}, \end{aligned} \tag{9.34}$$

where $\mathcal{L}_{QCD}$ is the Lagrangian (9.4) with N_f massless quarks and N_c colors, $d\mu[A]$ denotes the Yang-Mills field integration quasi-measure including gauge fixing and Faddeev-Popov determinant for the ghost fields (Karchev and Slavnov, 1985; Zahed and Brown, 1986). In accord with already known low energy QCD properties the following conditions are imposed:

a. Let $N_c \to \infty$, i.e. keep only the terms to leading order in N_c;
b. Assume that the chiral symmetry is spontaneously broken with pseudoscalar mesons π^a as Goldstone bosons;
c. Take the low-energy and long-wavelength limit in which one expects the Goldstone bosons to dominate in physics.

To introduce the chiral phase (order parameter of the quark medium) $\mathbf{U} = \exp\{i\pi^a(x)\lambda^a\}$, λ^a being the $U(N_f)$ generators, we perform a change of variables in the functional integral (9.34):

$$q \to q^U : q^U = \mathbf{U}^{-2} q_R + q_L; \quad \bar{q}^U = \bar{q}_R \mathbf{U}^2 + \bar{q}_L. \tag{9.35}$$

This is done in anticipation of the latter identification of $\pi^a(x)$ with the meson fields. The Jacobian of this substitution J after calculations gives

$$\begin{aligned} \ln J = &-\frac{iN_c}{48\pi^2} \int_0^1 d\alpha \int dx\, \epsilon^{\mu\nu\lambda\rho} \mathrm{Tr}\{L_\alpha L_\mu L_\nu L_\lambda L_\rho\} \\ &+ \frac{i\sqrt{2N_f}}{8\pi^2} \mathrm{Tr} \int dx\, G_{\mu\nu}(x) G_{\lambda\rho}(x) \epsilon^{\mu\nu\lambda\rho} \pi^0(x), \end{aligned} \tag{9.36}$$

where it was used the extension of $\mathbf{U}(x)$ to $\mathbf{U}(x,\alpha) = \exp\{i\alpha\pi^a(x)\lambda^a\}$, so that "extended" chiral currents are

$$L_\mu(x,\alpha) = \mathbf{U}^\dagger(x,\alpha)\partial_\mu \mathbf{U}(x,\alpha), \quad L_\alpha(x,\alpha) = \mathbf{U}^\dagger(x,\alpha)\partial_\alpha \mathbf{U}(x,\alpha). \tag{9.37}$$

The first term in (9.36) is **the anomalous Wess-Zumino term**. The corresponding part of the effective Lagrangian still "remembers" its quark origin through the appearance of the number of colors N_c, even if they have been eliminated as explicit degrees of freedom. The second term in (9.36) is just the Adler-Bell-Jackiw-Bardeen $U(1)$ anomaly with π^0 being the (N_f^2-1)-th pseudoscalar field. This term can be omitted if we keep only terms to the leading order in N_c. Thus the only remaining piece of $\mathcal{L}_{\text{eff}}$ obtained in this procedure is the Wess-Zumino term.

The next step involves the integration over the color (quark and gluon) degrees of freedom to obtain the generating functional given as an integral over the chiral fields

$$Z_{\text{QCD}} \text{ from (9.34)} \Rightarrow \int d\mu[\mathbf{U}] \exp\left\{ i \int dx\, \mathcal{L}_{\text{eff}}(\mathbf{U}) \right\}. \tag{9.38}$$

To make our story short it suffies to say that the main difficulty here lies in the integration over the gluon variables, since at the moment we are not able to sum up all diagrams containing closed gluon lines. Such a summation has not yet been performed even for planar diagrams. Nevertheless, using the functional integration methods in application to a gauge theory (Rajaraman, 1982; Faddeev and Slavnov, 1980), the large N diagrammatic analysis and assuming a spontaneous chiral symmetry breaking, Karchev and Slavnov (1985) obtained the following form for resulting effective Lagrangian

$$\mathcal{L}_{\text{eff}}(\mathbf{U}) = -\frac{F_\pi^2}{16}\mathrm{Tr}(L_\mu L^\mu) + \mathcal{L}_{WZ} + V_0(\pi^0), \tag{9.39}$$

where the pion decay constant $F_\pi^2 = N_c f_\pi^2$ is given in terms of the quark-gluon coupling parameter appearing in the Bethe-Salpeter amplitude for pseudoscalar Goldstone bosons. Since (9.39) clearly implements an expansion in the powers of the chiral currents L_μ, the Skyrme-type terms of order four and higher in L_μ are embodied in the functional $V_0(\pi^0)$. However, the exact determination of the potential $V_0(\pi^0)$ at least requires the summation of planar diagrams in a pure Yang-Mills theory.

Note, that in its full extent this problem is equivalent to that of the low-energy QCD solvability. Already at the early stage of recent studies (Witten, 1984) it became evident that the meson theory (equivalent to QCD) must be a rather complicated and its Lagrangian function must be dependent not only on pseudoscalar mesons, but as well on all low-lying meson multiplets. The modest results in the direct derivation of such a Lagrangian reflects the lack in a quantitative understanding of low-energy nonperturbative phenomena in QCD such as the spontaneous breaking of chiral symmetry, the condensation of quarks into mesons (bosonization), confinement of quarks and gluons and so on. Therefore a progress in that direction requires either further assumptions or developed "scenarios" for listed nonperturbative phenomena with subsequent tests for the reasonableness.

Among others here we should mention **the chiral bosonization method**, suggested in (Andrianov, 1985; Andrianov and Novozhilov, 1985). Leaving aside the interactions between quark flavor and color phases, it has been considered a system of quarks interacting with external color-singlet fields: vector V_μ, pseudovector A_μ, scalar S and pseudoscalar P. The Lagrangian density for thus described system is

$$\mathcal{L} = \bar{q}\not{D}q = \mathrm{i}\,\bar{q}\left[\gamma^\mu(\partial_\mu - V_\mu - \gamma_5 A_\mu) + \mathrm{i}\,S + \mathrm{i}\gamma_5 P\right] q, \tag{9.40}$$

where $\not{D}$ is the complete Dirac operator. It is assumed that the external fields are transformed in such a way that the Lagrangian (9.40) is invariant under gauge and local chiral transformations. If one ignores the quark masses, then the Dirac operator spectrum: $\not{D}q_k = K\,q_k$ is symmetrical and a gauge-invariant definition of the low energy region employs only the QCD cutoff parameter (or the chiral-breaking mass scale) Λ_{QCD}: $|K| < \Lambda_{\text{QCD}}$. For massive quarks it is necessary to introduce an additional parameter M - the assymetry of quark spectrum, and to define a low energy region by

$$|K - M| \leq \Lambda_{\text{QCD}}, \quad 0 \leq M \leq \Lambda_{\text{QCD}}. \tag{9.41}$$

The quark condensate $\langle \bar{q}q \rangle$ (the quark density overaged over the volume) is easily calculated and gives

$$\langle \bar{q}q \rangle = -\frac{N_c}{2\pi^2}\Big(\Lambda^2_{\rm QCD}\, M - \frac{M^3}{3}\Big). \tag{9.42}$$

Thus the Andrianov-Novozhilov bosonization method is based on the assumption, that in the low energy region of QCD there dominate the nonperturbative quark fluctuations, violating the chiral symmetry and leading to the formation of the quark condensate. Such an assumption makes it possible to integrate over fermion fields in $Z_{\rm QCD}$, taking into account the restrictions (9.41) and utilizing the finite mode regularization technique (Andrianov et al., 1986). For our exposition the most interesting in this approach is the available derivation of the 4th order terms in chiral currents L_μ for the effective Lagrangian

$$\mathcal{L}^{(4)}_{\rm eff}(\mathrm{U}) = \frac{N_c}{384\pi^2}\mathrm{Tr}\{[L_\mu, L_\nu]^2 - 2(L_\mu L^\mu)^2 + 4\partial_\mu\partial^\mu \mathrm{U}\partial_\mu\partial^\mu \mathrm{U}^\dagger\}. \tag{9.43}$$

Note that the complete Lagrangian also depends on external fields V_μ, A_μ, S and P, but we omit here the corresponding terms.

To be precise, recall, that an exact bosonization was proved only for two-dimensional QCD ('t Hooft, 1974b; Ebert and Pervushin, 1978; Gonzales and Redlich, 1984). In four dimensions an exact bosonization is precluded since the integration over gluon fields cannot be performed strictly due to the presence of gluon self-interaction. Nevertheless, if one restricts oneself to the simplest category of diagrams containing only the "two-body" correlations between quarks, then the integration in the right hand side of Eq. (9.34) over gluonic fields can be carried out exactly (Wrzecionko, 1988; Roberts and Cahill, 1987; Cahill, 1989). These direct calculations of the effective Lagrangian up to terms not higher than 4th order of chiral currents L_μ apart from the Wess-Zumino term give the following result:

$$\mathcal{L}_{\rm eff}(\mathrm{U}) = \alpha \mathrm{Tr}\,[L_\mu, L_\nu]^2 + \beta \mathrm{Tr}\,(L_\mu L^\mu) + \gamma \mathrm{Tr}\,[\partial_\mu \partial^\mu\, L_\nu, L^\nu]. \tag{9.44}$$

Here α, β and γ are coefficients all proportional to the number of colors N_c in the underlaying chromodynamics and can be expressed through just the running coupling constant α_s of QCD. Thus in such an approach we have an evident possibility to establish a relation between the phenomenological constants F_π and e in the Skyrme Lagrangian and the QCD fundamental parameters. Unfortunately, thus derived effective Lagrangian (9.44) contains the term (proportional to γ) with the higher derivatives of chiral currents. This is a rather depressive moment involving undesirable difficulties in a subsequent quantization procedure.

Lack of space combined with the lack of notable progress in the direct derivation of the meson Lagrangians equivalent to the low energy QCD prevent us from further reviewing of innumerable attempts in this field. We only draw the reader's attention that in much similar to expressions (9.39), (9.43) and (9.44) one can obtain effective Lagrangians not only from the underlying QCD functional (9.34) but as well from an intermediate quark model of the Nambu – Jona Lasinio type as it was advocated by Ebert and Reinhardt (1986). All listed results give us a right to conclude this expository section with the statement, that to get a more reliable effective chiral Lagrangians from the QCD first principles one needs more specific dynamical pictures of nonperturbative phenomena, more knowledge on the QCD vacuum structure together with more powerful and effective mathematical tools. Only after that an amusing concordance of the above listed results will get a likely explanation. These results also indicate that independently of the future form of a true effective meson Lagrangian (if any) the latter will at least contain the "kinetic" (quadratic in L_μ) term, the Wess-Zumino term and, probably a modified Skyrme term. Therefore all methods elaborated to study properties and possibilities of such a "minimal" effective meson theory are of great importance.

And the final note. At present it is generally accepted as a more reliable approach to construct (instead of to derive) a likely effective meson Lagrangian out of the fields associated with the low-lying meson multiplets, namely the pseudoscalar and vector (ρ, ω, A_1) multiplets. There is an intensive literature on this subject and we refer the reader as to some pioneering papers (Kaymakcalan, Rajeev and Schechter, 1984; Bando et al., 1985) and to comprehensive reviews (Meissner, 1988; Bando, Kugo and Yamawaki, 1988). We do not have here enough room to expose this very interesting and good working approach and find as an excuse that an interested reader will find all necessary information in the aforegiven references.

10. Skyrmion as a Fermion

"I would like to think that the fermion concept was just a good way of talking about the behaviour of some semi-classical construction, and that it was no more fundamental than renormalization."

Tony Skyrme

The fifth Skyrme's suggestion, as listed in Section 1.5, without any doubt appears to be the most attractive one. Recall, that one of the central ideas of Skyrme's approach was to discover a possible way to regard a bose-field theory originated soliton as a fermionic state. We have already mentioned that this idea entailed the new branch of study in nonlinear physics, called the Fermi-Bose transmutations. Localized structures with transmutations of spins and statistics has been studied in variety of nonlinear models, applied in gravitation ("geons") (Friedman and Sorkin, 1980; Sorkin, 1986), in condensed matter physics for model description of such novel effects as high temperature superconductivity ("anyons") (Arovas, Schrieffer, Wilczek and Zee, 1985; Fröhlich and Marchetti, 1988), the fractional quantum Hall effect ("holons") (Wilczek, 1982; Arovas, Wilczek and Schrieffer, 1984). We refer the reader to self-contained reviews and lecture notes on related questions (Mackenzie and Wilczek, 1988; Laughlin, 1988; Wilczek, 1990; Balachandran, Marmo, Skagerstam and Stern, 1991). Thus this Skyrme's conjecture appeared to be not only valid, but also productive.

10.1 The Finkelstein's Double-Valued Functionals

To demonstrate in what sense Skyrmions are fermions it is better to use the topological formalization of the distinction between boson and fermion states, provided in (Finkelstein, 1966; Finkelstein and Rubinstein, 1968). One may consult also (Ravenel and Zee, 1985; Schwarz, 1989) for more modern and accessible expositions. As a keystone in this topological approach it was used the quantum mechanical fact that a fermion wave function changes its sign or picks up a phase (π) when coordinate frame is adiabatically rotated by 2π radians. Under double 2π rotation the fermionic wave function must attain the initial value. In the configuration space $\mathbf{Map^o}(\mathbf{X};\Phi) \equiv \mathbf{M}$ of maps $\phi : \mathbf{X} \to \Phi$, where ϕ are classical fields at fixed moment of time and Φ is the field manifold of a classical theory, analogs of 2π rotations are closed paths or loops, starting and ending at a same field ϕ.

Recall that, being classical, a soliton field configuration must be single-valued under the action of the Poincaré transformations, in particular under spatial rotations. On the other hand, spinors are characterized by double-valuedness under 2π rotations. In homotopy language this fact can be expressed by means of the isomorphism for the fundamental group: $\pi_1(SO(3)) \simeq \mathbb{Z}_2$, which reflects the double connectivity of the $SO(3)$ manifold. There are two distinct classes of closed paths in $SO(3)$ (see Section 2.2), corresponding to two elements I and $-I$ of $\mathbb{Z}_2$, commonly associated with the integral and half-integral spin representations, respectively.

The homotopical classification of classical fields (as discussed in Section 2.2) and the topological charge $\mathbf{Q}$ by their definitions are independent of time. Therefore, they could be applied to classify quantum states. The set $\pi_o(\mathbf{M})$ of connection components of $\mathbf{M}$ can be identified with the set of homotopy classes of maps from $\mathbf{S}^3$ into Φ: $\{[\mathbf{S}^3, \Phi]\} = \pi_3(\Phi)$. It can be proven (Whitehead, 1966) that

$$\pi_k(\mathbf{M}) = \pi_{k+3}(\Phi), \tag{10.1}$$

therefore the elements of the fundamental group of the configuration space $\mathbf{M}$ might be associated to elements of the fourth homotopy group of Φ:

$$\pi_1(\mathbf{M}) = \pi_4(\Phi). \tag{10.2}$$

In the case of Skyrme model the field manifold $\Phi = SU(2)$, so that $\pi_4(\Phi) = \pi_4(SU(2)) = \mathbb{Z}_2$. It means that there are only two distinct types of loops in the Skyrme model configuration space: those that are trivial (or, in other words, contractable in a continuous way to a point on $\mathbf{M}$) and non-trivial (noncontractable) loops. A representation of a nontrivial path in the Hilbert space of quantum states should be an operator which changes the phase of wave functions by (π). As for trivial paths, the corresponding operators should not affect phases of quantum states. Thus we can construct two distinct quantization schemes for Skyrmions: fermionic and bosonic. To choose one of them means to impose a constraint on the Hilbert space of physically allowed states, which would select only desirable states. It is clear from the aforegiven, that Skyrmions in the original $SU(2)$ model **might be quantized as fermions,** if by one or another method we can exclude the possibility of trivial path representations in the Hilbert space. A possible form of this constraint was given by (Finkelstein, 1966; Finkelstein and Rubinstein, 1968) and later on by (Williams, 1970), as necessary and sufficient conditions for the existence of the, so-called, double-valued functionals or "spinorial structures" on field manifolds in the framework of nonlinear field theory.

To be more precise, let us consider the field $\phi(\boldsymbol{x}, t)$, $t_1 \leq t \leq t_2$, as a path p in $\mathbf{M}$ connecting the points $\phi_1(\boldsymbol{x}) = \phi(\boldsymbol{x}, t_1) \in \mathbf{M}$ and $\phi_2(\boldsymbol{x}) = \phi(\boldsymbol{x}, t_2) \in \mathbf{M}$. Denoting the corresponding quantum state vectors as $|\phi_i>$, the quantum transition amplitude from $|\phi_1>$ to $|\phi_2>$ can be written by means of Feynman's path integral (Schwartz, 1989) as

$$\langle\phi_2|\phi_1\rangle = \sum_{\alpha} \chi(\alpha) \int_{M_\alpha(\phi_1,\phi_2)} d\mu[\phi] \exp\left\{\frac{i}{\hbar}\int dx^4\, \mathcal{L}[\phi]\right\}. \tag{10.3}$$

Here the sum runs over the set of connection components $\{M_\alpha(\phi_1,\phi_2)\}$ of the space $\mathbf{M}(\phi_1,\phi_2)$ of paths connecting the points $\phi_1(\boldsymbol{x})$ and $\phi_2(\boldsymbol{x})$ (frequently named as *the space of histories of fields*). $\mathcal{L}[\phi]$ is the Lagrangian density of classical theory; $d\mu[\phi]$ is a quasi-measure in the functional space. Without any loss of generality the paths $\{\phi(\boldsymbol{x},t)\} = \{p\}$ might be considered as closed ones, i.e. $\phi_1 = \phi_2$, then the set of components $\{M_\alpha(\phi_1,\phi_2)\}$ can be identified with $\pi_1(\mathbf{M}) = \pi_{3+1}(\Phi)$. The coefficients $\chi(\alpha)$ in (10.3) obey the composition law[1]: $\chi(\alpha\beta) = \chi(\alpha)\chi(\beta)$ and thus determine a one-dimensional representation of $\pi_1(\mathbf{M})$. The calculation of $\chi(\alpha)$ in general case might be reduced to that in the case $\phi_1 = \phi_2$. The latter facts reflect that the space $\mathbf{M}(\phi_1,\phi_2)$ is either empty or homotopically equivalent to $\mathbf{M}(\phi_1,\phi_1)$.

Consider the point $\phi_1 \in [\mathbb{S}_3,\Phi]_i \equiv \Sigma_i$. Then in this homotopy class Σ_i there should be a closed path p such that the value of the functional (10.3) along the path (i.e. for $\phi_2 \in p$) changes continuously with ϕ_2 and does not attain its original value at the end point. At the same time, after going twice around the same path p, or equivalently after going around the path p^2 it must return to the initial value. In such a manner in the homotopy class Σ_i there should exist a nontrivial (a noncontractable to a point) closed path p, such that the path p^2 would be a trivial (a contractable) one. This picture reminds the familiar situation when the fermionic wave functions pick up a phase factor -1, the coordinate frame being adiabatically rotated by 2π radians. On the contrary, the bosonic wave functions (one can regard the latter as made up of two fermionic) does not change their sign under such a rotation.

Definition 10.1. *The functionals of the type (10.3) with the above described property are called* **the double-valued functionals** *due to (Finkelstein, 1966)*

Let us pause for a moment and discuss a little longer the appearance of the double-valued (multi-valued, in general) functionals in physical vocabulary. The meaning of this concept is at first sight imprecise and contradictive. However in various physical theories we frequently face the situation when the classical dynamical equations are gauge invariant, whereas the Lagrangian of the model is not. The corresponding action changes by the endpoint contributions and thus presents a typical example of a **multi-valued functional**. For classical physics, which is entirely determined by the dynamical equations of motion, this fact is irrelevant. Quantum theory makes greater use of the action because it enters the expression for the transition amplitude (see Eq. (10.3)). However it does so only via the exponential $\exp\{i/\hbar\cdot S\}$ and therefore it turns to be enough to require the gauge invariance of this exponential instead of that of the action S itself. To provide a consistent quantum description some

[1] This composition law follows directly from the group properties of the evolution operator $\exp\left\{-\frac{i}{\hbar}\hat{H}t\right\}$, where $\hat{H}$ is the Hamiltonian of the quantum system.

parameters (masses, coupling constants, charges etc.) that are unconstrained classically are forced to take on integer values (Jackiw, 1984 a,b). This phenomenon was named *the parameter quantization* in the field theory and bears a strong resemblance to Dirac's quantization of magnetic charge. We discuss the latter topic in full extent in the next section. For a mathematically rigorous definition of the multi-valued functionals one may consult with (Novikov, 1981; 1982). It is proved that this concept may be formulated exactly by introducing the idea of the covering system in close analogy to that of the universal covering space in topology (see Appendices B and E for more information).

In accord with Section 2.2, the construction which contains the information about the path structure of a homotopy class $[\mathbf{X}, \Phi]_i \equiv \Sigma_i$ is the first homotopy group $\pi_1(\Sigma_i)$. Due to the Whitehead theorem there exists the isomorphism

$$\pi_1(\Sigma_i) \simeq \pi_1(\Sigma_j), \quad i \neq j, \tag{10.4}$$

and therefore one can focus attention on the situation in any one homotopy class, namely in Σ_1. As demonstrated by Finkelstein (1966) **a necessary and sufficient condition for the existence of double-valued functionals** on Σ_1 is that the group $\pi_1(\Sigma_1)$ has an element of order 2, e.g.

$$\pi_1(\Sigma_1) \simeq \mathbb{Z}_2. \tag{10.5}$$

However the existence of double-valued functionals alone does not imply the existence of spinor structures. It must be shown, that the double-valuedness arises through following a path generated by a rotation of the soliton field configuration. Being continuous deformations complete rotations can only induce closed loops within Σ_1. If (10.5) holds, then in accord with the above discussion there exist two types of loops: the trivial (contractible) loops, corresponding to the integral spin and the nontrivial (noncontractible) loops, which can be linked with the half-integral spin. For such a double-connectivity of Σ_1 to be uniquely linked with the manifestation of spinor structure, it must be shown that $\pi_1(\Sigma_1) = \pi_1(SO(3))$, i.e. that the nontrivial loops of Σ_1 are in fact generated by actual 2π spatial rotation of the soliton configuration.

In summary, we say that a field theory with topologically nontrivial soliton solutions **admits spin** provided the following conditions hold[2] (Williams, 1970):

a. $\pi_1(\Sigma_1)$ has an element of order 2, or the isomorphism (10.5) holds.
b. Paths begining and ending at some $\phi_0 \in \Sigma_1$ and corresponding to rotation through 2π are noncontractable.

It is clear, that one can conceive those conditions as **a generalization of the Pauli exclusion principle** in the form, applicable in quantized nonlinear theory.

In order to evaluate $\pi_1(\Sigma_1)$ we can use the Whitehead theorem (10.4) in a particular form

[2] Condition (b) implies (a), but (a) is usually easier to verify and can be used as a test for the possibility of (b).

$$\pi_1(\Sigma_1) \simeq \pi_1(\Sigma_0), \tag{10.6}$$

therefore reducing the problem to the simpler one of seeking the path structure of the trivial sector Σ_0. One example of mapping belonging to Σ_0 is the constant mapping, which maps the whole of domain $\mathbf{X}$ into the fixed point $\phi_0 \in \Phi$:

$$\phi(\boldsymbol{x}) : \mathbf{X} \rightarrow \phi_0, \quad \text{for any} \quad \boldsymbol{x} \in \mathbf{X}. \tag{10.7}$$

Still further quotation from (Whitehead, 1966) shows that there is the homeomorphism between the space of closed loops in Σ_0 and the mapping space $\mathbf{Map}^{\mathbf{o}}(\mathbf{X}; \Phi)$ with suitable boundary conditions. The corresponding simple computations are reproduced in Appendix E. Thus the groups $\pi_1(\Sigma_0)$ and $\pi_4(\Phi)$ are isomorphic,

$$\pi_1(\Sigma_0) \simeq \pi_4(\Phi), \tag{10.8}$$

and one can use the tabulated results (see for example Table I in Appendix B) to find out whether a given theory passes the first test. In particular for the Skyrme model we have $\Phi = \mathbb{S}^3$, $\pi_4(\mathbb{S}^3) \simeq \mathbb{Z}_2$, whence $\pi_1(\Sigma_1) \simeq \mathbb{Z}_2$.

To verify the second condition it remains to prove that nontrivial closed paths in Σ_1 based at some point ϕ_0 actually correspond to a 2π spatial rotation. The details of the proof given by (Williams, 1970) for convenience of the reader are reproduced in Appendix E. In brief this proof consists in considering a closed path in Σ_1 as a continuous sequence of mappings of the type (2.5):

$$\phi^{(1)}(\lambda) = \phi_0(\lambda) + i\tau_a \cdot \phi^a(\lambda), \tag{10.9}$$

where λ is a parameter, $\phi_0(\lambda)$ and $\phi^a(\lambda)$ are defined through the inverse stereographic projection (2.4) components or spherical projective coordinates

$$\phi_0(\lambda) = \frac{\boldsymbol{r}^2 - r_0^2}{\boldsymbol{r}^2 + r_0^2}; \quad \phi^a(\lambda) = R^{ab}(\lambda) \frac{2\, r_0 x^b}{\boldsymbol{r}^2 + r_0^2}. \tag{10.10}$$

Here $R^{ab}(\lambda)$ is a rotation matrix. Thus we have a path in Σ_1 (the 1-soliton sector). To transfer from the 1-soliton to 0-soliton sector Σ_0 one can use as an appropriate automorphism the combined mapping $\phi_1^\dagger \cdot \phi_1$, as $\phi_1^\dagger$ creates an antisoliton. The essential element of Williams' proof comes from the observation that for $\lambda = 1/2$ a mapping $\psi : \mathbb{S}^4 \rightarrow \mathbb{S}^3$, associated with a path in Σ_0, is a suspension of the Hopf map $\mathbb{S}^3 \rightarrow \mathbb{S}^2$. Hence the path (10.9) corresponding to a 2π spatial rotation is nontrivial. This completes the Finkelstein-Williams proof of admittance of spin structures in the Skyrme model. Finally, it has also been shown for models of Skyrme's type that another defining property of a multi-fermion state, namely, the double-valuedness of the wavefunctional under fields interchange, implies double-valuedness under a 2π rotation (Finkelstein and Rubinstein, 1968).

Remark 10.1. The Finkelstein-Rubinstein-Williams scheme looks like rather abstract considerations of general character, which are model-independent. Notice, that in the above text we did not use any other attributes of Skyrme model, exept the choice of the field manifold Φ and the dimension of physical space-time. The procedure of exchange of particles which establishes the normal connection between spin and statistics of the $SU(2)$ Skyrmions is of the same general character and therefore does not care about the dynamical details of the model. It should be noted here due to (Balachandran, Gomm and Sorkin, 1987; Sorkin, 1988) that the original Finkelstein-Rubinstein-Williams scheme is limited in applications, as, for example, it fails to give an answer in the case of three and more flavors with $\Phi = SU(N)$ as a field manifold, for $N \geq 3$. All closed paths are trivial in this case, because of isomorphism $\pi_4(SU(N)) = 0$. This fact was first notified by E. Witten, who developed in (Witten, 1983a) a dynamical realization of Finkelsten-Rubinstein-Williams ideas.

The ideas behind such a dynamical approach to the Fermi-Bose transmutations go back to earliest treatments of the charge-monopole system. Therefore we find appropriate for pedagogical purposes to expose this problem to a certain extent.

10.2 The Charge-Monopole Multi-valued Action

The charge-monopole system, one of the classic problems in theoretical physics, gave rise to a number of new methods, approaches and even caused new branches of study. In his 1931 paper P. Dirac clearly showed that the quantum mechanics of an electric charge e in the given background field of a magnetic monopole of strength μ is consistent if and only if their product

$$e\mu = \frac{n\hbar}{2}, \quad n \in \mathbb{Z}. \tag{10.11}$$

Among others Tamm's result (Tamm, 1931) is remarkable in its relevance to the present subject. Starting with the Schrödinger equation for the charge-monopole system Tamm proved that if the condition (10.11) holds for $n = 1$, then the system possesses the half-integer angular momentum.

Let us start from the well-studied description of a charged particle on the surface of an unit sphere $\boldsymbol{r}^2 = 1$ through the Lagrangian density

$$\mathcal{L} = \tfrac{1}{2} m \dot{\boldsymbol{r}}^2 + \eta (\boldsymbol{r}^2 - 1), \tag{10.12}$$

where η is the Lagrange multiplier and the magnetic field is assumed to be trivial ($\boldsymbol{B} = 0$). The equation of motion (after the evaluation of an appropriate η) reads

$$m\ddot{x}_i + m x_i \cdot \sum_{i=1}^{3} \dot{x}_i^2 = 0 \tag{10.13}$$

and exhibits separately two descrete symmetries

$$P: t \to -t; \quad x_i \to -x_i. \tag{10.14}$$

Now, if we switch on the monopole magnetic field $\boldsymbol{B} = \mu \boldsymbol{r}/r^3$ the corresponding equation of motion

$$m\ddot{x}_i + m x_i \cdot \sum_{i=1}^{3} \dot{x}_i^2 = e\,(\dot{\boldsymbol{r}} \times \boldsymbol{B})_i = e\mu\epsilon_{ijk}\,\dot{x}_j \frac{x_k}{r^3} \tag{10.15}$$

where the Lorentz force in the right-hand side is added. It is easy to verify that Eq. (10.15) is invariant only under the combined operation of parity transformations (10.14). The other distinction consists in the problematical restoration of a Lagrangian, which leads to the equation (10.15) via the standard variational principle. There is no obvious term whose variation equals to the Lorentz force, since the simplest choice $\epsilon_{ijk}\dot{x}_i x_j x_k = 0$.

However this problem has a well-known solution. One can introduce a vector potential $\boldsymbol{A}$ for the monopole magnetic field such that

$$(\nabla \times \boldsymbol{A}) = \boldsymbol{B} = \frac{\mu \boldsymbol{r}}{r^3} \tag{10.16}$$

and to write down the appropriate action functional in the form

$$S = \int dt \left\{ \tfrac{1}{2} m \dot{\boldsymbol{r}}^2 + \eta\,(\boldsymbol{r}^2 - 1) + e\,(\dot{\boldsymbol{r}} \cdot \boldsymbol{A}) \right\}. \tag{10.17}$$

Although (10.17) yields (10.15) as the Euler-Lagrange equations it is not a well-defined expression. Indeed, there is a contradiction between the left- and right-hand side equalities in (10.16). Since for the monopole field $\nabla \cdot \boldsymbol{B} \neq 0$ while $\boldsymbol{B} = (\nabla \times \boldsymbol{A})$, the introduced vector potential $\boldsymbol{A}$ is singular in $\mathbb{R}^3$ (the Dirac string of singulariries) as it was first noted in (Dirac, 1931; 1948). Moreover, the equation of motion (10.15) is manifestly invariant under a gauge transformation

$$\boldsymbol{A}' = \boldsymbol{A} + \nabla \psi, \tag{10.18}$$

while the action (10.17) changes under (10.18) by the end-points terms:

$$S' = S + e \int (\nabla \psi \cdot d\boldsymbol{r}) = S + e \int d\psi. \tag{10.19}$$

Thus we have a typical example of a **multi-valued action**, and as discussed in the previous section it causes a troublesome situation in a quantum description of the charge-monopole system. Since the action (10.17) does enter a quantum mechanical transition amplitude (in Feynman form)

$$Z\,(t_2, t_1) = \int_{\boldsymbol{r}(t_1)}^{\boldsymbol{r}(t_2)} d\mu[\boldsymbol{r}(t)] \exp\left\{ \frac{i}{\hbar}\, S\,[\boldsymbol{r}(t)] \right\}, \tag{10.20}$$

the quantum mechanical treatment of the problem would be a gauge-dependent.

The way round this is the following. Although it is forbidden to define a non-singular $\boldsymbol{A}$ over the whole of any sphere about the origin, we can do this separately, for example on the northern and southern hemispheres. To implement this recipe consider a closed path γ of the charge on the sphere (see Fig. 10.1), putting $\boldsymbol{r}(t_1) = \boldsymbol{r}(t_2)$ in (10.20). According to (10.17) and (10.19) the troublesome term in (10.20) is

$$e \int_{t_1}^{t_2} dt\,(\dot{\boldsymbol{r}} \cdot \boldsymbol{A}) = e \oint_{\gamma} (\boldsymbol{A}(\boldsymbol{r}) \cdot d\boldsymbol{r}). \tag{10.21}$$

Fig. 10.1. A closed path γ of a charged particle on $\mathbf{S}^2$ with a Dirac monopole at the origin (not shown). The path γ devides the surface of $\mathbf{S}^2$ into two distinct discs D^+ and D^-, so that $\mathbf{S}^2 = D^+ \cup D^-$

By Stokes' theorem, the line integral of the vector potential $\boldsymbol{A}$ along γ can be converted into an integral of the magnetic field $\boldsymbol{B}$ over the surface bounded by γ. But it is possible iff the integrand in (10.21) is non-singular. To meet the latter requirement we can choose as following: when integrating over the disc D^+ we relegate the singularities of $\boldsymbol{A}$ to the other disc D^- and vice versa. Then we can write for the case of two different gauges

$$\begin{aligned} e \oint_{\gamma} (\boldsymbol{A}_1(\boldsymbol{r}) \cdot d\boldsymbol{r}) &= e \int_{D^+} (\mathrm{rot}\boldsymbol{A}_1(\boldsymbol{r}) \cdot d\boldsymbol{S}) = e \int_{D^+} (\boldsymbol{B}_1(\boldsymbol{r}) \cdot d\boldsymbol{S}), \\ e \oint_{\gamma} (\boldsymbol{A}_2(\boldsymbol{r}) \cdot d\boldsymbol{r}) &= -e \int_{D^-} (\mathrm{rot}\boldsymbol{A}_2(\boldsymbol{r}) \cdot d\boldsymbol{S}) = -e \int_{D^-} (\boldsymbol{B}_2(\boldsymbol{r}) \cdot d\boldsymbol{S}), \end{aligned} \tag{10.22}$$

where a minus sign comes in because γ bounds D^+ in a right-hand sense, and D^- in the left-hand sense. Making use of (10.22) we can calculate the change of action S when passing from one gauge to the other

$$\Delta S = e \oint_{D^+ \cup D^-} (d\boldsymbol{S} \cdot \boldsymbol{B}) = e \int_V d^3r\, \mathrm{div}\boldsymbol{B} = 4\pi e\mu, \tag{10.23}$$

where V is the volume bounded by the $\mathbf{S}^2$ surface. The natural requirement to get the same physics in both gauges or to make the results of (10.20) gauge-independent leads to the obvious condition

$$\exp\left\{\frac{\mathrm{i}e}{\hbar} \int_{D^+} (\boldsymbol{B}_1(\boldsymbol{r}) \cdot d\boldsymbol{S})\right\} = \exp\left\{-\frac{\mathrm{i}e}{\hbar} \int_{D^-} (\boldsymbol{B}_2(\boldsymbol{r}) \cdot d\boldsymbol{S})\right\}, \tag{10.24}$$

or equivalently due to (10.23)

$$\exp\left\{\frac{ie}{\hbar}\int_{D^{+}\cup D^{-}}(\boldsymbol{B}(\boldsymbol{r})\cdot d\boldsymbol{S})\right\}=\exp\left\{\frac{\mathrm{i}}{\hbar}\triangle S\right\}=1. \tag{10.25}$$

Hence we obtain **the Dirac quantization condition** for the product of electric and magnetic charges

$$\frac{1}{\hbar}\triangle S=2\pi n, \quad e\mu=\frac{n}{2}\hbar, \quad n\in\mathbb{Z}, \tag{10.26}$$

which is a clear example of the parameter quantization phenomena.

To conclude the section we fill like necessary to make some remarks of topological character.

Remark 10.2. To introduce a well-defined vector potential $\boldsymbol{A}$ for the monopole field we need at least two regions D^{+} and D^{-} covering the sphere $\mathbb{S}^2$. This reflects the well-known fact from topology that at least two charts are necessary to cover the $\mathbb{S}^2$ manifold. In the overlap region of these charts the potentials $\boldsymbol{A}_1$ and $\boldsymbol{A}_2$ are to differ at most by a gauge transformation, which might be considered as a requirement for a smooth connection between two charts.

Remark 10.3. A closed path γ in $\mathbb{S}^2$ one can regard as a mapping γ of a circle $\mathbb{S}^1$ into $\mathbb{S}^2$. The gauge transformations $U(t)=\exp\{\mathrm{i}\psi(\boldsymbol{r}(t))\}$ form the group $U(1)\simeq SO(2)$ with time-dependent elements, which might be considered as mappings of a circle $\mathbb{S}^1$ (a compactified time axis $\mathbb{R}^1$) into the group $SO(2)$. The connectivity of the space $\mathbf{Map}(\mathbb{S}^1,\mathbf{SO(2)})$ is defined through the fundamental group $\pi_1(SO(2))\simeq\mathbb{Z}$. Thus one can associate an integer-valued invariant $\mathbf{Q}=n$ (the Chern-Pontryagin index) to each of homotopy classes. In particular these integer numbers enter the Dirac quantization rule (10.11) and (10.26). And for closed paths the multi-valued action functional (10.17) might be written in the form

$$S'=S+e\int d\psi=S+2\pi\hbar\mathbf{Q}. \tag{10.27}$$

10.3 The Wess-Zumino Term and Witten's Realization of Skyrme's Suggestion

The lesson learned from the previous sections shows us a supplementary way of searching for an appropriate effective Lagrangian in order to describe the low-energy limit of QCD. First of all such a Lagrangian must have a smallest possible number of derivatives, and on the other hand it must exhibit the same symmetry properties as that of QCD Lagrangian (9.4) in the chiral limit. Let us try to meet these minimal requirements by the simplest $U(3)_L\otimes U(3)_R$ symmetric Weinberg-Gürsey Lagrangian (2.40):

$$L = -\frac{F_\pi^2}{16} \int d^3x \mathrm{Tr}(\mathbf{L}_\mu \mathbf{L}^\mu), \tag{10.28}$$

where from here on the chiral field $\mathbf{U}(\boldsymbol{x}, t)$ takes its values in the $SU(3)$ manifold, so that

$$\mathbf{U}(\boldsymbol{x}, t) = \exp\left\{\frac{2\mathrm{i}}{F_\pi} \lambda^a \pi^a(\boldsymbol{x}, t)\right\}, \quad a = \overline{1, 8}. \tag{10.29}$$

If the Goldstone bosons $\{\pi^a\}$ are treated as pseudoscalars in a sense that under parity transformations P pion fields transform as

$$P : \pi^a(\boldsymbol{x}, t) \ \rightarrow \ -\pi^a(-\boldsymbol{x}, t), \tag{10.30}$$

then the massless QCD is parity invariant. In terms of chiral fields $\mathbf{U}(\boldsymbol{x}, t)$ the latter operation corresponds to

$$P \cdot \mathbf{U}(\boldsymbol{x}, t) \cdot P^{-1} \ = \ \mathbf{U}^\dagger(-\boldsymbol{x}, t). \tag{10.31}$$

It is not difficult to present this QCD parity operation

$$P : \boldsymbol{x} \ \rightarrow \ -\boldsymbol{x}; \quad t \ \rightarrow \ t; \quad \mathbf{U} \ \rightarrow \ \mathbf{U}^\dagger = \mathbf{U}^{-1} \tag{10.32}$$

as a combination of two descrete symmetries, namely

$$P_0 : \boldsymbol{x} \ \rightarrow \ -\boldsymbol{x}, \ t \ \rightarrow \ t; \quad (-1)^{N_B} : \mathbf{U} \ \rightarrow \ \mathbf{U}^\dagger, \tag{10.33}$$

so that (10.32) is just

$$P = P_0 \cdot (-1)^{N_B}. \tag{10.34}$$

The notation $(-1)^{N_B}$ comes from the perturbative expansion of $\mathbf{U}$:

$$\mathbf{U}(\boldsymbol{x}, t) = 1 + \frac{2\mathrm{i}}{F_\pi} \sum_{a=1}^{8} \lambda^a \pi^a + \ldots, \tag{10.35}$$

which makes it evident that the operation $\mathbf{U} \ \rightarrow \ \mathbf{U}^\dagger$ is equivalent to $\pi^a \ \rightarrow \ -\pi^a$ and thus counts modulo two the number of bosons, N_B.

The problem, which reminds the charge-monopole story, is that the simplest Lagrangian (10.28) is invariant under P_0 and $(-1)^{N_B}$ separately, while QCD is invariant only under the combined operation P (10.32) or (10.34). The simplest processes that respect all symmetries of QCD but violate both P_0 and $(-1)^{N_B}$ one by one are

$$K^+K^- \ \rightarrow \ \pi^+\pi^0\pi^-; \quad \eta\pi^0 \ \rightarrow \ \pi^+\pi^0\pi^-. \tag{10.36}$$

Both processes are entailed by anomalies in QCD, and turn out to be kinematically excluded by the σ-model (10.28) since in each of them an even number of pseudoscalar mesons decay into an odd number, and vice versa.

10.3.1 The Wess-Zumino Term in Effective Chiral Lagrangian

Witten (1983a) proposed to handle with this problem in a manner similar to that in the charge-monopole system. This already known recipe tells us to add to (10.28) a higher-order $U(3)_L \otimes U(3)_R$-invariant term that breaks P_0 and $(-1)^{N_B}$ separately, while preserving their combination. We start with the Euler-Lagrange equation for (10.28)

$$\tfrac{1}{8} F_\pi^2 \, \partial_\mu \mathbf{L}^\mu = 0 \tag{10.37}$$

and modify it by adding a Lorentz-invariant term that has to violate P_0, consequently it must contain the totally anti-symmetric Levi-Civita symbol

$$\tfrac{1}{8} F_\pi^2 \, \partial_\mu \mathbf{L}^\mu + \eta \, \epsilon_{\mu\nu\lambda\rho} \mathbf{L}^\mu \mathbf{L}^\nu \mathbf{L}^\lambda \mathbf{L}^\rho = 0, \tag{10.38}$$

η is an (apparently) arbitrary constant. Indeed, the P_0-symmetry is violated since under $\boldsymbol{x} \rightarrow -\boldsymbol{x}$ we have $\epsilon_{\mu\nu\lambda\rho} \rightarrow -\epsilon^{\mu\nu\lambda\rho}$; $\partial_\mu \rightarrow \partial^\mu$ and $\mathbf{L}^\mu \rightarrow \mathbf{L}_\mu$, hence the transformed Eq. (10.38) reads

$$\tfrac{1}{8} F_\pi^2 \, \partial^\mu \mathbf{L}_\mu - \eta \, \epsilon^{\mu\nu\lambda\rho} \mathbf{L}_\mu \mathbf{L}_\nu \mathbf{L}_\lambda \mathbf{L}_\rho = 0. \tag{10.39}$$

With respect to the $(-1)^{N_B}$-operation we have $\pi^a \rightarrow -\pi^a$, so that $\mathbf{U}(x) \rightarrow \mathbf{U}^\dagger(x)$, $\mathbf{L}_\mu \rightarrow \mathbf{R}_\mu = \mathbf{U}\mathbf{L}_\mu\mathbf{U}^\dagger$, and taking into account the relation

$$\partial^\mu \mathbf{R}_\mu + \mathbf{U} \partial^\mu \mathbf{L}_\mu \mathbf{U}^\dagger = 0$$

from (10.39) we obtain (10.38). In other words, the redundant symmetries are lifted, while their combination P is not.

When trying to restore the Lagrangian by Eq. (10.38), which one needs to proceed to a quantum description, we find a trouble again. An obvious candidate might be

$$\mathrm{Tr}\,\{\epsilon^{\mu\nu\lambda\rho} \mathbf{L}_\mu \mathbf{L}_\nu \mathbf{L}_\lambda \mathbf{L}_\rho\} = 0, \tag{10.40}$$

but this vanishes identically in $(3+1)$ dimension due to the cyclic property of the trace and the skew symmetry of the Levi-Civita symbol. Still exploiting further the experience obtained in the previous section combined with the striking analogy of both problems, and following (Witten, 1983a), we can introduce a fifth fictitious dimension and regard our σ-model as an embedded into a five-dimensional space. The anomaly term in (10.38) one can treat as the analogue of the Lorentz force in (10.15), while the geometrical analogue of the one-dimensional closed path $\mathbb{S}^1$ on $\mathbb{S}^2$ is a four-dimensional quasi-sphere $\mathbb{S}^3 \otimes \mathbb{S}^1$ on $\mathbb{S}^3 \otimes \mathbb{S}^2$. This quasi-sphere one might consider as a compactification of the space-time $\mathbb{R}^3 \otimes \mathbb{R}^1$ due to required behaviour of the chiral field $\mathbf{U}(\boldsymbol{x}, t) \rightarrow \mathbb{1}$ as $|\boldsymbol{x}| \rightarrow \infty$, so that $\mathbb{R}^3 \cup \{\infty\} = \mathbb{S}^3$, and since $\mathbf{U}(\boldsymbol{x}, \infty) = \mathbf{U}(\boldsymbol{x}, -\infty)$ (closed path) we have $\mathbb{R}^1 \cup \{\infty\} = \mathbb{S}^1$. For the sake of simplicity, we can ignore for a moment such a proper treatment, and exchange the quasi-sphere $\mathbb{R}^3 \otimes \mathbb{R}^1$ for the sphere $\mathbb{S}^4$. The results do not change but in this way we can avoid in much the unjustified (for the present purposes) complexities in our exposition[3]. Therefore we can treat chiral fields, defined on $\mathbb{R}^3 \otimes \mathbb{R}^1 = \mathbf{S}^4$, as mappings

$$\mathbf{U}(x) : \mathbf{S}^4 \rightarrow SU(3), \tag{10.41}$$

while under the aforementioned extention with appropriate boundary conditions we have the $\mathbf{S}^5$-sphere as a new domain of chiral fields, i.e.

$$\mathbf{U}(x, s) : \mathbf{S}^5 \rightarrow SU(3). \tag{10.42}$$

Remark 10.4. To justify the above extention we need to require for maps from (10.41) to be homotopic to a constant map into some point on $SU(3)$. The latter condition comes from the following considerations. The $\mathbf{S}^4$ sphere is the boundary of two discs D_5^+ and D_5^- in $\mathbf{S}^5$. The picture we get is analogous to that in Fig. 10.1. Within one of the discs we can define a continuous family of spheres, starting with $\mathbf{S}^4$ (the image of physical space), which might be fibred into the bunch of $\mathbf{S}^3$ spheres, each of the latter - fibred into the bunch of $\mathbf{S}^2$ and so on up to $\mathbf{S}^0$. Thus we have a continuous sequence of spheres, each enclosing the next, starting with the physical space and ending at a single point (Goddard and Mansfield, 1986). The values of the field $\mathbf{U}$ on each sphere provide a homotopy between the values $\mathbf{U}$ taken on the physical space (or its image $\mathbf{S}^4$) and the constant map $\mathbf{U}_0$, which is its value at the single point.

In short the conclusion of Remark 10.4 might be expressed as a requirement for the homotopy group $\pi_4(\Phi)$ to be trivial and fortunately this is the case for $\Phi = SU(3)$:

$$\pi_4(SU(3)) = 0. \tag{10.43}$$

The extended maps (10.42) also fall into homotopy classes, so that

$$\pi_5(SU(3)) = \mathbb{Z}, \tag{10.44}$$

and one can construct a homotopic invariant, realizing the isomorphism (10.44), along the lines of Isham's algorithm (see, for example, Section 2.3). The result would be as follows

[3] To be precise, the image of space-time under a map $\mathbf{U}(\boldsymbol{x}, t)$ is homeomorphic to the image of $\mathbf{S}^4$ only if the whole boundary of space-time is mapped by $\mathbf{U}$ to the same point in $SU(3)$ manifold. On the other side, if $\mathbf{U}$ maps all the points $\{(\boldsymbol{x}, t = -\infty)\}$ to a single point in $SU(3)$, then $\mathbf{U}$ defines a map from $\mathbf{S}^3$ to $SU(3)$ in the trivial homotopy class. It means that for such maps $\mathbf{U}$ the topological charge is zero, and it does not correspond to a nontrivial solitonic configuration. If we wish to describe in the given framework a nontrivial configuration, the image of space-time that $\mathbf{U}$ generates in $SU(3)$ must be homeomorphic not to the image of $\mathbf{S}^4$ (Jain and Wadia, 1985), but rather to that of $\mathbf{S}^1 \otimes \mathbf{S}^3$. More general geometric treatment of Wess-Zumino terms one can find in (de Azcárraga, Izquierdo and Macfarlane, 1990; Krichever, Olshanetsky and Perelomov, 1986). In its turn we will use the indicated precise treatment in Chapter 11 in order to evaluate the Wess-Zumino Lagrangian for collective coordinates.

$$Q_5 = \int_{\mathbb{S}^5} d^5x\, J_5^0 = -\frac{\mathrm{i}\,\epsilon^{\mu\nu\lambda\rho\sigma}}{240\pi^2} \int_{\mathbb{S}^5} d^5x \mathrm{Tr}\left\{\mathbf{L}_\mu \mathbf{L}_\nu \mathbf{L}_\lambda \mathbf{L}_\rho \mathbf{L}_\sigma\right\}, \tag{10.45}$$

where we integrate over the whole sphere $\mathbb{S}^5$. The integrand in (10.45) is the 5-form

$$\omega_5 = -\frac{\mathrm{i}}{240\pi^2} \mathrm{Tr}\,[l^5], \tag{10.46}$$

where we use the 1-form $l = \mathbf{L}_\mu dx^\mu$ and omit the wedge product for convenience. If we take into account that $dl + l^2 = 0$, then ω_5 proves to be closed, since

$$d\omega_5 = -\frac{\mathrm{i}}{240\pi^2} \mathrm{Tr}\,[dl^5] = -\frac{\mathrm{i}}{48\pi^2} \mathrm{Tr}\,[l^6] = 0,$$

hence by Poincaré's lemma (see Appendix B) ω_5 is a locally exact form.

To construct the desired action in analogy with the charge-monopole one (with almost the same argumentation which we will not repeat here) rather than integrate over the whole of $\mathbb{S}^5$ in (10.45) we have to integrate over one of two complementary discs $D_5^\pm$ and write

$$\Gamma_{WZ} = \mu \int_{D_5^+} \omega_5 = -\mu \int_{D_5^-} \omega_5, \quad \mu \in \mathbb{R}. \tag{10.47}$$

The latter expression is known in literature under the name of **the Wess-Zumino action** (Wess and Zumino, 1971), which is manifestly $SU(3)_L \otimes SU(3)_R$-invariant and depends only on the space-time boundary of a disc $D_5^\pm$, namely, on $\partial D_5 \simeq (\mathbb{S}^3 \otimes \mathbb{S}^1)$, since ω_5 is locally exact. A small variation of the configuration (10.42) results in a variation of Γ_{WZ} which can be expressed entirely in terms of the variations of $\mathbf{U}$ from (10.41). This is because only the totally skew-symmetric part of $\delta\omega_5$ contributes to (10.47) and after a little algebra and by Stokes' theorem we have

$$\delta\Gamma_{WZ} = -\frac{\mathrm{i}\,\mu\epsilon^{\mu\nu\lambda\rho}}{48\pi^2} \int d^4x \mathrm{Tr}\left\{\mathbf{U}^{-1}\delta\mathbf{U}\mathbf{L}_\mu \mathbf{L}_\nu \mathbf{L}_\lambda \mathbf{L}_\rho\right\}, \tag{10.48}$$

where the boundary conditions $\mathbf{U}(x,0) = \mathbf{U}(x)$, $\mathbf{U}(x,1) = \mathbb{1}$ were taken into account. Summarizing, the modified σ-model action is

$$S_\pm = \frac{F_\pi^2}{16} \int d^4x \mathrm{Tr}\mathbf{L}_\mu^2 \pm \frac{-\mathrm{i}\,\mu\epsilon^{\mu\nu\lambda\rho\sigma}}{240\pi^2} \int_{D_5^\pm} d^5x \mathrm{Tr}\left\{\mathbf{L}_\mu \mathbf{L}_\nu \mathbf{L}_\lambda \mathbf{L}_\rho \mathbf{L}_\sigma\right\}, \tag{10.49}$$

that implies through the variational principle the equation of motion (10.38) with $\eta = \mathrm{i}\mu/48\pi^2$. The second important consequence is, that Γ_{WZ} does not depend on the details of extending the field $\mathbf{U}$ from a map from the physical space to one from five-dimensional space.

However there remains some arbitrariness in this procedure which is reflected by the sign ambiguity in (10.49). The minus sign is present because the physical space $\mathbb{S}^4$ bounds D_5^- with the opposite orientation. There is no way to choose between D_5^+ and D_5^- and this contour ambiguity in (10.49) can be resolved if one requires the generating functional and hence the exponential factor $\exp(\mathrm{i}\, S_{\pm})$ to be contour independent. Imposing the condition of consistency with quantum mechanics $\exp(\mathrm{i}\, S_{+}) = \exp(\mathrm{i}\, S_{-})$, which is the same as (10.23), we have

$$\mu\left\{\int_{D_5^+}\omega_5+\int_{D_5^-}\omega_5\right\}=\mu\int_{D_5^-\cup D_5^+}\omega_5=\mu\int_{\mathbf{S}^5}\omega_5=2\pi\mu\equiv 2\pi n, n\in\mathbb{Z}. \tag{10.50}$$

Therefore the parameter μ is forced to take on only integer values and once again we face the parameter quantization phenomenon. The relation (10.50) can be regarded as the analogue of the Dirac quantization condition (10.11) or (10.25). As it was demonstrated in (Witten, 1983b), the integer n in (10.50) can be identified with the number of colors N_c in QCD if we were to compare the QCD predictions for the various anomalous decays at low energy, such as $\pi^0 \to 2\gamma$, $2K \to 3\pi$, ..., with the ones based on (10.49). One might also compare the expression for the Wess-Zumino term in (9.36), which is the result of a direct derivation from QCD, with that in (10.49). So, the important conclusion is, that **the effective action (10.49) with the identification $\mu = N_c$ incorporates systematically the anomaly content of QCD and thus solves the aforestated problem.**

10.3.2 Spin and Statistics of Skyrmions

Now we are ready to understand simple but compelling Witten's arguments, as presented in (Witten, 1983b), that the statistics of states, corresponding to solitons in a chiral effective theory based on a $SU(N_c)$ gauge group, depends on the presence of the Wess-Zumino term. If the Wess-Zumino term is nontrivial, then the states with the unit topological charge $\mathbf{Q} = 1$ (Skyrmions) **must (not only can) be quantized as bosons for N_c even and as fermions for N_c odd.** In that very sense the Wess-Zumino term breaks the ambiguity in the choice of a quantization scheme, pointing to one of them in accordance with the number of colors in the underlying gauge theory.

To illustrate the above statement, write down following (Zahed and Brown, 1986) transition amplitude for the Skyrmion at rest (to leading order in $\hbar$)

$$\langle Sk(T)|Sk(0)\rangle = \exp\left(-\frac{\mathrm{i}}{\hbar}M_{Sk}T\right)[1+O(\hbar)], \tag{10.51}$$

where M_{Sk} is the Skyrmion mass, and T is the compactified length in the time direction $\mathbf{S}^1$. Now, if we rotate adiabatically the Skyrmion over 2π around the third axis, then quantum mechanics tells us that to leading order in $\hbar$, we have

$$\frac{\langle Sk(T)|Sk(0)\rangle_{2\pi}}{\langle Sk(T)|Sk(0)\rangle} = \exp\left(-\frac{\mathrm{i}}{\hbar}2\pi J\right)[1+O(\hbar)], \tag{10.52}$$

where $J = J_3$ is the angular momentum (spin in this context) of the Skyrmion. To evaluate J we have to embed the $SU(2)$ hedgehog configuration (4.33) in a $SU(N_f)$-valued chiral field with $N_f \geq 3$ (otherwise the Wess-Zumino term would not contribute, as it vanishes for two flavors). Thus we deal with the field of the form

$$\mathbf{U}(\boldsymbol{x}) = \begin{pmatrix} \exp\left[\mathrm{i}\tau_a x_a/r \cdot \Theta(r)\right] & \vdots & 0 \\ & \vdots & 0 \\ \cdots\cdots\cdots\cdots\cdots & & \cdots \\ 0 \quad\quad 0 & \vdots & 1 \end{pmatrix}. \tag{10.53}$$

Using the equivalence of rotations in space to isorotations for the hedgehog configurations, expressed in Eq. (8.1), we describe the rotated Skyrmion by

$$\mathbf{U}(\boldsymbol{x},t) = A(t) \cdot \mathbf{U}(\boldsymbol{x}) \cdot A^{\dagger}(t), \tag{10.54}$$

where $A(t)$ is an $SU(N_f)$ matrix of a time-dependent 2π rotation around a spatial axis. The further evaluation of J, which requires a further specification of $A(t)$ in (10.54), are given in Appendix B.5. Here we proceed, observing that to leading order the only term, which will contribute to the phase factor value (10.52) from the adiabatic rotation (10.54), is the Wess-Zumino term. Indeed this is the only term linear in the time derivatives ($\mathbf{L}_0 = \mathbf{U}^{\dagger}\partial_0\mathbf{U}$) in the action (10.49). Any other combination of L's involves at least two time-derivatives, and yields a higher order corrections in $\hbar$. Thus we need to calculate the classical action

$$S_{cl} \;=\; -M_{Sk}T \;+\; \frac{(-\mathrm{i})\,N_c}{240\pi^2}\int_{D_5^+}\mathrm{Tr}\left[l^5\right] \tag{10.55}$$

on rotated configurations (10.54) and after the intermediate algebra from Appendix B.5, we have

$$S_{cl} \;=\; -M_{Sk}T \;+\; N_c\mathbf{Q}\pi, \tag{10.56}$$

where $\mathbf{Q}$ is the topological charge of the configuration (10.53). Thus for the adiabatically rotated $SU(N_f)$ Skyrmion we get

$$\frac{\langle Sk(T)|Sk(0)\rangle_{2\pi}}{\langle Sk(T)|Sk(0)\rangle} = \exp\left(\mathrm{i}\,\pi N_c\mathbf{Q}\right)[1 + O(\hbar)], \tag{10.57}$$

and comparison with (10.52) shows, that up to an integer the spin of the Skyrmion is determined by

$$J \;=\; \mathbf{Q}\,\frac{N_c}{2}. \tag{10.58}$$

So, in accord with the previous statement, for $\mathbf{Q} = 1$ **the Skyrmion is a fermion if N_c is odd and a boson otherwise.** In other words, the amplitude for the rotated Skyrmion acquires an additional phase $\exp\{iN_c\Gamma_{WZ}\} = (-1)^{N_cQ}$ in comparison with the amplitude of an unrotated Skyrmion. It is not difficult to note that this is still another implementation of the Pauli exclusion principle. In its turn this approach could not be applied in the $SU(2)$ case, as the Wess-Zumino action vanishes for two flavors.

A unified approach, applicable in the both aformentioned situations, was developed in the paper (Balachandran, Gomm and Sorkin, 1987). It was demonstrated that one can regard the $\mathbb{Z}_2$ ambiguity in the $SU(2)$ model as a discrete Wess-Zumino term. That means in general that the Skyrme model with any number of flavors might be considered as the theory containing the Wess-Zumino term (in its ordinary or extended version). For a construction of wave functions in the presence of the Wess-Zumino term it was implemented modernized Dirac's idea, used in his treatment of monopoles (Dirac, 1948). The wave functions or state vectors of a quantum system, which are determined only up to a phase, might be regarded as not functions on the configuration space $\mathbf{M}$, but rather as functions on a $U(1)$ bundle $\widehat{\mathbf{M}}$ over $\mathbf{M}$. The bundle $\widehat{\mathbf{M}}$ is obtained by associating a circle $\mathbf{S}^1$ to each point of $\mathbf{M}$. The bundle $\widehat{\mathbf{M}}$ is a trivial when circles or "fibers" $\mathbf{S}^1$ are glued together in the way that $\widehat{\mathbf{M}} = \mathbf{M} \otimes \mathbf{S}^1$ and nontrivial or twisted in any other case. The latter possibility gives rise to a nontrivial Wess-Zumino term. The further development of this idea demonstrated that symmetries of quantum systems are not in general the symmetries of underlying classical action. The group $\widehat{G}$, which acts on $\widehat{\mathbf{M}}$, was constructed as a central extension of classical symmetry group G, and it was shown that in general $\widehat{G}$ may not even contain G as a subgroup. The symmetries of quantum system are transformations from either the group $\widehat{G}$ itself, or its subgroup. It is because of this change in the group that the states might have half odd integral spin and change sign under 2π rotation, in spite of the fact that the symmetry group G of the underlying classical theory does not admit representations of half odd integral spin. For example, let us revisit once more the charge-monopole system. The group of classical symmetries, which operates on the configuration space $\mathbf{M}$ is $G = SO(3)$, while the quantum group $\widehat{G}$, acting on wave functions of quantized system is a central extention of $SO(3)$ by $U(1)$. There are precisely two central extensions of $SO(3)$ by $U(1)$, namely $\widehat{G}_1 = SO(3) \otimes U(1)$ and $\widehat{G}_2 = U(2) = [SU(2) \otimes U(1)]/\mathbb{Z}_2$. In the framework of this formalism, the already known results sound in the form: if $e\mu$ is an integer, the group $\widehat{G}_1$ occurs and states have integral spin; if $e\mu$ is one half of an odd integer, the quantum group $\widehat{G}_2$ takes place and states change sign under 2π rotation. However, we have no enough place to discuss this promising approach any longer and therefore refer the reader to the full length exposition in (Balachandran, Marmo, Skagerstam and Stern, 1991)

The normal form of spin-statistics correlation for Skyrmions (for any number of flavors) has been established in (Sorkin, 1988). As the consequence it was shown that this correlation will exist whenever the underlying theory incorporates the possibility of pair creation and annihilation. This is the present case in the problem of spin and statistics of Skyrmions.

As far as it is clear from results listed here, the ambiguity of quantization procedure in the case of $SU(2)$ Skyrme model might be improved only in a rather tricky way in a now-popular spirit of the "Cheshire cat." The conventional Wess-Zumino term, which uniquely picks out one of two possible quantization schemes, does not exist for this case but, as was done in

(Balachandran, Gomm and Sorkin, 1987) it is possible to convince oneself that the same role is appropriate to its discrete analog or the "smile", which only remains in the two flavor case. To do that it is necessary to consider the $SU(2)$ model as a restriction of more general $SU(N_f)$ theory.

11. Quantized $SU(3)$ Skyrmions and Their Interactions

"I am therefore inclined to believe that the description of quantum mechanism ... has to be regarded as incomplete and indirect description of reality, to be replaced at some later date by a more complete and direct one."

Albert Einstein

Taking into account that among possible readers of this book, there would be people, mostly interested in practical implementations of Skyrme's approach we add a brief account of relevant topics in this Chapter. It might rather be considered as a guide in innumerable literature, appeared on the subject during the last decade. In advance, we apologize to the authors, who will not find their contributions to the field in our references. We mostly refer to the papers, which help us to understand the situation with the Skyrme model applications, and hope that our own experience, exposed here, might be useful for some readers. Among other sources, we mostly learnt from (Adkins and Nappi, 1985; Balachandran, 1986; Balachandran, Marmo, Skagerstam and Stern, 1991; Yabu and Ando, 1988; Zahed and Brown, 1986)

11.1 The $SU(3)$ Generalized Lagrangian in Terms of Collective Coordinates

We assume that there are at least three flavors. For $N_f = 3$, where the field $\mathbf{U}$ takes values in $SU(3)$, we use generalized classical Skyrme's ansatz by writing

$$\mathbf{U}_c(\boldsymbol{x}) = \begin{pmatrix} \exp\left(\mathrm{i}\tau_a x_a \Theta(r)/r\right) & 0 \\ & 0 \\ 0 \quad 0 & 1 \end{pmatrix}. \tag{11.1}$$

In the manner described in Section 8.1 we introduce collective coordinates $A(t)$ (where $A(t)$ is now an element of $SU(3)$).

11.1.1 The $SU(3)$ Skyrme "Collective" Lagrangian

Inserting the so-called Witten's ansatz $\mathbf{U}(\boldsymbol{x},t) = A(t)\cdot\mathbf{U}_c(\boldsymbol{x})\cdot A(t)^{-1}$ (where $A(t)\in SU(3)$ is the collective coordinate) into the Skyrme Lagrangian (8.4) and integrating in $\boldsymbol{x}$ we find

$$L_{\mathrm{SK}}(A,\dot{A}) = \tfrac{1}{4} I_{\alpha\beta}(\mathbf{U}_c)\mathrm{Tr}\,[\lambda^\alpha\,\mathrm{A}^{-1}\dot{\mathrm{A}}]\mathrm{Tr}\,[\lambda^\beta\,\mathrm{A}^{-1}\dot{\mathrm{A}}], \tag{11.2}$$

where the "moment of inertia" tensor

$$I_{\alpha\beta} = -\frac{F_\pi^2}{8} I^{(2)}_{\alpha\beta}(\mathbf{U}_c) - \frac{1}{16e^2} I^{(4)}_{\alpha\beta}(\mathbf{U}_c), \tag{11.3}$$

is conveniently presented in two pieces

$$I^{(2)}_{\alpha\beta} = \int d^3x\,\Big[2\delta_{\alpha\beta} - \tfrac{1}{2}\,\mathrm{Tr}\,[\lambda_\alpha(\mathbf{U}_c\lambda_\beta\mathbf{U}_c^\dagger + \mathbf{U}_c^\dagger\lambda_\beta\mathbf{U}_c)]\Big], \tag{11.3a}$$

$$I^{(4)}_{\alpha\beta} = \int d^3x\,\mathrm{Tr}\,\Big\{[\lambda_\alpha - \mathbf{U}_c^\dagger\lambda_\alpha\mathbf{U}_c\,,\,\mathbf{U}_c^\dagger\partial_i\mathbf{U}_c][(\alpha\to\beta)]\Big\}. \tag{11.3b}$$

After a tedious but straightforward calculations (not reproduced here) from (11.2) and (11.3) one comes to the following expression:

$$L_{\mathrm{SK}} = -E(\mathbf{U}_c) - \frac{a(\mathbf{U}_c)}{2}\,[A^{-1}\dot{A}]_i^2 - \frac{b(\mathbf{U}_c)}{2}\,[A^{-1}\dot{A}]_a^2, \tag{11.4}$$

with

$$[B]_i \equiv \mathrm{Tr}\,(\lambda_i B),$$

$$a(\mathbf{U}_c) = \frac{2}{3}\frac{\pi}{F_\pi e^3}\int_0^\infty dx\,\sin^2\Theta\,\{x^2 + 4[\sin^2\Theta + (x\Theta')^2]\}, \tag{11.4a}$$

$$b(\mathbf{U}_c) = \frac{\pi}{2}\frac{1}{F_\pi e^3}\int_0^\infty dx\,(1-\cos\Theta)\{x^2 + [2\sin^2\Theta + (x\Theta')^2]\}. \tag{11.4b}$$

Here the index i takes values 1, 2, 3; a takes values from 4 to 7 and $E(\mathbf{U}_c)$ is the classical energy of the $SU(3)$ Skyrmion. In Eqs. (11.4a,b) we made use of the dimensionless variable $x = eF_\pi r$. Numerically, if one assumes that $\mathbf{U}_c$ is the $SU(3)$ minimal energy configuration, we have

$$a(\mathbf{U}_c) \simeq \frac{447}{F_\pi e^3},\quad b(\mathbf{U}_c) \simeq \frac{156}{F_\pi e^3}. \tag{11.5}$$

Thus we stored up the first part of the Lagrangian to be quantized by canonical methods.

11.1.2 The Wess-Zumino Term for Collective Coordinates

As we have seen in two previous chapters, a Lagrangian pretending to account for the anomalies in QCD with necessity involves a term L_{WZ} in addition to L_{SK} (see Section 9.3 and 10.2 for available justification of the Wess-Zumino term from the physical point of view). Therefore the next inevitable point is to calculate the Wess-Zumino term for the introduced collective coordinates. First of all we have to understand how to pass from the Wess-Zumino action S_{WZ}, as given in Eq. (10.49), to the Wess-Zumino Lagrangian L_{WZ}, which we need to perform the quantization procedure (in accord with lines of Chapter 8). The idea of this passage (Balachandran, Lizzi, Rodgers and Stern, 1985) in brief consists in the following. If one approximates the chiral field $\mathbf{U}$ by Witten's ansatz $A(t)\mathbf{U}_c(\boldsymbol{x})A^\dagger(t)$, then in accord with Section 10.3 the image of space-time this ansatz generates in $SU(3)$ manifold is topologically the same as the image of $\mathbf{S}^1 \otimes \mathbf{S}^3$, with $\mathbf{U}_c$ being the map from $\mathbf{S}^3$ (compactified space) and $A(t)$ being the map from $\mathbf{S}^1$ (compactified time). We assume that in this case one may think of $\mathbf{S}^1$ as the boundary of a 2-dimensional disc D_2 and extend $A(t)$ to the whole of D_2. Points on D_2 are parametrized by two variables t and s, where t still has the significance of time (labels the boundary $\mathbf{S}^1$ of D_2) and s is a radial variable. Thus extended

$$\mathbf{U}(\boldsymbol{x},t,s) = A(t,s)\mathbf{U}_c(\boldsymbol{x})A^\dagger(t,s) \tag{11.6}$$

becomes a map from $D_2 \otimes \mathbf{S}^3$ into $SU(3)$, that reduces to Witten's ansatz at the domain boundary $\mathbf{S}^1 \otimes \mathbf{S}^3$. The image of $D_2 \otimes \mathbf{S}^3$ under the map (11.6) is the region of integration for the 5-form defined in (10.46). The corresponding integral (the Wess-Zumino action) is independent (modulo 2π) of the particular choice of the extension $A(t,s)$, and depends only on $A(t)$.

Summarizing, on extended ansatz (11.6) S_{WZ} from (10.49) can be written as follows:

$$\begin{aligned} S_{\text{WZ}}(\mathbf{U}) &= -\frac{\mathrm{i}N_c}{240\pi^2}\int_{D_2} ds\,dt \int_{\mathbf{S}^3} \operatorname{Tr}\left[\mathbf{U}(\boldsymbol{x},t,s)^{-1}d\mathbf{U}(\boldsymbol{x},t,s)\right]^5 \\ &= S_{\text{WZ}}(\mathbf{U}_c) + \int_{D_2}\int_{\mathbf{S}^3} dW_4. \end{aligned} \tag{11.7}$$

Here $W = W(A,\mathbf{U}_c)$ is a 4-form. Using Stokes' theorem, we can then write:

$$\int_{D_2}\int_{\mathbf{S}^3} dW_4 = \left\{\int_{\partial D_2^{(1)}} - \int_{\partial D_2^{(2)}}\right\}\int_{\mathbf{S}^3} W_4 + \int_{t_1}^{t_2}\int_{\mathbf{S}^3} W_4|_{s=1}, \tag{11.8}$$

where the boundaries $\partial D_2^{(1),(2)}$ are defined (Balachandran, 1986) through

$$\begin{aligned} \partial D_2 &= \partial D_2^{(1)} \cup \partial D_2^{(2)} \cup \partial D_2^{(3)}, \\ \partial D_2^{(1)} &= \{(s,t_1)|0 \le s \le 1\}, \\ \partial D_2^{(2)} &= \{(s,t_2)|0 \le s \le 1\}, \\ \partial D_2^{(3)} &= \{(1,t)|t_1 \le t \le t_2\}, \end{aligned} \tag{11.9}$$

so that t_1 and t_2 denote the initial and final times in the action principle. [The minus sign in front of the $\partial D_2^{(2)}$ integral arises because it was assumed that the integrals over $\partial D_2^{(1)}$ are done in the increasing s direction]. In obtaining the Euler-Lagrange equations, we do not vary $S_{\mathrm{WZ}}(\mathbf{U}_c)$ or the integrals over $\partial D_2^{(1),(2)}$ at all. Thus, these terms can be discarded, and S_{WZ} can be replaced by a simplified action with an associated Lagrangian L_{WZ} where

$$\int_{t_1}^{t_2} dt\, L_{\mathrm{WZ}} = \int_{t_1}^{t_2} \int_{S^3} W_4|_{s=1}. \tag{11.10}$$

To implement this recipe in detail, we write the Wess-Zumino term (11.7) using the differential form $l = \mathbf{U}^\dagger d\mathbf{U}$, as in (10.46) with $l^5 \equiv l \wedge l \wedge l \wedge l \wedge l$. Introducing another left-invariant Maurer-Cartan forms

$$a = A^\dagger dA, \quad u = \mathbf{U}_c^\dagger d\mathbf{U}_c, \tag{11.11}$$

we can readily deduce the simplified action [in accord with the announcement (11.7)] involving only a four dimensional integral:

$$\begin{aligned} S_{\mathrm{WZ}}(\mathbf{U}) = S_{\mathrm{WZ}}(\mathbf{U}_c) - \frac{\mathrm{i}N_c}{48\pi^2} \int_{t_1}^{t_2} \int_{S^3} \mathrm{Tr} \Big\{ & u^3\, a - a^3\, u - \frac{1}{2}(u\, a)^2 \\ & + \mathbf{U}_c\, (u-a)^3\, \mathbf{U}_c^\dagger\, a - a^3\, \mathbf{U}_c\, (u-a)\, \mathbf{U}_c^\dagger \\ & - \frac{1}{2}[a\, \mathbf{U}_c\, (u-a)\, \mathbf{U}_c^\dagger]^2 \Big\}. \end{aligned} \tag{11.12}$$

This result is valid even if A has a spatial dependence. When A depends only on t, and $\mathbf{U}_c$ depends only on $\boldsymbol{x}$, there is a further drastic simplification. Since $a^2 = (A^{-1}\dot{A})^2\, dt \wedge dt$, $a^3 = (A^{-1}\dot{A})^3\, dt \wedge dt \wedge dt$ are zero due to the screw symmetry of the wedge product, and $S_{\mathrm{WZ}}(\mathbf{U}_c)$ is zero by its definition[1], the simplified effective action is

$$\int_{t_1}^{t_2} dt\, L_{\mathrm{WZ}} = -\frac{\mathrm{i}N_c}{48\pi^2} \int_{t_1}^{t_2} \int_{S^3} \mathrm{Tr}\, u^3 (a + \mathbf{U}_c^\dagger a \mathbf{U}_c) \tag{11.13}$$

and the required effective Lagrangian is

$$L_{\mathrm{WZ}} = -\frac{\mathrm{i}N_c}{48\pi^2} \int_{S^3} \mathrm{Tr}\, u^3 (A^{-1}\dot{A} + \mathbf{U}_c^\dagger A^{-1}\dot{A}\mathbf{U}_c). \tag{11.14}$$

For the (11.1) configuration, this expression can be further simplified, as $u = \mathbf{U}_c^\dagger d\mathbf{U}_c$ has vanishing entries along the third row and column. For h being an element of the isospin subgroup

$$h = \begin{pmatrix} \tilde{h} & \begin{matrix} 0 \\ 0 \end{matrix} \\ \begin{matrix} 0 & 0 \end{matrix} & 1 \end{pmatrix}, \quad \tilde{h} \in SU(2), \tag{11.15}$$

[1] Indeed, the action $S_{\mathrm{WZ}}(\mathbf{U}_c)$ is defined by a 5-form on the 3-manifold $SU(2)$, which might be only trivial (see Appendix B.3 for details).

and taking into account the rotational invariance of the measure d^3x we have

$$h\left[\int u^3(\boldsymbol{x})\right]h^\dagger=\int u^3[\boldsymbol{R}(\boldsymbol{h})\boldsymbol{x}]=\int u^3(\boldsymbol{x}),\tag{11.16}$$

where it was used the fact

$$\tilde{h}\cdot\tau_i\cdot\tilde{h}^\dagger=\tau_j\cdot R_{ji}(\tilde{h}),\quad R(\tilde{h})\in SO(3).\tag{11.17}$$

Thus, the nonvanishing 2×2 block of the integral of u^3 commutes with $\tilde{h}$, and by Schur's lemma, it is, therefore, a multiple of the identity:

$$\int u^3=\text{const}\times\begin{pmatrix}1&&\\&1&\\&&0\end{pmatrix}.\tag{11.18}$$

It is possible to evaluate the latter constant in terms of the topological charge $\mathbf{Q}(\mathbf{U}_c)$ by taking the trace to finally find

$$\int u^3=12\pi^2\mathbf{Q}(\mathbf{U}_c)\times\begin{pmatrix}1&&\\&1&\\&&0\end{pmatrix}.\tag{11.19}$$

On the other hand, the second term in L_{WZ} involves the integral

$$\int_{S^3}U_cu^3\mathbf{U}_c^\dagger=\int_{S^3}(d\mathbf{U}_c\mathbf{U}_c^\dagger)^3=\int_{S^3}u^3.\tag{11.20}$$

The passage from the second to the third term in the Eq. (11.20) involves a calculation similar to the preceding one. Finally, for the Wess-Zumino Lagrangian in terms of collective coordinates we find

$$L_{\mathrm{WZ}}=-\frac{\mathrm{i}}{2}\sqrt{\frac{1}{3}}N_c\mathbf{Q}(\mathbf{U}_c)\,[A^{-1}\dot{A}]_8.\tag{11.21}$$

11.1.3 The Symmetry Breaking Term

As stated in Chapter 9, while in the $SU(2)$ Skyrme model the chiral limit is a good approximation due to smallness of current u- and d-quark masses, this is not the case for the $SU(3)$ Skyrme model. The magnitude of the s-quark current mass is comparable with the QCD cutoff Λ_{QCD}, therefore we have to take into consideration the $SU(3)$ breaking effects. It is also known [see for example (Adkins and Nappi, 1984; Chemtob, 1985; Praszalowicz, 1989)] that the $SU(3)$ symmetry breaking treatment in the standard collective coordinate approach leads to some problems, and the main one is the wrong value for F_π, which has to be assumed to bring the model predictions in agreement with experimental data. There exist three main approaches to treat the symmetry breaking:

1° **The perturbative approach** (Praczalowicz, 1985; Chemtob, 1985), based on the assumption that the $SU(3)$ *symmetry breaking is not very large*, and the Lagrangian can be devided into the symmetry preserving part and the symmetry breaking one. Then the former is treated by the $SU(3)$ collective-coordinate method and the latter is evaluated by the perturbation method. The principal shortcomings of this approach are the following: a) while the mass ordering of the baryons within multiplets is correctly reproduced, the magnitudes of the splittings are too small in comparison with experimental values; b) to reproduce the average values of the baryon masses in multiplets $\underline{8}$ and $\underline{10}$, this method requires a quite small value of $F_\pi(\simeq 46\,\mathrm{MeV})$.

2° **The bound state approach** (Callan and Klebanov, 1985; Callan, Hornbostel and Klebanov, 1988; Nowak, Rho and Scoccola, 1988; Niman and Riska, 1990), is based on the assumption that *the symmetry breaking is large*, and that a hyperon is a bound state of the $SU(2)$ Skyrmion and a kaon. Therefore the fields are devided into two parts: the $SU(2)$ part which describes π-meson degrees of freedom, and the other part which does that for K-mesons. Hyperon masses are calculated in two steps: a) by expanding the K-meson part in the $SU(2)$ Skyrmion background one derives equations of motion for radial kaon densities; b) by rotating the Skyrmion and introducing $SU(2)$ collective coordinates one calculates mass splitting within the baryon multiplet. The first shortcoming is the wrong mass ordering, obtained in such an approach, for example $m_\Sigma > m_{\Sigma^*}$. On the other hand, the approach results in the same F_π value as in the $SU(2)$ Skyrme model ($F_\pi = 108\,\mathrm{MeV}$) to fit N and Δ masses, and the masses calculated in this way satisfy the relations derived by the naive quark model, e.g. $m_\Delta - m_N = m_{\Sigma^*} + (1/2)m_\Sigma - (3/2)m_\Lambda$.

3° **The rigid rotator approach** (Yabu and Ando, 1988) is based on the assumption that the only essential zero modes are the $SU(2)$ rotations. The motives are as following. If the diagonal $SU(3)$ symmetry is not broken, then $\mathbf{U}(\boldsymbol{x}; A) = A \cdot \mathbf{U}_c \cdot A^\dagger$ with any constant $A \in SU(3)$ is a classical solution and the diagonal symmetry gives rise to the zero modes of the (11.1) configuration. Since this symmetry is broken $\mathbf{U}(\boldsymbol{x}; A)$ is not a solution except that A belongs to the upper $SU(2)$ subgroup. Therefore the zero modes of this solution are determined by the upper $SU(2)$ diagonal rotations. Let the symmetry breaking effect be of an intermediate strength (*the finite symmetry breaking*), then the deformation of solution (11.1) is small and $\mathbf{U}(\boldsymbol{x}; A)$ is not very far from the actual configuration for any A. This gives the arguments to treat $SU(3)$ diagonal rotation in terms of collective coordinates as excitations about strangeness freedom without zero modes, i.e. the strange zero modes are assumed to be almost frozen. There is a claim, that the perturbative and the bound state approaches listed before are two extreme approximations to this rigid rotator approach. Therefore we find it appropriate to keep closer to this one approach in the present exposition, since the previous two one can obtain in the weak and in the strong symmetry breaking limits, respectively.

A natural term to account for not only the $SU_L(3) \otimes SU_R(3)$ symmetry breaking but also for the diagonal $SU(3)$ breaking is

$$L_{\mathrm{SB}} = \int d^3x \left\{ \frac{F_\pi^2}{32} \left(m_\pi^2 + m_\eta^2\right) \mathrm{Tr}\left(\mathbf{U} + \mathbf{U}^\dagger - 2\right) + \frac{\sqrt{3}\,F_\pi^2}{24} \left(m_\pi^2 - m_K^2\right) \mathrm{Tr}\left(\lambda_8(\mathbf{U} + \mathbf{U}^\dagger)\right) \right\}. \tag{11.22}$$

Its form is taken in such way that it reduces to the mass terms of the π, K, η mesons, when we expand

$$\mathbf{U} = \exp\left[\frac{2\mathrm{i}}{F_\pi} \sum_{\alpha=1}^{8} \lambda_\alpha \varphi_\alpha\right]$$

around the classical vacuum $\mathbf{U} = \mathbb{1}$:

$$L_{\mathrm{SB}} = \int d^3x \left\{ \tfrac{1}{2}\, m_\pi^2 \sum_{i=1}^{3} \varphi_i^2 + \tfrac{1}{2}\, m_K^2 \sum_{a=4}^{7} \varphi_a^2 + \tfrac{1}{2}\, m_\eta^2\, \varphi_8^2 + O(\varphi^4) \right\}. \tag{11.23}$$

Expanding the total action to the second order in the octet meson fields, one also recovers the familiar Gell Mann – Okubo quadratic mass rule

$$m_\pi^2 + 3\, m_\eta^2 - 4\, m_K^2 = 0, \tag{11.24}$$

which holds experimentally within 3%.

We are now ready to proceed with the semiclassical approximation. After substituting $\mathbf{U}(\boldsymbol{x},t) = A(t)\mathbf{U}_c(\boldsymbol{x})A(t)^\dagger$ into L_{SB} and performing the spatial integration, we find that

$$L_{\mathrm{SB}}\,(A, \dot{A}) = -\tfrac{1}{2}\, c(\mathbf{U}_c)\, [1 - D^{(8)}(A)], \tag{11.25}$$

where it is denoted

$$D^{(8)}(A) = \tfrac{1}{2}\, \mathrm{Tr}\left(\lambda_8 A^\dagger \lambda_8 A\right) \tag{11.26}$$

and for the $SU(3)$ Skyrmion (11.1)

$$c(\mathbf{U}_c) \doteq \frac{4\,\pi}{3\, F_\pi e^3} \left(m_K^2 - m_\pi^2\right) \int_0^\infty dx\, x^2 [1 - \cos\Theta(x)]. \tag{11.27}$$

As a result of the all above consideration we can finally write down the total Lagrangian for the $SU(3)$ Skyrmion in terms of collective coordinates:

$$L\,(A, \dot{A}) = L_{Sk} + L_{\mathrm{WZ}} + L_{\mathrm{SB}}, \tag{11.28}$$

with the terms specified in formulae (11.4), (11.21) and (11.25), respectively.

11.2 Quantization in the Presence of the Wess-Zumino Term

As we have already considered the general scheme of the $SU(2)$ Skyrmion quantization in Section 8.2, we fill free here to afford a so-called telegraphic style of exposition. On the other hand, it gives us more place to focus on specific features which arise due to the presence of the Wess-Zumino term.

11.2.1 Canonical Quantization

Since $A \in SU(3)$, it can be locally parametrized by eight variables $\xi = (\xi_1, \xi_2, \ldots, \xi_8)$, which coordinate the $SU(3)$ manifold. Thus, A in the Lagrangian is $A[\xi(t)]$ and $\dot{A}$ is $(\partial A/\partial\xi_\alpha)\dot{\xi}_\alpha$, while the momentum conjugate to ξ_α is

$$P_\alpha = \frac{\partial L}{\partial \dot{\xi}_\alpha} = -a(\mathbf{U}_c)\,[A^\dagger \dot{A}]_i \left[A^\dagger \frac{\partial A}{\partial \xi_\alpha}\right]_i - b(\mathbf{U}_c)\,[A^\dagger \dot{A}]_a \left[A^\dagger \frac{\partial A}{\partial \xi_\alpha}\right]_a - \frac{\mathrm{i}}{2\sqrt{3}} N_c \mathbf{Q}(\mathbf{U}_c) \left[A^\dagger \frac{\partial A}{\partial \xi_\alpha}\right]_8 . \tag{11.29}$$

As it was specified above i takes values $1, 2, 3$, a takes values $4, 5, 6, 7$ and α takes values from 1 to 8. Here the $SU(3)$ "angular velocities" are defined by

$$[A^\dagger \dot{A}]_\alpha = \mathrm{Tr}\,(\lambda_\alpha A^\dagger \dot{A}). \tag{11.30}$$

Remark 11.1. As an appropriate coordinate system $\{\xi_i\}$ for $SU(3)$ manifold [following, for example, (Yabu and Ando, 1988)] one can choose the $SU(3)$ Euler angles:

$$A[\xi(t)] = \exp\left(-\mathrm{i}\frac{\xi_1}{2}\lambda_3\right)\cdot\exp\left(-\mathrm{i}\frac{\xi_2}{2}\lambda_2\right)\cdot\exp\left(-\mathrm{i}\frac{\xi_3}{2}\lambda_3\right)\cdot\exp(-\mathrm{i}\xi_4\lambda_4) \times \exp\left(-\mathrm{i}\frac{\xi_5}{\sqrt{3}}\lambda_8\right)\cdot\exp\left(-\mathrm{i}\frac{\xi_6}{2}\lambda_3\right)\cdot\exp\left(-\mathrm{i}\frac{\xi_7}{2}\lambda_2\right)\cdot\exp\left(-\mathrm{i}\frac{\xi_8}{2}\lambda_3\right). \tag{11.31}$$

with the ranges

$$0 \le \xi_1, \xi_3, \xi_6, \xi_8 \le 2\pi; \quad 0 \le \xi_4 \le \frac{\pi}{2}; \qquad 0 \le \xi_2, \xi_7 \le \pi; \quad 0 \le \xi_5 \le 3\pi. \tag{11.31a}$$

Such a choice (as it becomes clear after some algebra) leads to a separation of variables and to an exact diagonalization of the semi-classical Hamiltonian, as it allows one to write the quadratic Casimir operators $C_2(SU_R(2))$ and $C_2(SU_L(3))$ in the form separate about Euler angles.

From general theory (Balachandran, Marmo, Scagerstam and Stern, 1983; 1991) one knows that there exists a nonsingular matrix $N(\xi) = [N_{\alpha\beta}(\xi)]$ such that

$$\mathrm{i}\, A\lambda_\alpha = \frac{\partial A}{\partial \xi_\beta} N_{\beta\alpha}. \tag{11.32}$$

The reproduction of the proof for the sake of completeness is given in Appendix D.3. This fact gives us the possibility to introduce "angular momentum" variables

$$\mathcal{M}_\alpha = -P_\beta\, N_{\beta\alpha}, \tag{11.33}$$

and using the standard canonical Poisson brackets

$$\{\xi_\alpha, \xi_\beta\} = \{P_\alpha, P_\beta\} = 0, \quad \{\xi_\alpha, P_\beta\} = \delta_{\alpha\beta}, \tag{11.34}$$

we find

$$\begin{aligned} &\{A_{\alpha\beta}, A_{\gamma\delta}\} = 0, \\ &\{\mathcal{M}_\alpha, A_{\beta\gamma}\} = i[A\lambda_\alpha]_{\beta\gamma}, \\ &\{\mathcal{M}_\alpha, \mathcal{M}_\beta\} = -2\, f_{\alpha\beta\gamma}\, \mathcal{M}_\gamma, \end{aligned} \tag{11.35}$$

where $f_{\alpha\beta\gamma}$ are the $SU(3)$ stuctural constants. The expression for $\mathcal{M}_\alpha$ in terms of collective coordinates and velocities follows from (11.29) and (11.33):

$$\mathcal{M}_\alpha = 2\,\mathrm{i}\, a(\mathbf{U}_c)\, [A^\dagger \dot{A}]_i\, \delta_{i\alpha} + 2\,\mathrm{i}\, b(\mathbf{U}_c)\, [A^\dagger \dot{A}]_a\, \delta_{a\alpha} + \frac{N_c}{\sqrt{3}} \mathbf{Q}(\mathbf{U}_c)\, \delta_{\alpha 8}. \tag{11.36}$$

The Hamiltonian of the system is obtained via the general rule

$$H = P_\alpha \dot{\xi}_\alpha - L(A, \dot{A}), \tag{11.37}$$

and, after the substitution for P_α and $L(A, \dot{A})$ from Eqs. (11.29) and (11.28), respectively, reduces to a diagonalized form

$$H = \frac{\mathcal{M}_i\, \mathcal{M}_i}{8\, a(\mathbf{U}_c)} + \frac{\mathcal{M}_a\, \mathcal{M}_a}{8\, b(\mathbf{U}_c)} + E(\mathbf{U}_c) + \frac{1}{2}\, c(\mathbf{U}_c)\, [1 - D^{(8)}(A)]. \tag{11.38}$$

When we pass on to quantum mechanics the Poisson brackets (11.35) are replaced by commutators following standard Dirac's prescription. Denoting a quantum operator which corresponds to any classical variable F by $\hat{F}$, we then write down the commutation relations as

$$\begin{aligned} &[\hat{A}_{\alpha\beta}, \hat{A}_{\gamma\delta}] = 0, \\ &[\hat{\mathcal{M}}_\alpha, \hat{A}_{\beta\gamma}] = -[\hat{A}\lambda_\alpha]_{\beta\gamma}, \\ &[\hat{\mathcal{M}}_\alpha, \hat{\mathcal{M}}_\beta] = -2\,\mathrm{i}\, f_{\alpha\beta\gamma} \hat{\mathcal{M}}_\gamma. \end{aligned} \tag{11.39}$$

A forthcoming quantization consists as usual in finding an irreducible representation of the appropriate algebra of observables. But (fortunately or not) this is not the whole story, when one deals with the system with a nontrivial Wess-Zumino term (Rabinovici, Schwimmer and Yankielowicz, 1984), and from now on we are going to discuss the peculiarities inserted by its presence. To do that in a proper way one needs a more deep insight in the symmetries of the system under study.

11.2.2 Symmetries, Constraint and Spectrum

As we shall see in a moment, the total Lagrangian $L(A, \dot{A})$ from (11.28) has a *gauge* symmetry. That is, there is a *time dependent* group of symmetries under which L changes by a total derivative:

$$L \to L + \frac{d\psi}{dt}. \tag{11.40}$$

This is a well studied situation [see, for example, (Sudarshan and Mukunda, 1974)], which falls under

Proposition 11.1. Whenever a Lagrangian theory possesses a time dependent group of symmetries, it does not lead to conservation laws, but rather to constraints on the phase space in classical theory and to constraints on the allowed states in quantum theory.

The best known example of this kind is the gauge invariance in Electrodynamics with the associated constraint represented by the Gauss law. It will be clear later on, that for the Skyrme model we should expect actually only one constraint because the gauge group is obtained by gauging a certain $U(1)$ group. This constraint will be related to the Wess-Zumino term and will uniquely pick out one single quantization scheme, removing the ambiguities listed in Section 8.1. Exactly the Wess-Zumino term alters the precise nature of the constraint and influences the choice of the quantization scheme.

Let us now show how the gauge symmetry appears. First of all notice that here we make use of collective coordinates $A(t) \in SU(3)$, which are defined up to a right hypercharge multiplication. Indeed, if instead of $A(t)$ we substitute into Witten's ansatz

$$A'(t) = A(t) \cdot h(t), \quad \text{with} \quad h(t) = \exp\{\mathrm{i}\,\beta(t) Y\} \in U_R(1), \tag{11.41}$$

where $\beta(t)$ is a real-valued function and $Y = \mathrm{diag}\,[1, 1, -2] = (\sqrt{3})\lambda_8$ then both matrices $A(t)$ and $A'(t)$ lead to the same perturbations of the $SU(3)$ Skyrmion, as

$$A'(t) \cdot \mathbf{U}_c(\boldsymbol{x}) \cdot A'^{\dagger}(t) = A(t) \cdot h(t) \cdot \mathbf{U}_c(\boldsymbol{x}) \cdot h^{\dagger}(t) \cdot A'^{\dagger}(t) = \mathbf{U}(\boldsymbol{x}, t). \tag{11.42}$$

Here it was used the obvious fact, that $h(t)$ from (11.41) commutes with $\mathbf{U}_c(\boldsymbol{x})$ given by Eq. (11.1). Thus the space of collective coordinates is not the $SU(3)$ as a whole, but rather the 7-dimensional manifold $SU(3)/U_R(1)$, and since we have described the dynamical system (in the frame of semi-classical approximation) with variables whose number exceeds the number of actual degrees of freedom (eight instead of seven), therefore we have earned a local $U_R(1)$ symmetry[2].

[2] This $U_R(1)$ symmetry might be considered either as a redundant one with the associated right hypercharge Y_R deprived of any physical meaning, or alternatively one can regard the symmetry of collective coordinates as $SU(3)$ and treat the freedom of $U_R(1)$ as a constraint. Due to its easiness the latter approach is usually adopted.

To verify the latter statement we need the auxiliary relations, obtained with a little algebra from (11.41), e.g.

$$[A^\dagger \dot{A}]_\alpha \to [A^\dagger \dot{A}]_\alpha + \mathrm{i}\sqrt{3}\,\mathrm{Tr}\,(\lambda_\alpha \lambda_8)\,\dot{\beta}(t) = [A^\dagger \dot{A}]_\alpha \quad \text{for} \quad \alpha \neq 8; \tag{11.43a}$$

$$[A^\dagger \dot{A}]_8 \to [A^\dagger \dot{A}]_8 + 2\mathrm{i}\sqrt{3}\,\dot{\beta}(t), \tag{11.43b}$$

and on using them it is straightforward to find out that L_{SK} from Eq. (11.4) and L_{SB} from (11.25) remain unaltered by transformations (11.41) from $U_R(1)$ group, while

$$\begin{aligned} L_{\mathrm{WZ}}(A, \dot{A}) &= -\frac{\mathrm{i}}{2}\sqrt{\tfrac{1}{3}} N_c \mathbf{Q}(\mathbf{U}_c)[A^{-1}\dot{A}]_8 \\ &\to -\frac{\mathrm{i}}{2}\sqrt{\tfrac{1}{3}} N_c \mathbf{Q}(\mathbf{U}_c)[A^{-1}\dot{A}]_8 + N_c \mathbf{Q}(\mathbf{U}_c)\dot{\beta}(t) \\ &= L_{\mathrm{WZ}}(A, \dot{A}) + N_c \mathbf{Q}(\mathbf{U}_c)\dot{\beta}(t). \end{aligned} \tag{11.44}$$

Therefore the total Lagrangian (11.28) has the property (11.40). Since $\mathcal{M}_8$ from Eq. (11.36) is the infinitesimal generator of the finite transformation (11.41) it is just the coefficient of $-\sqrt{3}\dot{\beta}(t)$ in (11.44). Thus for $L(A, \dot{A})$ to be invariant under transformations (11.41) or (11.43) we have to impose **the constraint**:

$$\chi = \mathcal{M}_8 + \frac{N_c}{\sqrt{3}}\mathbf{Q} = 0. \tag{11.45}$$

Thus, we have a constrained Hamiltonian system and are to modify our Hamiltonian (11.38) to

$$H = \frac{\mathcal{M}_i \mathcal{M}_i}{8\,a(\mathbf{U}_c)} + \frac{\mathcal{M}_a \mathcal{M}_a}{8\,b(\mathbf{U}_c)} + E(\mathbf{U}_c) + \frac{1}{2}\,c(\mathbf{U}_c)\,[1 - D^{(8)}(A)] + \kappa\chi, \tag{11.46}$$

where κ is a Lagrange multiplier. Since $\{\chi, H\} = 0$, the constraint (11.45) is a primary one and there are no secondary constraints due to Dirac's classification.

Now we are ready to proceed further with the quantization procedure fixed in the previous section, bearing in mind the constraint (11.45) together with the relation $\hat{A}^\dagger \hat{A} = 1$ and the requirement $\hat{\mathcal{M}}_\alpha = \hat{\mathcal{M}}_\alpha^\dagger$. It is appropriate to perform the procedure in two steps. In the first step, we ignore the constraint $\hat{\chi}$ and realize the algebra (11.39) on a Hilbert space $\mathcal{H}$. In the second step, we impose the constraints on states, thereby reducing $\mathcal{H}$ to a subspace $\mathcal{H}_{SU(3)}$. The observables $\hat{F}$ [that have to be gauge invariant operators] are required to commute with the first class constraint $\hat{\chi}$. They have then the property of leaving $\mathcal{H}_{SU(3)}$ invariant, $\hat{F}\mathcal{H}_{SU(3)} \subseteq \mathcal{H}_{SU(3)}$.

The Hilbert space $\mathcal{H}$ we start with is the space of functions on $SU(3)$ with the scalar product

$$(\psi, \phi) = \int_{SU(3)} d\mu(A)\,\psi(A)^\star \phi(A), \tag{11.47}$$

where $d\mu(A)$ is the Haar measure on $SU(3)$ and $\psi, \phi \in \mathcal{H}$. On this space, the operator A is realized by the rule

$$[\hat{A}_{\alpha\beta}\psi](A) = A_{\alpha\beta}\psi(A), \tag{11.48}$$

while the action of $\hat{\mathcal{M}}_\alpha$ is defined by the relation

$$[e^{\mathrm{i}\theta_\alpha \hat{\mathcal{M}}_\alpha}\psi](A) = \psi(Ae^{-\mathrm{i}\theta_\alpha\lambda_\alpha}), \tag{11.49}$$

for all constants θ_α. Realizations of operators in the form (11.48) and (11.49) one can substantiate from the symmetries of the Lagrangian $L(A, \dot{A})$. Therefore let us list the symmetries and corresponding quantum numbers, which will label wave functions describing the quantum states (Jain and Wadia, 1985; Yabu and Ando, 1988).

1° **Isorotations** $SU_L(2) \subset SU_L(3) : (I, I_3)$

$$A \to V_L A, \quad V_L \in SU_L(2). \tag{11.50a}$$

2° **Hypercharge** $U_L(1) \subset SU_L(3) : Y$

$$A \to \exp\{\mathrm{i}\theta_L\lambda_8\}A, \quad \exp\{\mathrm{i}\theta_L\lambda_8\} \in U_L(1). \tag{11.50b}$$

3° **Spatial rotations** $SU_R(2) \subset SU_R(3) : (J, J_3)$

$$A \to AV_R^\dagger, \quad V_R \in SU_R(2). \tag{11.50c}$$

4° **Right hypercharge** $U_R(1) \subset SU_R(3) : Y_R$

$$A \to A\exp\{-\mathrm{i}\theta_R\lambda_8\}, \quad \exp\{\mathrm{i}\theta_R\lambda_8\} \in U_R(1). \tag{11.50d}$$

The latter symmetry, being a local one, was already taken into account through the constraint (11.45). It reduces the space $\mathcal{H}$ to the subspace $\mathcal{H}_{SU(3)}$ of functions which fulfill

$$\psi(A\,e^{-\mathrm{i}\theta_8\lambda_8}) = e^{-\mathrm{i}N_c\mathbf{Q}(\Gamma_c)\theta_8/\sqrt{3}}\psi(A). \tag{11.51}$$

In an equivalent form the subspace $\mathcal{H}_{SU(3)}$ of $\mathcal{H}$ can be defined by

$$\mathcal{H}_{SU(3)} = \Big\{\psi \in \mathcal{H}\,|\,\psi(A\,e^{\mathrm{i}\beta Y}) = \psi(A)\exp\big[\mathrm{i}\beta N_c\,\mathbf{Q}(\mathbf{U}_c)\big]\Big\}. \tag{11.52}$$

Since $\exp\{\mathrm{i}\,2\,\pi Y\} = 1$ and $\mathbf{Q}(\mathbf{U}_c)$ can take only integer values, this equation shows that N_c must be an integer. This consideration one might regard as an alternative approach to obtain Witten's result from the previous chapter (Witten, 1983).

Remark 11.2. For the two flavor model, the twofold quantization ambiguity was due to the fact that the physical states could belong to either one of the two inequivalent representations of the group $\mathbb{Z}_2 = \{(\mathbb{1}, -\mathbb{1})\}$. That is, they could be either even or odd functions of A. Now let us regard the two flavor model as an appropriate restriction of the three flavor model. Since

$\exp i\pi Y = \mathrm{diag}[-1,-1,1]$ the restriction of the three flavor gauge group to two flavors contains $\mathbb{Z}_2$. The gauge constraint for three flavors uniquely determines the value of $\exp[i\pi Y]$ on the physical states. Consequently, the quantization ambiguity of the two flavor model disappears when it is regarded as the restriction of the three flavor model.

For a more detailed study of the system spectrum of states one requires to set up a basis in the space $\mathcal{H}_{SU(3)}$. In order to implement this we can make use of the analogy with the symmetrical top (Landau and Lifshitz, 1977). The latter is a rigid body whose moment of inertia tensor is such that two of three principal moments of inertia are equal to each other and different from the third one. The collective coordinates A for the symmetric top are $SU(2)$ matrices, constructed out of Euler angles, that describe the position of the top. The corresponding Hamiltonian is symmetric against transformations $A \to V\,A\,W$, where V is an $SU(2)$ matrix that depends upon a general space rotation: left multiplication by V is a symmetry because of isotropy of space, and W is an element of a $U(1)$ subgroup of $SU(2)$ and is of the type $\exp\{i\theta\tau_3\}$, right multiplication by W is a symmetry because of the nature of the top. Then energy eigenfunctions as functions of the Euler angles (α,β,γ) are given [up to normalization] by the matrix elements $D^J_{m'm}(\alpha,\beta,\gamma)$ of the matrix $D^J(\alpha,\beta,\gamma)$, which is the $(2J+1)$-dimensional irreducible representation of the rotation group. The index m in the wave function $D^J_{m'm}(\alpha,\beta,\gamma)$ is the eigenvalue of the third component of ordinary (space-fixed) angular momentum and the index m' is the eigenvalue of the generator of right $U(1)$ symmetry. $D^J_{m'm}(\alpha,\beta,\gamma)$ is the simultaneous eigenfunction of the latter two operators, $\hat{\mathbf{J}}^2$ and the Hamiltonian.

To demonstrate the analogy in an explicit form, notice that for the system (11.28) or (11.46) the transformation $A \to V\,A\,W$ is also a symmetry, where now V belongs to $SU(3)$ and W lie in the $SU(2)$ or $U(1)$ subgroup of $SU(3)$ mentioned in (11.50). Whence replacing $SU(2)$ by $SU(3)$ and the right $U(1)$ symmetry by $SU(2)\otimes U(1)$ we are allowed to construct the eigenfunctions as matrix elements of $SU(3)$. Unitary irreducible representations of $SU(3)$, denoted $D^{(pq)}(A)$ or simply by (p,q), are realized in the space of traceless tensors with p lower and q upper indices and therefore labelled by two integers p and q. The singlet for example is $(0,0)$, the octet is $(1,1)$ and the decuplet is $(3,0)$. The eigenfunctions here will be matrix elements $D^{(pq)}_{\alpha\beta}(A)$ of the representation matrices $D^{(pq)}(A)$, where the collective indices α and β are again the eigenvalues of symmetry generators that could be simultaneously diagonalized with the Hamiltonian (11.46). According to the list of symmetries (11.50) in the $SU(3)$ case generators $\hat{\mathbf{I}}^2$, $\hat{I}_3$ and the left hypercharge $\hat{Y}_L$ can be simulteneously diagonalized so that α or β stands for three indices (I, I_3, Y) collectively. Since the index m for the symmetric top corresponds to the *left symmetry generator*, we are to relate the index β with the $SU(3)$ left symmetry, which is clearly the flavor symmetry and whence $\beta \equiv (I, I_3, Y)$ gives the *flavor quantum numbers* of the state $D^{(pq)}_{\alpha\beta}$. In the same manner $\alpha \equiv (I', I'_3, Y')$ has to be related with eigenvalues of the diagonal generators of the *right* $SU(2)$ and $U_R(1)$ symmetries in the state $D^{(pq)}_{\alpha\beta}$. Still in accord with (11.50) we have to identify (I', I'_3) with the *spin quantum numbers* (J, J_3) of the state $D^{(pq)}_{\alpha\beta}$ and Y' with the right hypercharge Y_R, when we speak about a general situation.

Summarizing, one can state (in adopted quantum mechanical notations) that a basis for the Hilbert space $\mathcal{H}$ is the collection of all matrix elements

$$\langle I, I_3, Y|D^{(p,q)}(A)|J, J_3, Y'\rangle \equiv D^{(p,q)}_{(J,J_3,Y')(I,I_3,Y)}(A), \tag{11.53}$$

where (p, q) and all the remaining indices take on all possible values. When passing to a basis for the Hilbert space $\mathcal{H}_{SU(3)}$ we are to take into account that $U_R(1)$ is a local symmetry, which imposes the constraint (11.45) or equivalently (11.51) on the physical state space. It follows that $\mathcal{H}_{SU(3)}$ is spanned by the functions

$$\langle I, I_3, Y|D^{(p,q)}(A)|J, J_3, (1/3)N_c\mathbf{Q}\rangle = D^{(p,q)}_{(J,J_3,(1/3)N_c\mathbf{Q})(I,I_3,Y)}(A), \tag{11.54}$$

or equivalently by the functions, which are symmetric under the $U_R(1)$ transformations:

$$D^{(p,q)}_{(J,J_3,Y')(I,I_3,Y)}(A\, e^{-i\theta_8\lambda_8})D^{(p,q)}_{(J,J_3,Y')(I,I_3,Y)}(A) \cdot e^{-i\theta_8\sqrt{3}Y'}. \tag{11.55}$$

Let us now specialise to the case $N_c = 3$, $\mathbf{Q}(\mathbf{U}_c) = 1$ so that $Y' = N_c\mathbf{Q}(U_c)/3 = 1$. With this restriction on Y', the allowed representations $D^{(pq)}(A)$ are those for which it is fulfilled the following condition

$$p + 2q = 0(\text{mod}3). \tag{11.56}$$

The lowest allowed multiplets are the octet $(1,1)$ and decuplet $(3,0)$. The preceding discussion shows that the wave functions from these multiplets transform as octets and decuplets under flavor $SU(3)$. The restriction $Y' = 1$ forces J to be $1/2$ for the octet $\underline{8} = (1.1)$ and $3/2$ for the decuplet $\underline{10} = (3.0)$, so that we have an octet of spin 1/2 states and a decuplet of spin 3/2 states. The former is identified with the spin 1/2 baryon octet and the latter with the spin 3/2 baryon decuplet. Thus with $N = 3$, the quantization of the $SU(3)$ Skyrmion reproduces the known $SU(3)$ representations and spins of baryons. At last, note that the 2π rotation $\exp\{i(1/2)2\pi\lambda_3\}$ around the third isospin axis is equal to $\exp\{i\pi\sqrt{3}\lambda_8\}$. Therefore by (11.55) all the wave functions in $\mathcal{H}_{SU(3)}$ change sign under 2π rotation. That is all states in $\mathcal{H}_{SU(3)}$ are fermionic, in accord with Witten's result (Witten, 1983).

Thus we convinced ourselves, that after all the addition of the Wess-Zumino term to the SU(3) generalized Skyrme model provides a qualitatively regular classification for baryon states. The next natural step is to consider quantitative results for observables, obtained in the framework of the approach.

11.2.3 Static Observables and Mass Formulae

In order to make some contacts with experiment here we just mention how the most basic phenomenological results of the $SU(3)$ Skyrme model can be derived. Without going into detail of calculations and sweeping away any attempt at a review of the extensive literature on phenomenological aspects of the Skyrme approach, which appeared during the 1980s decade, we focus on the principal ideas only.

Static Observables. Various static observables of baryons and hyperons such as charge densities and charge radii, magnetic moments and magnetic charge radii, axial coupling constant g_A and so on are associated through standard relations with the left and right Noether currents, defined as in Eqs. (1.45–47) and topological (baryon) current (1.38). For $L = L_{\rm SK} + L_{\rm WZ} + L_{\rm SB}$ the Noether currents will have the form

$$\begin{aligned} \mathbf{N}^a_{\mu,L} = & -\frac{\mathrm{i}F_\pi^2}{8}\,\mathrm{Tr}\left(\frac{\lambda^a}{2}\mathbf{R}_\mu\right) + \frac{\mathrm{i}}{8\,e^2}\,\mathrm{Tr}\left(\left[\frac{\lambda^a}{2},\mathbf{R}_\nu\right][\mathbf{R}_\mu,\mathbf{R}^\nu]\right) \\ & - \frac{N_c}{48\,\pi^2}\,\epsilon_{\mu\nu\lambda\rho}\mathrm{Tr}\left(\frac{\lambda^a}{2}\mathbf{R}^\nu\mathbf{R}^\rho\mathbf{R}^\lambda\right); \quad a = \overline{1,8}, \end{aligned} \tag{11.57}$$

where $\mathbf{R}_\nu = \mathbf{U}\partial_\nu\mathbf{U}^\dagger$ and the right currents $\mathbf{N}^a_{\mu,R}$ are given by interchanging $\mathbf{R}_\nu$ with $\mathbf{L}_\nu$. The vector and axial vector currents are obtained as

$$\mathbf{V}^a_\mu = \tfrac{1}{2}\left(\mathbf{N}^a_{\mu,L} + \mathbf{N}^a_{\mu,R}\right): \quad \mathbf{A}^a_\mu = \tfrac{1}{2}\left(\mathbf{N}^a_{\mu,R} - \mathbf{N}^a_{\mu,R}\right), \tag{11.58}$$

while the electro-magnetic current is to be

$$\mathbf{V}^{\rm EM}_\mu = \mathbf{V}^3_\mu + \frac{1}{\sqrt{3}}\mathbf{V}^8_\mu. \tag{11.59}$$

Then static observables are defined in terms of current components, e.g.: An electric charge radius.

$$\left(r_{\rm E}\right)^2 = \int d^3x\, r^2 \mathbf{V}_0(r). \tag{11.60a}$$

A magnetic moment:

$$\mu_k = \tfrac{1}{2}\int d^3x\,[\boldsymbol{r}\times\mathbf{V}]_k. \tag{11.60b}$$

A magnetic charge radius:

$$\left(r_{\rm M}\right)^2_k = \tfrac{1}{2}\int d^3x\, r^2[\boldsymbol{r}\times\mathbf{V}]_k. \tag{11.60c}$$

At the next step we are to obtain these currents in terms of collective coordinates by putting Witten's ansatz in (11.57) and manipulating it and subsequent expressions from (11.60) to be functions of $A^\dagger\dot{A}$. Then using the wave functions for the baryonic states, described in the previous subsection one can calculate matrix elements of operators, representing various observables. This is a general scheme, which does not contain anything except standard quantum mechanical calculations. Therefore we are not going to reproduce these straight but a bit tedious calculations here, and refer the reader to systematic and instructive papers (Adkins and Nappi, 1985; Kanazawa, 1987). The predicted values for most of static observables are consistent with experiment at the average level of 30%.

The Mass Formulae. In the standard manner (Guadagnini, 1984) the mass formulae are obtained as eigenvalues of the Hamiltonian (11.46), which should be previously diagonalized. Recall, that in the absence of the symmetry breaking term, i.e. the $c(\mathbf{U})_c = 0$ case, the Hamiltonian is diagonalized for the $SU(3)$ D-functions (11.54), which can be expressed in terms of the $SU(3)$ Euler angles. By writing

$$\sum_{a=4}^{7} \hat{\mathcal{M}}_a \hat{\mathcal{M}}_a = \sum_{\alpha=1}^{8} \hat{\mathcal{M}}_\alpha \hat{\mathcal{M}}_\alpha - \sum_{i=1}^{3} \hat{\mathcal{M}}_i \hat{\mathcal{M}}_i - \hat{\mathcal{M}}_8^2, \tag{11.61}$$

and taking into account the constraint (11.45) with $\mathbf{Q} = 1, N_c = 3$ one can present the Hamiltonian (11.46) in the form:

$$\begin{aligned} H = E(\mathbf{U}_c) &+ \frac{1}{8}\left(\frac{1}{a(\mathbf{U}_c)} - \frac{1}{b(\mathbf{U}_c)}\right) C_2(SU_R(2)) - \frac{3}{8\,b(\mathbf{U}_c)} \\ &+ \frac{1}{2\,b(\mathbf{U}_c)} C_2(SU_L(3)) + \frac{1}{8} c(\mathbf{U}_c)\,[1 - D^{(8)}(A)], \end{aligned} \tag{11.62}$$

where

$$C_2(SU_L(3)) \equiv \sum_{\alpha=1}^{8} \hat{\mathcal{M}}_\alpha \hat{\mathcal{M}}_\alpha; \quad C_2(SU_R(2)) \equiv \sum_{i=1}^{3} \hat{\mathcal{M}}_i \hat{\mathcal{M}}_i$$

are the quadratic Casimir operators of $SU_L(3)$ and $SU_R(2)$, respectively. Taking the eigenvalues of the Casimir operators for the wave functions (11.54) we obtain the mass formulae [the static energy eigenvalues for (11.62) without $c(\mathbf{U}_c)$-dependent term]

$$\begin{aligned} M_{J,Y'}^{(p,q)} = M_{cl}(\mathbf{U}_c) &+ \left(\frac{1}{a(\mathbf{U}_c)} - \frac{1}{b(\mathbf{U}_c)}\right) \frac{1}{2} J(J+1) \\ &+ \frac{1}{8\,b(\mathbf{U}_c)} \left\{\frac{4}{3}[p^2 + q^2 + 3(p+q) + pq] - 3Y'^2\right\}, \end{aligned} \tag{11.63}$$

where due to the constraint (11.45) $Y' = 1$. Thus for the average masses of the octet $\underline{8}$ with $J = 1/2$ we have

$$M_{\underline{8}} = M_{cl}(\mathbf{U}_c) + \frac{3}{8\,a(\mathbf{U}_c)} + \frac{3}{4\,b(\mathbf{U}_c)}, \tag{11.64a}$$

and for that of the decuplet $\underline{\mathbf{10}}$ with $J = 3/2$

$$M_{\underline{10}} = M_{cl}(\mathbf{U}_c) + \frac{15}{8\,a(\mathbf{U}_c)} + \frac{3}{4\,b(\mathbf{U}_c)}. \tag{11.64b}$$

Thus, as it was expected, in the chiral limit all the particles within the same multiplet have the same mass.

To obtain the experimentally evident mass splitting within the multiplets as displayed in Fig. 11.1 we need to take into account the symmetry breaking term.

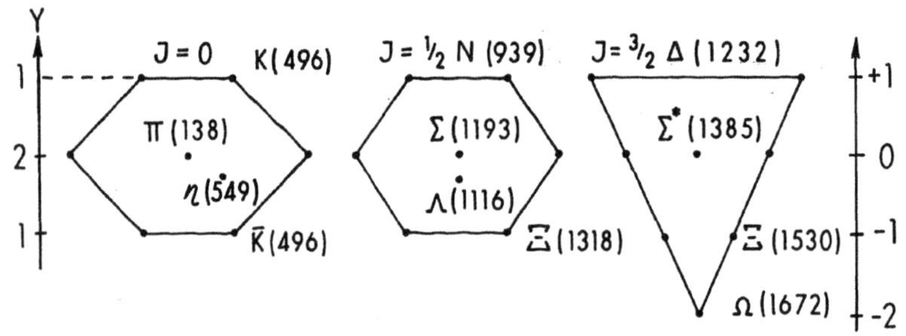

Fig. 11.1. The $SU(3)$ classification of hadrons. The experimental data for the hadron masses is given in brackets

Here arises the problem of diagonalization of the Hamiltonian (11.46) in the $c(\mathbf{U}_c) \neq 0$ case, which was solved by (Yabu and Ando, 1988). They found that in the $SU(3)$ Euler angles (11.31) the symmetry breaking term depends only on one of them, namely:

$$\tfrac{1}{2}\, c(\mathbf{U}_c)\, [1 - D^{(8)}(A)] = \tfrac{3}{4}\, c(\mathbf{U}_c)\, \sin^2 \xi_4. \tag{11.65}$$

For further details we refer the reader to the mentioned original paper. Here we just state that the mass splitting is correctly reproduced in this approach and leads to the following improved mass formulae

$$\begin{aligned} M^{(p,q)} &= M_{cl}(\mathbf{U}_c) + \frac{1}{2}\left(\frac{1}{a(\mathbf{U}_c)} - \frac{1}{b(\mathbf{U}_c)}\right) J(J+1) \\ &\quad + \frac{1}{2b(\mathbf{U}_c)}(\mathcal{E}_{\mathrm{SB}} - \mathcal{E}_0). \end{aligned} \tag{11.66}$$

Here $\mathcal{E}_{\mathrm{SB}}$ is the dimensionless eigenvalue of the following eigenequation:

$$\left[C_2(SU_L(3)) + 6\, c(\mathbf{U}_c)\, b \sin \xi_4\right] D_{\mathrm{SB}}^{(p,q)} = 4\, \mathcal{E}_{\mathrm{SB}}\, D_{\mathrm{SB}}^{(p,q)}, \tag{11.67}$$

where $D_{\mathrm{SB}}^{(p,q)}$ are the states which diagonalize the Hamiltonian (11.46) in the $c(\mathbf{U}_c) \neq 0$ case. $\mathcal{E}_0$ is the vacuum-like fluctuation eigenvalue, which is zero in no symmetry breaking limit and it becomes large with the increase of the symmetry breaking. It was also shown that the $SU(3)$ Skyrme model smoothly reduces to the $SU(2)$ original model when the strength of the symmetry breaking $6\,c(\mathbf{U}_c)\,b$ becomes large. Regarding M_{cl}, a, b and c in (11.66) as fitting parameters Yabu and Ando (1988) claimed that the best fitted mass spectrum obtainable in the chiral $SU(3)$ soliton models (that have at most second order time derivative terms) is as displayed in Fig. 11.2.

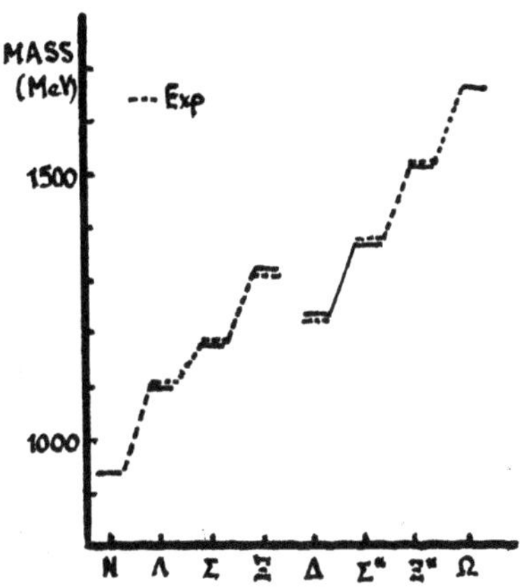

Fig. 11.2. The best-fit result for the masses of the $\underline{8}$ and $\underline{10}$ multiplets of baryons. The solid line refers to the Yabu-Ando result, experimental values are presented by the dashed lines

As a result of preceding discussion one should acknowledge that within 30% accuracy the Skyrme model seems to give reasonable account of the static properties of the light baryons. Beyond any doubt by one or another method it is possible to improve the achieved accuracy. However before embarking on a systematic way of improving the present calculations in the framework of this model or that of modifying the model itself, it is relevant to extend the analysis to dynamical issues. That is why one requires to test the ability of the Skyrme approach in describing the principal processes where baryons (in the image of Skyrmions) take part.

11.3 Skyrmions' Interactions: Nuclear Forces and Nuclear Matter

As a prelude remind the reader that the point of departure in Skyrme's search, resulted in the soliton model of baryons, were problems related to unsatisfactory description of nuclear matter (see Chapter 1). It is curious to see what a new knowledge on the subject was obtained as a result of the Skyrme approach development and its more thorough study. With this in mind we give here a brief outlook of the present status in the field of $SU(2)$ Skyrmion dynamics.

The presented material is an extended version of a part of our previous review (Makhankov, Rybakov and Sanyuk, 1992).

11.3.1 The Skyrmion-Skyrmion Interaction and Nuclear Forces

To calculate the Skyrmion-Skyrmion interaction it is generally assumed the product-ansatz form [see Eq. (7.3) and the relevant discussion of Skyrmion-Skyrmion interaction mostly from topological point of view in Section 7.1] for the two-baryon configuration:

$$\mathbf{U}_{12} = \mathbf{U}_0^A(\boldsymbol{r} - \boldsymbol{r}_1) \cdot \mathbf{U}_0^B(\boldsymbol{r} - \boldsymbol{r}_2), \tag{11.68}$$

where $\mathbf{U}_0^A = A \cdot \mathbf{U}_0 \cdot A^\dagger$, $\mathbf{U}_0^B = B \cdot \mathbf{U}_0 \cdot B^\dagger$; and $A, B \in SU(2)$. From the energy value computed for this configuration it is substracted the energy of the two widely separated Skyrmions, i.e. $2\,E_1$. The obtained result is declared to be the Skyrmion-Skyrmion potential $V(\boldsymbol{R}; A, B)$, where $\boldsymbol{R} = \boldsymbol{r}_1 - \boldsymbol{r}_2$. In a quantum description the matrices A and B are random and one ought to average over them. It turns out that the result would be the same, if one average some products of Pauli matrices $\underline{\sigma}_i$ and $\underline{\tau}_j$, $i, j = 1, 2$ over the standard spin-isospin states of nucleon $|N>$, where $\underline{\sigma}_i$ and $\underline{\tau}_j$ are spin and isospin operators of individual nucleons. In particular, the following relation holds

$$\langle N'|\mathrm{Tr}(\tau_\alpha A^\dagger \tau_\beta A)|N\rangle = -\tfrac{2}{3}\,\langle N'|\sigma_\alpha \tau_\beta|N\rangle, \tag{11.69}$$

and by taking it into account one leads to the NN-potential

$$V^{NN} = V_c + (\underline{\tau_1} \cdot \underline{\tau_2})\big[(\underline{\sigma_1} \cdot \underline{\sigma_2})\, V_{\sigma\tau} + S_{12} V_{T\tau}\big]. \tag{11.70}$$

Here $V_c, V_{\sigma\tau}, V_{T\tau}$ correspond, respectively, to the central, spin-isospin and tensor parts contributions of the NN-potential and are functions of the separation R, and S_{12} is the conventional tensor operator $S_{12} = 3\,(\underline{n} \cdot \underline{\sigma_1})\,(\underline{n} \cdot \underline{\sigma_2}) - (\underline{\sigma_1} \cdot \underline{\sigma_2})$, with $\underline{n} = \underline{R}/R$.

A comparison of (11.70) with the Paris potential (Jackson, Jackson, and Pasquier, 1985), which represents the most satisfactory compromise between theoretical and phenomenological potentials currently available, shows a good enough agreement at large distances ($R > 2\,\mathrm{fm}$) and a qualitative agreement at intermediate ones ($1\,\mathrm{fm} < R < 2\,\mathrm{fm}$). While the central potential V_c has the desired strongly repulsive core, it lacks the intermediate-range attraction, which is small in comparison with the nucleon mass, but plays a crucial role in nuclear physics, forming the binding energy of nuclei. In the standard boson-exchange treatment this attractive component of V_c is explained through an isoscalar $\pi\pi$-exchange (an exchange by an effective σ-meson). This problem in the intermediate-range attraction (to our belief, mostly related with the shortcomings of the product-ansatz approximation) in the framework of the Skyrme model is generally assumed to be due to missing contributions from quantum fluctuations. Recently, however, it has been demonstrated in (Gomm, Jain, Johnson and Schechter, 1986a; 1986b) that to obtain the correct QCD scaling behaviour for the energy-momentum tensor in the Skyrme model one should introduce an explicit scalar meson field in the model. Such an additional scalar field immediately leads to an intermediate-range attraction of sufficient strength (Yabu, Schwesinger and Holzwarth, 1989).

Remark, that in a relatively simple way one can extract from (11.70) the one-pion-exchange potential, which is related with the Skyrmion structure in the asymptotic ($r \to \infty$) region

$$\Theta(r) \simeq \frac{3\lambda f_{\pi NN}}{8\pi m_\pi} (1 + m_\pi r) \frac{\exp(-m_\pi r)}{r^2}, \tag{11.71}$$

where $f_{\pi NN}$ is a numerical constant, $\lambda = 2/F_\pi$. By making use of (11.71) one obtains the following potential

$$V_\pi^{NN} \simeq \frac{m_\pi f_{\pi NN}^2}{12\pi} \left[Y_2(m_\pi r) S_{12} + Y_0(m_\pi r) (\underline{\sigma_1} \cdot \underline{\sigma_2})\right] (\underline{\tau_1} \cdot \underline{\tau_2}), \tag{11.72}$$

where the Yukawa functions are defined as

$$Y_0(x) = \frac{e^{-x}}{x}; \quad Y_2(x) = \left(1 + \frac{3}{x} + \frac{3}{x^2}\right) Y_0(x). \tag{11.73}$$

Thus, the one-pion-exchange interaction evidently contributes into the components $V_{\sigma\tau}$ and $V_{T\tau}$. One should say, that a satisfactory agreement with the Paris potential is not only due to the one-pion-exchange. For example, at $R < 1.5\,\mathrm{fm}$ the contribution of V_π^{NN} is less then 30%. Therefore the Skyrme model provides a qualitatively correct account for the multi-pion-exchange processes, as well.

But the phenomenological Paris potential besides components (11.70) contains other terms of considerable importance. That is, for example, the spin-orbit interaction, which has the form

$$\left[V_{LS} + (\underline{\tau_1} \cdot \underline{\tau_2}) V_{LS\tau}\right] (\boldsymbol{L} \cdot \boldsymbol{S}), \tag{11.74}$$

where now $\boldsymbol{S} = (\underline{\sigma_1} + \underline{\sigma_2})/2$, $\boldsymbol{L} = -\mathrm{i}(\boldsymbol{R} \times \nabla_R)$. To derive the spin-orbit interaction in the Skyrme model one has to go beyond the static approximation. That is equivalent to an account of the kinetical nonstationary part of the Hamiltonian, related with $\mathrm{Tr}\, \mathrm{L}_0^2$ [see, for example, Eq. (1.42)] (Niman and Riska, 1990).

Derived in this way the isospin-dependent component of the spin-orbit interaction V_{LS}' does in fact agree at long distances with the corresponding component of the Paris potential. In contrast to this, the isospin-independent component has the incorrect sign. A natural way to overcome the problem of the sign appeares to be the inclusion of the sixth-order term $\mathcal{L}_{(6)}$, as given in Eq. (2.44). With this term one obtains an isospin-independent spin-orbit interaction which is negative (i.e. has the correct sign), and which at short and intermediate distances dominates over the contribution from $\mathcal{L}_{(4)}$ (Riska and Schwesinger, 1989). This is so far the strongest phenomenological evidence for the necessity of including a sixth-order term in the field derivatives in the Skyrme model.

11.3.2 The Meson-Baryon Interaction

The problem of constructing a consistent quantum description of meson-baryon processes presents a crucial test of validity of the Skyrme approach to be a reliable tool to study elementary particle processes. Since this approach is closely related with current algebra and phenomenological Lagrangians method (it embodies the nonlinear σ-model as itself) it is very likely that the Skyrmion dynamics will go through current algebra requirements and will not alter the so-called soft-pion threshold theorems. The trouble comes from the fact, that to study soft-pion Skyrmion dynamics we need to consider the model beyond the tree level approximation. But the model as it stands is not renormalizable, since it is both infrared and ultravialet divergent [see discussion in (Zahed and Brown, 1986)]. Nevertheless, it has been demonstrated, that the Skyrme model does yield a finite S-matrix. Moreover, an analysis of πN-interaction in the framework of the Skyrme model has revealed hitherto unknown phenomenologically useful relations, which are in a good agreement with experiment.

To make one step at a time, recall that to describe the πN-scattering in the Skyrme model the pion field has to be coupled to the Skyrmion. Formally, the simplest method is the symmetric ansatz, proposed by Schnitzer (1984) when the hedgehog configuration (4.33) is modified for

$$U(\boldsymbol{x}) = \exp\left(\frac{\mathrm{i}}{2F_\pi}(\underline{\tau}\cdot\underline{\pi})\right)\exp\left(\mathrm{i}(\underline{\boldsymbol{n}}\cdot\underline{\tau})\,\Theta(r)\right)\exp\left(\frac{\mathrm{i}}{2F_\pi}(\underline{\tau}\cdot\underline{\pi})\right), \quad (11.75)$$

where $\Theta(r)$ describes the Skyrmion; $\underline{\pi}(\boldsymbol{r})$ is the pion field and $\underline{\boldsymbol{n}} = \underline{\boldsymbol{r}}/r$. In order to obtain an effective interaction Lagrangian one performes a power series expansion of the Lagrangian density (1.39) in powers of π. Up to the second order the result reads

$$\begin{aligned}\mathcal{L}_{\text{soft}} &= \mathcal{L}_{SK} + \frac{1}{2}\,\partial_\mu\underline{\pi}\cdot\partial^\mu\underline{\pi} + \frac{1}{2F_\pi}\,\partial_\mu\underline{\pi}\cdot\boldsymbol{A}^\mu \\ &\quad + \frac{1}{4F_\pi^2}\,(\partial_\mu\underline{\pi}\times\underline{\pi})\cdot\boldsymbol{V}^\mu\ldots,\end{aligned} \quad (11.76)$$

where $\boldsymbol{V}^\mu$ and $\boldsymbol{A}^\mu$ are corresponding vector and axial vector Skyrmion currents. The obtained expression (11.76) up to some negligible distinctions has the same form as the phenomenologically successful Weinberg's current algebra Lagrangian (Weinberg, 1967). On the basis of the observed similarity it has been approved that the soft-pion threshold theorems (the Goldberger-Treiman relations, the Adler-Weisberger relations and so on) together with other current algebra requirements hold true in the Skyrme model.

Remark 11.3. The underlying physical idea merits to be quoted from (Holzwarth and Schwesinger, 1986), as it has some roots in the initial Skyrme's motives. From the study of various extended objects like elastic bodies or droplets of fluids, it is known that once the normal modes for small classical oscillations around a stable minimum are determined, their amplitudes may serve as the dynamical coordinates to be subject to quantization (collective coordinates). What is perhaps less known is the fact, that if the extended object possesses a diffuse surface the normal modes may display an oscillatory behaviour as $|\boldsymbol{x}| \to \infty$ and thus correspond to scattering states of the underlying basic field quanta at large distances from the object. It is precisely this situation we have in the Skyrme model. The baryon is a lump (or vortex) with a diffuse surface in terms of underlying meson fields. At large distances the fluctuations of this lump will correspond to the field quanta, the mesons. Indeed, since the classical soliton field interpolates to the classical vacuum once we get away from the center of soliton, it follows that small oscillations about the soliton become small oscillations about the vacuum when not in the neighborhood of the soliton. In accord with Skyrme's ideas, realised in his "Mesonic Fluid" model, small oscillations about the vacuum are to be interpreted as pions. Whence we get a visual picture of the scattering process: $\pi + N \to \pi + N$. One may hope therefore to obtain a full account of πN-scattering just by determining the normal modes of the Skyrmion with the suitable boundary conditions.

The other way of introducing the pion field is due to (Mattis and Karliner, 1985; Mattis and Peskin, 1985; Karliner and Mattis, 1986) when the hedgehog (4.33) is extended to

$$U(\boldsymbol{x}) = \exp\left\{ \mathrm{i}(\underline{n} \cdot \underline{\tau})\, \Theta(r) + \frac{2\mathrm{i}}{F_\pi} \underline{\tau} \cdot \underline{\pi})(\boldsymbol{r}) \right\}, \quad \underline{n} = \frac{\boldsymbol{r}}{r}. \tag{11.77}$$

Now upon the same expansion, one does not obtain the Lagrangian (11.76), but instead the πN-interaction is effectively described by the action in the form:

$$S \simeq \int d^4x\, \pi^a(\boldsymbol{x}, t) \hat{D}_{ab}(\mathbf{U}_c) \pi^b(\boldsymbol{x}, t), \tag{11.78}$$

where $\hat{D}_{ab}$ is a linear differential operator that depends on the hedgehog profile, and commutes with the so-called K-spin:

$$\left[\boldsymbol{K}, \hat{D}_{ab}(\mathbf{U}_c)\right] = 0. \tag{11.79}$$

By definition $\boldsymbol{K} = \boldsymbol{J} + \boldsymbol{I}$, where $\boldsymbol{J}$ is the total angular momentum of mesons and $\boldsymbol{I}$ is the total isospin. The Eq. (11.79) manifests the availability of the so-called K-symmetry in πN-scattering (Mattis and Mukerjee, 1988), which is relevant to the G_1-invariance of the Skyrmion configuration (as discussed in Section 4.2), whence one can relate the K-symmetry with the invariance under the $SU(2)_K$ group. It is useful to stress here the analogy with the central potential scattering in quantum mechanics, when the orbital angular momentum is conserved.

Notice that thus defined the K-symmetry holds in the adiabatic approximation only, that is at a relatively high energy of incoming particles in the absence of resonances with the low-laying excitations of Skyrmions. Roughly speaking, the scattering time has to be small if compare with the period of rotation: Skyrmion rotates through a negligible angle during the scattering process. To exploit the K-symmetry, one can expand the π-field in the action (11.78) in K-spherical harmonics (elements of irreducible representation of $SU(2)_K$). This leads to a set of normal-mode equations for the radial field functions, which asymptotically describe free propagating pions. The asymptotic solutions of this set, expressed in terms of spherical Bessel and Neumann functions, give the possibility to define the S-matrix [for details we refer the reader to the original papers (Mattis and Karliner, 1985; Mattis and Peskin, 1985; Karliner and Mattis, 1986)]. The final result may be stated as follows: ***Given the pion energy ω, the scattering process is completely specified by the total isospin I and angular momentum J of the πN system, and the relative incoming and outgoing pion angular momenta L and L'.*** The corresponding S-matrix in this approximation (from now on we consider 0^- multiplets if otherwise is not stated) can be written in terms of the reduced matrix elements $\mathbf{S}_{KLL'}$, for pion scattering off a Skyrmion:

$$\begin{aligned} \mathbf{S}_{L's'LsIJ}(\omega) &= (-1)^{s'-s}\left[(2s+1)(2s'+1)\right]^{1/2} \\ &\times \sum_K (2K+1) \begin{Bmatrix} K & I & J \\ s' & L' & 1 \end{Bmatrix} \begin{Bmatrix} K & I & J \\ s & L & 1 \end{Bmatrix} \mathbf{S}_{KL'L}. \end{aligned} \tag{11.80}$$

Here s and s' are the values of $I = J$-spins for incoming and outgoing baryons, namely $s = 1/2$ for N and $s = 3/2$ for Δ; the K values range in accord with the restriction $|L-1| \le K \le L+1$, $|L'-1| \le K \le L'+1$; and $\left\{\begin{smallmatrix} j_1 & j_2 & j_3 \\ j_4 & j_5 & j_6 \end{smallmatrix}\right\}$ are standard Wigner's $6-j$ symbols. The predictions of πN-scattering parameters in such a treatment (with the exception of S-waves) have been remarkably successful, i.e. predicted partial wave amplitudes agree qualitatively very well with empirically observed amplitudes.

Moreover, as it was announced above, this analysis implies two linear relations between the S-matrix elements:

$$\begin{aligned} \mathbf{S}_L\Big(I=\frac{3}{2}, J=L-\frac{1}{2}\Big) &= \frac{L-1}{4L+2}\,\mathbf{S}_L\Big(\frac{1}{2}, L-\frac{1}{2}\Big) \\ &+ \frac{3(L+1)}{4L+2}\,\mathbf{S}_L\Big(\frac{1}{2}, L+\frac{1}{2}\Big), \end{aligned} \tag{11.81a}$$

$$\begin{aligned} \mathbf{S}_L\Big(I=\frac{3}{2}, J=L+\frac{1}{2}\Big) &= \frac{L+2}{4L+2}\,\mathbf{S}_L\Big(\frac{1}{2}, L+\frac{1}{2}\Big) \\ &+ \frac{3L}{4L+2}\,\mathbf{S}_L\Big(\frac{1}{2}, L-\frac{1}{2}\Big). \end{aligned} \tag{11.81b}$$

In particular one gets from (11.81a) a familiar relation

$$\mathbf{S}_1\Big(\frac{1}{2}, \frac{3}{2}\Big) = \mathbf{S}_1\Big(\frac{3}{2}, \frac{1}{2}\Big),$$

which follows from the Chew-Low static model. Remark, that **the Mattis-Karliner-Peskin relations** (11.81) do not work very well for S, P or D channels for which the ignored coupling effects between rotational and the vibrational small oscillation modes seem to be important. While the detailed construction of the phase shifts gives a fairly good agreement with the experimental data in F and higher channels. Among other achievements, these relations were subsequently generalized to the case of arbitrary meson spin and isospin (Mattis, 1986), as well as to the $SU(3)$ case (Braaten and Cai, 1989). For further progress in this direction we refer the reader to a very instructive paper (Mattis and Braaten, 1989) and to references therein.

11.3.3 The Skyrme Model and Nuclear Matter

The earliest attempt to describe a dense nuclear matter by means of effective meson fields were made by Skyrme himself (Skyrme, 1962). Introducing as above in Eqs. (5.7) vectors $\underline{\mathbf{X}}, \underline{\mathbf{Y}}$, and $\underline{\mathbf{Z}}$ we can write for the energy density $u = E/V$ and for the baryon density $n = N/V = |\mathbf{Q}|/V$ (here V is the occupied volume) the following expressions:

$$u = \frac{1}{2\lambda^2}(\boldsymbol{X}^2 + \boldsymbol{Y}^2 + \boldsymbol{Z}^2) + \epsilon^2\left[(\boldsymbol{X}\times\boldsymbol{Y})^2 + (\boldsymbol{Y}\times\boldsymbol{Z})^2 + (\boldsymbol{Z}\times\boldsymbol{X})^2\right], \quad (11.82)$$

$$n = \frac{1}{2\pi^2}|\boldsymbol{X}\cdot(\boldsymbol{Y}\times\boldsymbol{Z})|. \quad (11.83)$$

By recalling that a geometric average does not exceed an arithmetic average, one comes to inequalities:

$$\tfrac{1}{3}(\boldsymbol{X}^2 + \boldsymbol{Y}^2 + \boldsymbol{Z}^2) \geq (|\boldsymbol{X}|\,|\boldsymbol{Y}|\,|\boldsymbol{Z}|)^{2/3} \geq |\boldsymbol{X}\cdot(\boldsymbol{Y}\times\boldsymbol{Z})|^{2/3}, \quad (11.84a)$$

$$\tfrac{1}{3}\left[(\boldsymbol{X}\times\boldsymbol{Y})^2 + (\boldsymbol{Y}\times\boldsymbol{Z})^2 + (\boldsymbol{Z}\times\boldsymbol{X})^2\right] \geq |\boldsymbol{X}\cdot(\boldsymbol{Y}\times\boldsymbol{Z})|^{4/3}. \quad (11.84b)$$

By making use of these inequalities from (11.82) and (11.83) one deduces the estimate on the energy density of nuclear matter through the baryon density n:

$$u \geq \frac{1}{2\lambda^2}\left(2\pi^2 n\right)^{2/3} + \epsilon^2\left(2\pi^2 n\right)^{4/3}. \quad (11.85)$$

In the dense matter limit the second term in (11.85) becomes a dominant one, whence

$$u > \epsilon^2\left(2\pi^2 n\right)^{4/3} \simeq 160\,\epsilon^2 n^{4/3}. \quad (11.86)$$

It is curious to compare Skyrme's estimate (11.86) with that obtained in the bag approximation of the quark-gluon plasma theory (Pethick, 1985), namely

$$u \simeq B + \frac{9}{4}\pi^{2/3}\left(1 + \frac{2\alpha_s}{3\pi}\right)n^{4/3}, \quad (11.87)$$

where B is a quark bag constant and α_s is the running QCD coupling constant.

It is evident that estimates (11.86) and (11.87) are consistent with each other. This fact tells us that the Skyrme model to some extent accounts for high-energy hadron physics as well. Thus the range of applicability of the model might be extended beyond the low-energy regime and by force of this improving its status as a bridge between the QCD and the traditional nuclear physics description. Still another conclusion made by Kutschera and Pethick (1985) on the basis of the observed accordance of estimates is that in a dense medium of Skyrmions there are significant many-body effects that tend to cancel the long-range part of the two-Skyrmion interaction. Therefore they claimed that this "screening effect" limits the interaction between Skyrmions to nearest neighbors in dense matter.

Numerical calculations presented in (Castillejo, Jones, Jackson, Verbaarschot and Jackson, 1989) show, that if one considers a set of isolated Skyrmions then under an increasing extent of condensation they would form a structure similar to a face-centered (cubic) lattice with a spacing a and symmetries in the form:

$$\mathbf{U}\left(x+\frac{a}{2}, y+\frac{a}{2}, z\right) = \tau_3\, \mathbf{U}(x,y,z)\, \tau_3; \;\ldots, \tag{11.88}$$

where dots stand for transformations obtained by circular permutations in (11.88). The symmetry (11.88) means that situated at the center of face the Skyrmion in isospace is turned on the angle π around the outward normal to the face. This ensures maximum attraction between Skyrmions (remind the discussing in Section 7.1). Under further condensation Skyrmions expand themselves in size while losing their individuality. Therefore there happens a phase transition to a condensed medium. Denoting $\mathbf{U} = \sigma + \mathrm{i}\,(\underline{\pi}\cdot\underline{\tau})$, then before this phase transition it holds to be valid the parabolic approximation to the mean-field value, i.e.

$$\langle\sigma\rangle \simeq (a - a_{cr})^2, \tag{11.89}$$

thus indicating that the above phase transition has to be of the second kind. The field configuration as itself obeys the additional symmetry at the phase transition point:

$$\begin{aligned}
\sigma\left(x+\frac{a}{2}, y, z\right) &= -\sigma\,(x,y,z),\\
\pi_1\left(x+\frac{a}{2}, y, z\right) &= -\pi_1\,(x,y,z),\\
\pi_2\left(x+\frac{a}{2}, y, z\right) &= \pi_2\,(x,y,z),\\
\pi_3\left(x+\frac{a}{2}, y, z\right) &= \pi_3\,(x,y,z).
\end{aligned} \tag{11.90}$$

Finally, at a further decrease of the spacing $a < a_{cr}$, the energy per an individual Skyrmion continues to decrease, attending its minimal value at $a = a_0$, where the field configuration is approximately described by the following functions:

$$\begin{aligned}
&\sigma = \cos kx \cos ky \cos kz, \quad k = \frac{2\pi}{a}\\
&\pi_1 = \sin kx\,\left(1 - \tfrac{1}{2}\sin^2 ky - \tfrac{1}{2}\sin^2 kz + \tfrac{1}{3}\sin^2 ky \sin^2 kz\right)^{1/2}, \ldots
\end{aligned} \tag{11.91}$$

and so on by circular permutations. Thus, the surfaces defined by $\sigma = 0$ are orthogonal to the planes with $[x, y, z = (a/4) + m(a/2)]$, $m \in \mathbb{Z}$. Remark, that this leads to a considerable abatement of the mass per Skyrmion (with a factor $\approx (5/6)$) if compare with that for a free one.

An interesting interpretation of this effect was suggested by (Manton and Ruback, 1986; Manton, 1987) who considered the Skyrme model as defined not on the flat space $\mathbb{R}^3$, but instead on $\mathbf{S}^3$ sphere of some radius R. Since the metric element on $\mathbf{S}^3$ in terms of spherical coordinates (μ, ϑ, α) is

$$d\dot{s}^2 = R^2[d\mu^2 + \sin^2\mu\,(d\vartheta^2 + \sin^2\vartheta\, d\alpha^2)]; \quad \mu \in [0, \pi], \tag{11.92}$$

and the volume element

$$dV = R^3 \sin^2\mu \sin\vartheta d\mu d\vartheta d\alpha, \tag{11.93}$$

the transition from $\mathbb{R}^3$- to $\mathbf{S}^3$-model can be achieved by the following change of variables: $r \rightarrow R\sin\mu$, $dr \rightarrow d\mu$ in the energy functional (6.1). The result is

$$\begin{aligned} E_{\mathbf{S}^3} = \frac{2\sqrt{2}\pi\epsilon}{\lambda} \int_0^\pi d\mu \sin^2\mu \Bigg\{ L\left[\left(\frac{d\Theta}{d\mu}\right)^2 + \frac{2\sin^2\Theta}{\sin^2\mu}\right] \\ + \frac{1}{L}\frac{\sin^2\Theta}{\sin^2\mu}\left[2\left(\frac{d\Theta}{d\mu}\right)^2 + \frac{\sin^2\Theta}{\sin^2\mu}\right]\Bigg\}, \end{aligned} \tag{11.94}$$

where we put $L = R/\lambda\epsilon\sqrt{2}$. In standard manner minimization of (11.94) with respect to the radial profile function $\Theta(\mu)$ leads to the Euler-Lagrange equation

$$\begin{aligned} \frac{d^2\Theta}{d\mu^2} + \frac{\sin 2\mu}{\sin^2\mu}\frac{d\Theta}{d\mu} - \frac{\sin 2\Theta}{\sin^2\mu} + \frac{2}{L^2}\frac{\sin^2\Theta}{\sin^2\mu}\frac{d^2\Theta}{d\mu^2} \\ + \frac{\sin 2\Theta}{L^2 \sin^2\mu} \times \left\{\left(\frac{d\Theta}{d\mu}\right)^2 - \frac{\sin^2\Theta}{\sin^2\mu}\right\} = 0, \end{aligned} \tag{11.95}$$

which admits the simple uniform solution

$$\Theta(\mu) = \pi - \mu. \tag{11.96}$$

The energy associated with (11.96) is

$$E = 3\sqrt{2}\pi^2 \frac{\epsilon}{\lambda}\left(L + \frac{1}{L}\right). \tag{11.97}$$

For $L = 1$, the solution (11.96) realizes the absolute minimum of the energy and moreover it saturates the topological lower bound

$$E \geq 6\sqrt{2}\pi^2 \frac{\epsilon}{\lambda}|\mathbf{Q}| \tag{11.98}$$

for $\mathbf{Q} = 1$, contrary to the situation in the original Skyrme model (see Section 3.3). Note that for $L > 1$ there are configurations with lower energy (Loss, 1987; Manton, 1987; Wirzba and Bang, 1990).

It is straightforward to see that the solution (11.96) corresponds to an uniform distribution of matter on $\mathbb{S}^3$, that is to a condensed phase. As it has been shown in (Manton and Samols, 1990; Wirzba and Bang, 1990) this solution is stable for $L < \sqrt{2}$, i.e. for $R = R_{\text{cr}} \equiv 2\,\lambda\epsilon$, there appears to be the phase transition and since the bifurcation at $L = \sqrt{2}$ is of standard "pitchfork" type it corresponds to the phase transition of the second kind. Therefore in the Skyrme-Manton model the radius R of the spatial sphere $\mathbb{S}^3$ should depend on the baryon density: $R \sim n^{-1/3}$. If $L > \sqrt{2}$, the condensed phase becomes unstable one and in the limit $L \to \infty$ one recovers the usual flat space isolated Skyrmions. In the case of large L it is possible to get L-dependent corrections to the Skyrmion mass (Manton and Samols, 1990) in the form

$$E \simeq 6\sqrt{2}\pi^2 \frac{\epsilon}{\lambda}\left(1.231445 - 0.419\,\frac{1}{L^2} + \frac{1}{4\,L^3}\right), \tag{11.99}$$

which implies **the Skyrmion mass abatement effect**, correlated with the nuclear matter density.

There is a hope, that such a scheme might be applicable to a description of the following hypothetical quantum effect. In some extreme conditions of high temperature or density, or in highly curved universe, baryonic matter may turn into quark plasma: the phase where the quarks are unconfined and the chiral symmetry is restored. The analysis of the flat Skyrmion symmetries has been already performed in detail (see Chapters 4, 9 and 11), therefore we focus here on the $\mathbb{S}^3$-model symmetries. The $\mathbb{S}^3$ Hamiltonian has the symmetry group $SO(4)_S \otimes SO(4)_{Ch}$, where $SO(4)_S$ is the isometry group of the spatial $\mathbb{S}^3$ and $SO(4)_{Ch} = SU_L(2) \otimes SU_R(2)$ is the chiral symmtry group of the field manifold $\mathbb{S}^3 = SU(2)$. Now when the spatial sphere $\mathbb{S}^3$ has the radius L greater then $\sqrt{2}\,(\approx 0.8\text{fm})$ the configuration is similar to the flat space Skyrmion with all its symmetric features. [Remember from Sections 4.2 and 9.3.1 that the chiral symmetry has to be broken to its diagonal subgroup.] On the spatial sphere of smaller radius the Skyrmion is presented by the identity map (11.96) and hence no point either in space or in flavor space is special. The unbroken group is the diagonal $SO(4)$ subgroup of the full symmetry group [one can regard it at equal footing as either chiral $SO(4)$ or spatial one]. As a consequence quantum states of pions interacting with a Skyrmion are classified by a larger group than in the case of flat space. Thus one may say that there is **a chiral symmetry restoration effect**, the quantum effect which definitely does not happen in the absence of a Skyrmion (Manton, 1987).

At present there are in progress numerous attempts to apply the promising Skyrme approach to various aspects of the dense nuclear matter [see (Walhout, 1990) and references therein]. The calculations show that even the basic Skyrme model, through a relatively small amount of computational efforts and almost without adjustable parameters, predicts the behaviour one expects in the dense matter regime and produces numbers which are accurate to the same degree as those it produces for few-baryon systems. Among attractive advantages of the solitonic picture of dense matter the principal are the following:

a. The aproach starts with a Lorentz-invariant Lagrangian;
b. Many-body effects are incorporated at an equal footing with few-body events;
c. The approach treats nuclear interactions and structure on the same level.

In the light of initial Skyrme's ideas, beyond doubt, a reliable description of dense nuclear matter would be the best tribute to the memory of the author of the model.

Concluding Remarks

Completing the book, authors console themselves with a hope, that they managed to convince the reader in a miracle efficiency of topological methods, especially when applied to particle physics. Tony Skyrme is now widely recognized as a pioneer in the field, being able to disclose a wide spectrum of topology resources in description of still so unfamiliar microcosm events. His deep intuition, bright mathematical talent and far reaching physical ideas allowed him to discover the crucial role of topological stability in nuclear matter. This important observation became a cornerstone in construction of the nuclear model, bearing at present Skyrme's name and providing one with an amazingly simple description of fundamental properties of baryons. While being historically the first known example of nonlinear chiral model, the Skyrme model substantially predates and in much supplies the results obtained later on in the framework of Quantum Chromodynamics. As it turned out, the well-known distinction between fermionic and bosonic states in quantum theory has the evident topological background. The analysis of interrelations between QCD on one side and the chiral models on the other, performed by t'Hooft and Witten, showed the necessity to supply the Skyrme action with the multivalued topological term, known as the Wess-Zumino term. The subsequent semi-classical quantization via Bogolubov's method of the $SU(3)$ generalized Skyrme-Witten model leads to a natural explanation of the observed baryon mass spectrum, in particular picking out two principal multiplets, i.e. the octet and the decouplet of baryons.

There are already some predicted and observed physical phenomena, which gained the most reliable explanation in the framework of the Skyrme model. Among them are the Callan-Rubakov effect of catalysis of the proton decay in the presence of a monopole, and the so-called "proton-spin" puzzle. [Unfortunately, their clear exposition occurs to be beyond the scope of the present book.] In general, it is already possible to conclude, that the image of baryon as chiral soliton made already a good deal as in clarifying so well in systematization of previously isolated pieces of information on the nuclear interactions. Moreover in this approach one was able to obtain a satisfactory description of meson-baryon and baryon-baryon processes.

We completely agree with other authors and researchers that it is too early to make any definite conclusion, concerning possibilities and outlook of the Skyrme model approach, since we have only started to appreciate its importance in nuclear and particle physics. It will not be a surprise if such an approach would be found as appropriate in description of condensed matter phenomena, in general. Therefore to conclude we would rather quote once more Skyrme's original paper: "... **there are a number of encouraging resemblances between the consequences of the theory and the phenomena of strongly interacting particles; quantitative comparison is hardly significant however, until the quantized theory is understood.... Whether or not this theory has any direct connection with the description of the observed strongly interacting particles the principles involved seem important enough for further study."** (Skyrme, 1962). More than thirty years have passed since these words were written, but they still remain to be actual and instructive.

Part IV

Appendices

Appendix A

Chiral Symmetry

In Section 1.2 it was already discussed the discovery in 1940s of Θ- and τ-mesons and the reasons why they were christened by the name "strange." Here we briefly remind some more details of the story how this "$\Theta - \tau$" paradox was resolved, and list some algebraic and geometric aspects of the chiral symmetry. Having analysed experimental data in 1956 T.D. Lee and C.C. Yang found out that the parity conservation law in weak interactions had never been checked experimentally and they proposed the way of such a verification. The experiments of C.S. Wu's team performed in 1957 proved beyond any doubts that the parity conservation law breaks down in weak interactions and that the "$\Theta - \tau$" paradox should be resolved in favour of one particle which gained the name K-meson. This naturally yielded a question of the extention of the isotopical internal symmetry group of hadrons in a manner, that in addition to three $SU(2)$ isotopical rotation generators $\mathbf{I}_i$ which do not change parity of states, the extended symmetry group should include three extra generators of transformations mixing states of various parity. Some ways of such an extention had already been studied by A. Pais (1953, 1954) and was known as a **chiral symmetry** of hadrons. An explicit form of the chiral transformation, which mixes states of different parity (γ_5-transformation), was proposed by B. Touschek (1957) and in (Watanabe, 1957) it was shown, that the γ_5-operator is the simplest one among those anticommuting with the parity operator.

The term of "chirality" was invented long ago (see footnote 1 in Chapter 1) and as a matter of fact the chiral symmetry might be related with a well-known fact from geometry that there are only two possible orientations in the n-dimensional Euclidean space: the left and the right ones. The word "chira" means "a hand" in Greak whereby appeared the name of symmetry.

In modern literature the chiral symmetry means that there is an invariance of the left-handed and right-handed spinor fields under isotopical transformations separately. In the case of massless particles this symmetry gives rise to a conserved characteristics, which is named **helicity**. This quantity is determined by the double spin projection on the particle momentum: $(\boldsymbol{\sigma} \cdot \boldsymbol{p})/|\boldsymbol{p}|$, such that when the directions of spin and momentum coincide the helicity is **right** (R) and **left** (L) in the opposite case. Then the group of transformations acting independently onto left and right states, denoted as $SU_L(2) \otimes SU_R(2)$ is called **the chiral group**. In general situation, instead of the $SU(2)$ isotopic group another internal symmetry group G can occur, then chiral transformations would be given by elements of the group $G_L \otimes G_R$.

The intensive implementation of the chiral symmetry ideas has been launched by Feynman and Gell Mann (1958), where it has been issued the idea of the universal $V-A$ structure of charged weak currents. In terms of chiral symmetry all the leptons enter weak currents with their left helicity states and the antiparticles with their right states. Further developments lead to appearence of such fundamental branches in particle physics as Gell Mann's Current Algebra, Phenomenological (or Effective) Lagrangians Method, Chiral σ-Model approach and so on. On the other hand, an extension of the isotopic symmetry in order to include strange particles had led Gell Mann and Neeman to the unitary $SU(3)$ symmetry of hadronic world, to the idea of quarks and thereby to creation of QCD fundamentals. A comprehensive discussion on these topics the reader will find in the monographs (De Alfaro, Fubini, Furlan and Rossetti, 1973; Lee, 1971; Treiman, Jackiw and Gross, 1972; Volkov and Pervushin, 1978) and in the review article (Gasiorovicz and Geffen, 1969) actual up to now. In what follows we provide the reader with initial information, necessery for understanding the body matter.

A.1 Algebraic Aspects of Chiral Symmetry

Let us discuss a little bit further the chiral algebra (1.6) - (1.8) and give some of its features in an explicit realization. Note, that the generators of isorotations $\mathbf{I}_i$ commute with the parity operator $\mathbf{P}$, and the chiral rotations generators $\mathbf{K}_j$ anticommute:

$$[\mathbf{P}, \mathbf{I}_i] = 0; \quad [\mathbf{P}, \mathbf{K}_i]_+ = 0. \tag{A.1}$$

Let us set also the commutation relations between the $SU(2) \otimes SU(2)$ generators in the vector representation and the components of the 4-isovector $(\phi_0, \underline{\phi})$:

$$[\mathbf{I}_i, \phi_j] = i\epsilon_{ijk}\phi_k; \quad [\mathbf{I}_i, \phi_0] = 0; \tag{A.2}$$

$$[\mathbf{K}_i, \phi_j] = -i\delta_{ij}\phi_0; \quad [\mathbf{K}_j, \phi_0] = i\phi_j, \tag{A.3}$$

with $\mathbf{I}_i$ and $\mathbf{K}_i$ being integrals of motion i.e. functionals of fields and canonical momenta, defined in accord with the Noether theorem. Here we treat the bracket [,] as the Poisson bracket multiplied by (−i) in classical theory and as the commutator in quantum theory.

To illustrate we give here a detailed calculation of the first relation from (A.2) in the classical theory of massless isovector fields as defined by the Lagrangian:

$$\mathcal{L}_\phi = \frac{1}{2}\,\partial_\mu\underline{\phi}\partial^\mu\underline{\phi}. \tag{A.4}$$

The canonical momentum

$$\pi = \frac{\partial\mathcal{L}_\phi}{\partial\dot{\underline{\phi}}} = \dot{\underline{\phi}} \tag{A.5}$$

and the isospin current

$$\mathbf{I}_i = \int d^3x\pi_k[\underline{\omega}_i, \underline{\phi}]_k = \int d^3x\pi_k\epsilon_{klm}\delta_{il}\phi_m$$
$$= \epsilon_{kim}\int d^3x\pi_k\cdot\phi_m, \tag{A.6}$$

give the following Poisson bracket

$$[\mathbf{I}_i, \phi_j] = \int d^3x'\Big(\frac{\delta\mathbf{I}_i}{\delta\phi_s}\frac{\delta\phi_j}{\delta\pi_s} - \frac{\delta\mathbf{I}_i}{\delta\pi_s}\frac{\delta\phi_j}{\delta\phi_s}\Big)$$
$$= \int d^3x'\big(\epsilon_{sim}\phi_m\delta_{js}\delta(\boldsymbol{r}-\boldsymbol{r}')\big) = \epsilon_{jim}\phi_m. \tag{A.7}$$

In the same manner one can calculate the brackets for left and right chiral generators defined in (1.7):

$$[\mathbf{L}_j, \phi_0] = -\frac{\mathrm{i}}{2}\phi_j; \quad [\mathbf{L}_i, \phi_j] = \frac{\mathrm{i}}{2}(\delta_{ij}\phi_0 + \epsilon_{ijk}\phi_k); \tag{A.8}$$

$$[\mathbf{R}_j, \phi_0] = \frac{\mathrm{i}}{2}\phi_j; \quad [\mathbf{R}_i, \phi_j] = -\frac{\mathrm{i}}{2}(\delta_{ij}\phi_0 - \epsilon_{ijk}\phi_k). \tag{A.9}$$

The relations (A.8) and (A.9) can be rewritten in a compact form, using the quaternionic representation of the 4-isovector

$$\mathbf{U}(\boldsymbol{x},t) = \mathbb{1}\,\phi_0(\boldsymbol{x},t) + \mathrm{i}\,\underline{\tau}\,\underline{\phi}(\boldsymbol{x},t), \tag{A.10}$$

where $\underline{\tau}$ are isospin Pauli matrices. Then (A.8) and (A.9) assume the form

$$[\mathbf{L}_j, \mathbf{U}] = -\tfrac{1}{2}\tau_j\mathbf{U}; \quad [\mathbf{R}_j, \mathbf{U}] = \tfrac{1}{2}\mathbf{U}\tau_j. \tag{A.11}$$

For isospinor fields ψ it proves to be convenient preliminary to divide them into the left and the right components

$$\psi_L = \tfrac{1}{2}(1-\gamma_5)\psi; \quad \psi_R = \tfrac{1}{2}(1+\gamma_5)\psi, \tag{A.12}$$

transforming under the fundamental representations $(1/2, 0)$ and $(0, 1/2)$ of the $SU(2)$ group. Acting by left and right generators, we have

$$[\mathbf{L}_i, \psi_L] = -\tfrac{1}{2}\tau_i\psi_L; \quad [\mathbf{R}_i, \psi_L] = 0; \tag{A.13}$$

$$[\mathbf{R}_i, \phi_R] = -\tfrac{1}{2}\tau_i\psi_R; \quad [\mathbf{L}_i, \psi_R] = 0. \tag{A.14}$$

When constructing **a nonlinear realization** of the $SU(2)_L \otimes SU(2)_R$ chiral symmetry we are to put the following constraint:

$$\phi_0{}^2 + \underline{\phi}^2 = 1, \tag{A.15}$$

on the 4-isovector components $\phi_\alpha = (\phi_0, \underline{\phi})$, thereby only three components remain independent and according to (Weinberg, 1968) under chiral boosts they are transformed in the following manner:

$$[\mathbf{K}_i, \phi_j] = -\mathrm{i}\,\delta_{ij}f(\underline{\phi}^2) - \mathrm{i}\phi_i\phi_j g(\underline{\phi}^2). \tag{A.16}$$

Here $f(\underline{\phi}^2)$ is an arbitrary regular function, and $g(\underline{\phi}^2)$ is given through $f(\underline{\phi}^2)$ due to Jacobi identity for commutators:

$$g(\underline{\phi}^2) = \frac{1 + 2\, f(\underline{\phi}^2) f'(\underline{\phi}^2)}{f(\underline{\phi}^2) - 2\, \underline{\phi}^2 f'(\underline{\phi}^2)}, \tag{A.17}$$

where $f'(\underline{\phi}^2)$ denotes derivative with respect to $\underline{\phi}^2$. The relations (1.6) and (A.2) remain unaltered, and spinor fields commute with $\mathbf{K}_i$ as

$$[\mathbf{K}_i, \psi] = \frac{1}{2} \frac{\epsilon_{ijk} \tau_j \phi_k}{f(\underline{\phi}^2) - (f^2 + \underline{\phi}^2)^{1/2}} \psi. \tag{A.18}$$

Conserved Noether currents, related charges as well as some other dynamical characteristics one constructs in the well-studied manner, partly demonstrated in Section 1.4.3. It is not difficult to write analogous formulae for a more general unitary chiral group $SU(N)_L \otimes SU(N)_R$ and even for $G_L \otimes G_R$ symmetry. The recipe is simple: instead of generators and structural constants of the $SU(2)_L \otimes SU(2)_R$ group one ought to use the generators and structural constants of the group of interest and appropriately change the number of fundumental fields [for more details see e.g. (Isham, 1969; Barnes and Isham, 1970; Barnes, Dondi and Sarkar, 1972)].

A.2 Geometrical Aspects of Chiral Symmetry

When constructing nonlinear realizations of the internal symmetry groups it is convenient to relate this problem to geometry of curved spaces. This connection goes back to Cartan (1929), who proposed also the moving frame method to construct nonlinear realizations of symmetry groups as representations of groups in Riemannian spaces. From (A.15) we see that the vector $(\phi_0, \boldsymbol{\phi})$ lies on the 3-dimensional sphere $\mathbb{S}^3$ in 4-dimensional Euclidean space. On the other hand a sphere $\mathbb{S}^3$ can be regarded as the manifold of the isogroup $SU(2)$ whose elements are parametrized through the vector $(\phi_0, \boldsymbol{\phi})$ in accord with the formula (A.10). Therefore it is natural to consider as the field variable the field $\mathbf{U}(\boldsymbol{x}, t)$ with its values lying on the $SU(2)$ manifold. This field is called **the principal chiral field**. In general the term principal chiral fields is refered to fields with values on the manifold of a compact group G or on the coset space G/H with H being the stationary subgroup.

Evidently, there exists a close relation between such a topological property of a manifold as **orientability**, which allows one to tell the left coordinate system on it from the right one, the left screws and spirals from those right and so on, and a possibility to define on it a chiral group $G_L \otimes G_R$ action. Anyway, on the orientable $SU(2)$ manifold the chiral group $SU(2)_L \otimes SU(2)_R$ acts via the left and right shifts

$$\delta_L \mathbf{U} = \delta\xi \cdot \mathbf{U},\ \delta_R \mathbf{U} = \mathbf{U} \cdot \delta\eta,\ \delta_L \mathbf{U}^{-1} = -\mathbf{U}^{-1} \cdot \delta\xi,\ \delta_R \mathbf{U}^{-1} = -\delta\eta \cdot \mathbf{U}^{-1}, \tag{A.19}$$

where $\mathbf{U} \in SU(2)$, $\delta\xi$ and $\delta\eta$ are infenitesimal elements of the $SU(2)$ algebra. On the $SU(2)$ group there is the only Riemannian metric and connection, which are invariant with respect to transformations from $SU(2)_L \otimes SU(2)_R$. The explicit form of the metric might be given, for instance, by the left-invariant form on $SU(2)$:

$$\mathbf{l}(\mathbf{U}, d\mathbf{U}) = \mathbf{U}^{-1} d\mathbf{U}, \tag{A.20}$$

where $d\mathbf{U}$ is a matrix constructed of the differential of the matrix elements of $\mathbf{U}$. The left invariance of the form (A.20) $\delta_L \mathbf{l} = 0$ follows from (A.19) and under the right shifts we have

$$\delta_R \mathbf{l} = -\delta\eta \cdot \mathbf{U}^{-1} \cdot d\mathbf{U} + \mathbf{U}^{-1} \cdot d\mathbf{U}\delta\eta = [\mathbf{l}, \delta\eta]. \tag{A.21}$$

A similar result is obtained via the right-invariant form on $SU(2)$:

$$\mathbf{r}(\mathbf{U}, d\mathbf{U}) = d\mathbf{U}\mathbf{U}^{-1}; \quad \delta_L \mathbf{r} = [\delta\xi, \mathbf{r}]; \quad \delta_R \mathbf{r} = 0.$$

By use of the parametrization (A.10), taking into account the constraint (A.15) and the well-known relation for the Pauli matrices

$$\tau_a \tau_b = \mathrm{i}\epsilon_{abc}\tau_c + \delta_{ab}$$

one obtains the following expression for the form (A.20)

$$\begin{aligned} \mathbf{l}(\mathbf{U}, d\mathbf{U}) &= \mathbf{U}^{-1} d\mathbf{U} = (\phi_0 - \mathrm{i}\tau_a\phi^a)(d\phi_0 + \mathrm{i}\tau_b d\phi^b) \\ &= \mathrm{i}\tau^a \left[\left(\delta_{ab} + \frac{\phi_a \phi_i}{\phi_0{}^2} \right) \phi_0 + \epsilon_{abi}\phi^b \right] d\phi^i \\ &= \mathrm{i}\tau^a w^a, \end{aligned} \tag{A.22}$$

where $a, i = 1, 2, 3;$ $w^a = \xi_i^a d\phi^i$ are the basis left-invariant 1-forms, which determine a left-invariant reference frame on the $SU(2)$ manifold through the known theorem on existence of n linearly independent left-(right-)invariant 1-forms on a Lie group of rank n [see (Zhelobenko, 1970) p. 41]. Furthermore, we are given with the quadratic Cartan-Killing form of differentials

$$ds^2 = g_{ij}(\phi) d\phi^i d\phi^j = \xi_i^a \xi_j^a d\phi^i d\phi^j = -\frac{1}{2} \operatorname{Tr} \mathbf{l}^2, \tag{A.23}$$

which is invariant against the left as well as the right shifts (A.19). Whereby the explicit form for the metric tensor of the $SU(2)$ manifold follows as

$$g_{ij}(\phi) = \xi_i^a \xi_j^a = \delta_{ij} + \frac{\phi_i \phi_j}{\phi_0{}^2} = \delta_{ij} + \frac{\phi_i \phi_j}{1 - \phi_i{}^2}. \tag{A.24}$$

Recall, that equations (A.22) and (A.24) have been used in the body text.

In similar way we define (through the form $\mathbf{l}(\mathbf{U}, d\mathbf{U})$) the Riemannian connection which enters the covariant derivative of vector fields on the $SU(2)$:

$$D_A \mathbf{l}(B) = A \cdot \mathbf{l}(B) + \tfrac{1}{2} [\mathbf{l}(A), \mathbf{l}(B)], \tag{A.25}$$

with A being a vector field which determines the direction of this derivative, and $\mathbf{l}(B)$ is the form (A.20) as calculated for the field B. For detailed calculations we refer the reader to (Slavnov and Faddeev, 1971; Weinberg, 1968; Meetz, 1969). It should be also underlined, that the action of the chiral $SU(2)_L \times SU(2)_R$ group by means of the left and right shifts (A.19) on the manifold $\mathbb{S}^3 \simeq SU(2) \otimes SU(2)/SU(2)$ is related to the fact that there are defined two kinds of the absolute parallelism on "remarkable" spheres $\mathbb{S}^0, \mathbb{S}^1, \mathbb{S}^3$ and $\mathbb{S}^7$ [Adams' lemma, see e.g.(Schwartz, 1968)], which can be parametrized respectively by real, complex, quaternion and octonion numbers.

We now possess a minimal body of geometrical knowledge necessary to construct chiral invariant Lagrangians. Instead of the forms $\mathbf{l}$ and $\mathbf{r}$ we would rather use their counterparts, i.e. vector fields

$$\mathbf{L}_\mu = \mathbf{U}^{-1}(x)\partial_\mu \mathbf{U}(x); \quad \mathbf{R}_\mu = \mathbf{U}(x)\partial_\mu \mathbf{U}^{-1}(x), \tag{A.26}$$

which at any given μ are the elements of the $SU(2)$ Lie algebra. The left-invariant vector field $\mathbf{L}_\mu$ is transformed under the left shifts $\mathbf{U}(x) \to V \cdot \mathbf{U}(x)$ and under the right shifts $\mathbf{U}(x) \to \mathbf{U}(x) \cdot V$, respectively

$$\begin{aligned} \mathbf{L}_\mu &\to (VU)^{-1}\partial_\mu(VU) = U^{-1}V^{-1}V\partial_\mu U = \mathbf{L}_\mu; \\ \mathbf{L}_\mu &\to (UV)^{-1}\partial_\mu(UV) = V^{-1}U^{-1}\partial_\mu U \cdot V = V^{-1} \cdot \mathbf{L}_\mu V, \end{aligned} \tag{A.27}$$

with $V \in SU(2)$ being a global isotopical turn ($V \cdot V^{-1} = \mathbb{1}, \partial_\mu V = 0$). In order to reconstruct a chiral field $\mathbf{U}(x)$ (as a group element) via a vector field $\mathbf{L}_\mu$ (as an element of algebra) it is necessary and sufficient that the condition

$$\partial_\mu \mathbf{L}_\nu - \partial_\nu \mathbf{L}_\mu + [\mathbf{L}_\mu, \mathbf{L}_\nu] = 0 \tag{A.28}$$

holds, which is just the Maurer-Cartan structure equation or in another respect the zero curvature condition. The field $\mathbf{L}_\mu$ in the parametrization (A.10) is given as the standard expansion via the basis τ_a-matrices

$$\mathbf{L}_\mu = i\tau_a \mathbf{L}^a_\mu, \quad \mathbf{L}^a_\mu = \xi^a_i \frac{\partial \phi^i}{\partial x^\mu}; \quad a = 1, 2, 3. \tag{A.29}$$

In addition to the form of simplest chiral Lagrangian given in Section 2.4.1 [see Eq. (2.40)], by use of the metric $g_{ij}(\phi)$ from (A.24) it is possible to obtain one more expression

$$\mathcal{L}_{(2)} = \frac{1}{2}\partial_\mu \phi^i \partial^\mu \phi^i + \frac{1}{2}\frac{\phi_i \phi_j \partial_\mu \phi^i \partial^\mu \phi^i}{(1 - {\phi_i}^2)}. \tag{A.30}$$

In its turn the Weinberg chiral Lagrangian (Weinberg, 1967)

$$\mathcal{L}_{(2)} = \frac{1}{2}\frac{\partial_\mu \pi^i \partial^\mu \pi^i}{(1 - \pi^2)}; \quad \pi^2 = \pi_i \pi^i \tag{A.31}$$

one obtains from (A.30), when instead of the ϕ_i it has been used components of stereographic projection of $\mathbb{S}^3$ onto Euclidean space $\mathbb{R}^3$ [see Eq. (2.4) in Chapter 2].

Construction of chiral invariants of higher order in terms of $\mathbf{L}_\mu$ turns out to be not so easy problem and there were suggested various ways for its solution (Coleman, Wess and Zumino, 1969; Callan, Coleman, Wess and Zumino, 1969; Barnes, Dondi and Sarkar, 1972; Barnes, Ketley, Nicole and O'Donnel, 1977; Volkov and Pervushin, 1978; Weinberg, 1968). The simplest algorithm (workable for sure in the general case) uses the Gürsey matrices (Chang and Gürsey, 1967)

$$\hat{U} = \left(\frac{1+\gamma_5}{2}\right) U + \left(\frac{1-\gamma_5}{2}\right) U^{-1}, \tag{A.32}$$

with the parametrization for U as in Section 1.4. Then all the $SU(2) \otimes SU(2)$ invariant Lagrangians are given by combining the bloks $\gamma_\mu \hat{U}$ and their derivatives, while $SU(2)$- and Lorentz-invariance are achieved through taking traces of resulted combinations and summing over free indices $\mu, \nu, \ldots$.

Appendix B

A Concise Account of Algebraic Topology

First we need to clarify for what reason algebraic topology methods began to penetrate into particle theory. In 1950s there were obtained experimental data, which clearly acknowledged the extended sizes and the internal structure of hadrons, and to this end made it unsound to regard these particles as point-like objects. Therefore an appropriate mathematical body got necessary to describe extended objects in field theory. In the meantime, it became evident that the standard approaches, such as based on the perturbation theory, and proved to be good-working in linear field theories are almost useless in this respect [see, for example, discussion in (Filkenstein and Misner, 1959; Felsager, 1981; Schwarz, 1989)]. There were found to be unworkable the accustomed for physicist methods of analytical and differential geometries of euclidean spaces. Roughly speaking, this 'old machinery' still provided one with some tools for local studies of nonlinear fields, but shows to be invalid when one handles with global problems. One cannot use any more a global parametrization of a field, which is the keystone of any linear field theory. Instead, only a local parametrization can be introduced for nonlinear fields (by a one-to-one correspondence with a domain of Euclidean space $\mathbb{R}^n$), and one has to deal with several coordinate systems to get an appropriate description of structures in nonlinear fields. Thus one comes to the necessity to use the elements of algebraic topology, and we find it appropriate to expose here an elementary introduction to this subject.

B.1 Smooth Manifolds

If to construct a field theory in a way, that field functions take their values not in $\mathbb{R}^n$, but instead on a field manifold Φ, then it turns out that this scheme includes all linear theories of the Maxwell electrodynamics type, as well as essentially nonlinear ones, appropriate for description of extended objects.

Recall, that **a manifold** is a topological space which has the same structure as $\mathbb{R}^n$ in the neighbourhood of each point. More precisely a manifold is defined by setting of a finite or numerable set of overlapping **charts** (U_i, φ_i), where by U_i it is denoted the neighbourhood of a point $\phi \in \Phi$, and φ_i stands for a homeomorphism of U_i on the region $V_i \in \mathbb{R}^n$ (see Fig. B.1).

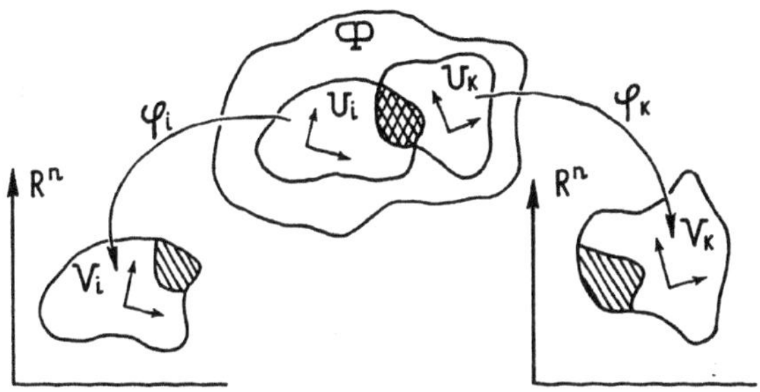

Fig. B.1. A fragment of an atlas on Φ as made of two neighbourhoods U_i and U_k of a point $\phi \in \Phi$, which overlap in the shaded region. Coordinate homeomorphisms φ_i and φ_k respect the topology of Φ, while their inverse maps provide U_i and U_k with coordinate systems (indicated by arrows) as duplicated from open Euclidean subsets $V_i, V_k \subset \mathbb{R}^n$

The family of these charts $\cup_i(U_i, \varphi_i)$, which covers the whole Φ, is called **the atlas**[1]. The functions φ_i^{-1} set a local parametrization (a coordinate system) in U_i, and in a overlap region $U_i \cap U_j$ there should be defined transition functions (the change of coordinates):

$$\varphi_{ik} = \varphi_i \cdot \varphi_k^{-1}; \quad \varphi_{ik} \cdot \varphi_k = \varphi_i \cdot \varphi_k^{-1} \cdot \varphi_k = \varphi_i. \tag{B.1}$$

Through the setting of an atlas on the manifold Φ one defines **a smooth structure**, which together with Φ itself form a concept of **a smooth manifold**. The smoothness of such a manifold is determined by the smoothness of the transition functions (B.1), i.e. if $\varphi_{ik} \in \mathbf{C}^m$, where the latter is the class of m times differentiable functions, then one says that such a manifold is of class $\mathbf{C}^m$. Different smooth structures are possible on manifolds, and among them are Riemannian geometry, conformal geometry, geometry of absolute parallelism, complex and symplectic geometries and so on. By use of local coordinates one can apply conventional analysis in studying smooth manifolds, i.e. to define functions, to construct tangent vectors, differential operators and more.

In this context the spaces of real $\mathbb{R}^n$ and complex $\mathbb{C}^n$ numbers are *trivial manifolds* since their atlases might consist from an only chart (the global Cartesian system of coordinates). A simple nontrivial example of manifolds is a sphere $\mathbf{S}^n$, more complex ones are manifolds of groups $SO(n)$ and $SU(n)$, the real and complex projective spaces $\mathbb{R}\mathbf{P}^n$ and $\mathbb{C}\mathbf{P}^n$. These manifolds and their products are usually chosen to be field manifolds, for example, of Quantum Chromodynamics, of the Weinberg-Salam model, of nonlinear σ-model etc. There are no global coordinate system without singularities for such manifolds and their atlases consist of at least two charts.

Here we give a simple example of the atlas of the sphere $\mathbf{S}^n$ embeded in $\mathbb{R}^{n+1}$ and defined by the equation

[1] This terminology has been introduced by K.F. Gauss, when he studied the problem of cartography (map-making) of the Earth surface.

$$\sum_{i=1}^{n+1} x_i^2 = 1. \tag{B.2}$$

A coordinate system on $\mathbf{S}^n$ can be introduced if we put out one of its points, then one-to-one correspondence between the rest of $\mathbf{S}^n$ and $\mathbb{R}^n$ is given by the stereographic projection. For example, let us put out the north pole N with the coordinates $x_N = \{0, \ldots, 1\}$, then each point x of the $\mathbf{S}^n \backslash \{N\}$ is related to the point z_N with the coordinates $z_N = \{z_N^1, \ldots, z_N^n\}$ of the equatorial plane ($x_{n+1} = 0$) by means of the ray emerging from the pole N and passing through the point x and z_N (see Fig. B.2).

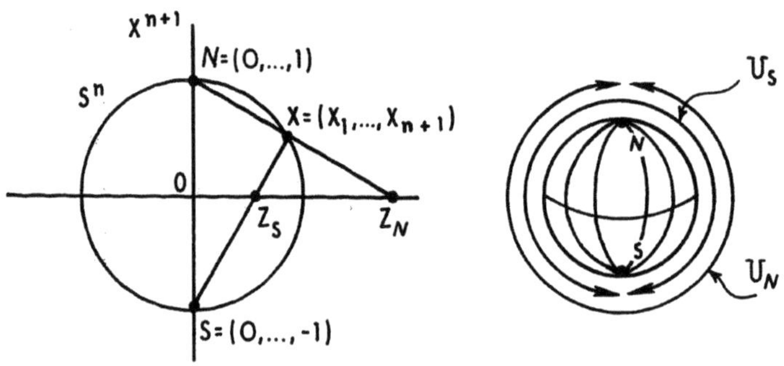

Fig. B.2. The stereographic projection of a sphere $\mathbf{S}^n$ onto the Euclidean space $\mathbb{R}^n$ (the left hand side). The sphere with the cut out Nothern pole $N = (0, \ldots, 1)$ proves to be in a one-to-one correspondence with $\mathbb{R}^n$. In the right hand side picture there is schematically displayed the atlas of two charts, covering the sphere $\mathbf{S}^n$

The coordinate homeomorphism is then given by

$$\varphi_N : \mathbf{S}^n \setminus \{N\} \to \mathbb{R}^n : z_N^i = \frac{x_i}{1 - x_{n+1}}; \quad i = \overline{1, n}, \tag{B.3a}$$

i.e. we have the chart (U_N, φ_N) with local coordinates $\{z_N^1, \ldots, z_N^n\}$. Treating in the same way the south pole we have

$$\varphi_S : \mathbf{S}^n \backslash \{S\} \to \mathbb{R}^n : z_S^i = \frac{x_i}{1 + x_{n+1}}; \quad i = \overline{1, n}, \tag{B.3b}$$

defining the coordinates in the chart (U_S, φ_S). Then the atlas $\{U_N, U_S\}$ covers the whole sphere and in the overlapping region of the charts $U_N \cap U_S$ the transition functions are as follows:

$$\varphi_N \cdot \varphi_S^{-1} : z_N^i = \frac{x_i(z_S)}{1 - x_{n+1}(z_S)} = z_S^i |z_S|^{-2}; \tag{B.4a}$$

$$\varphi_S \cdot \varphi_N^{-1} : z_S^i = \frac{x_i(z_N)}{1 + x_{n+1}(z_N)} = z_N^i |z_N|^{-2}, \tag{B.4b}$$

where for φ_S^{-1}:

$$x_i = \frac{2\, z_S^i}{z_S^2 + 1}; \quad x_{n+1} = \frac{z_S^2 - 1}{z_S^2 + 1}; \quad z_S^2 = \sum_{i=1}^{n} (z_S^i)^2;$$

and φ_N^{-1} one obtains from φ_S^{-1} by the exchange $(S \to N)$.

Jacobians of these transition functions are definitely positive, which implies **the orientability** of $\mathbf{S}^n$. In general, a manifold Φ is orientable if the Jacobian of transition functions

$$J_{\alpha\beta} = \det \left(\frac{\partial x_\alpha^i}{\partial x_\beta^k} \right) > 0 \tag{B.5}$$

in all overlap regions of charts $(U_\alpha, \varphi_\alpha)$, (U_β, φ_β). When $J < 0$, the corresponding overlapping charts are endowed with the coordinate system of the opposite orientation (e.g. the left and right triads for $n = 3$). Orientability of a manifold is a topological invariant and remains unaltered under all possible continuous transformations[2] .

The notion of **connectedness** is another important invariant of a manifold. We say a topological space is **connected**, if it cannot be presented as a union of two (or more) closed non-overlapping and non-empty sets. For manifolds it looks like more applicable the notion of **a pathwise- or arcwise-connectedness**, when any two points on manifold can be linked by a path.

Definition B.1. *The* **path** *is a continuous mapping of the segment* $[0, 1] \in \mathbb{R}^1$ *into the manifold* Φ.

Note, that these two notions of connectedness are identical for the manifolds of physical interest. If a manifold Φ is orientable and connected, then it can be oriented in two ways [see, (Dubrovin, Novikov and Fomenko; 1984) Section 1], i.e. there are two orientations.

In the main text we have been frequently exploiting properties of the Lie groups $SU(2)$, $SO(3)$ and $SU(3)$ and therefore it is of use to give here some details of the structure of these groups as manifolds. As far as the $SO(n)$ and $SU(n)$ groups are the matrix groups, they are topological spaces with defined on continuous operations, such as multiplication and taking the inverse element. In the corresponding parameter space $\mathbb{R}^{n^2}$ of $n \times n$ matrix transformations one can regard the group $SO(n)$ as a surface determined by the equation

$$A \cdot A^T = \mathbb{1}, \quad \det A = 1. \tag{B.6}$$

Hence, due to equation (B.6), only $n(n-1)/2$ parameters out of the whole number n^2, are independet ones. In this sence the $SO(n)$ group is a $n(n-1)/2$-dimensional manifold with each element $A \in SO(n)$ determined by a set of

[2] Here we leave aside such exotic manifolds as, for instance, the Alexander 'wild sphere'.

$m = n(n-1)/2$ parameters $(a^1, \ldots, a^m)$. In the same way the group $SU(n)$ is a surface in $\mathbb{C}^{n^2} \simeq \mathbb{R}^{2n^2}$ given by the equation

$$U \cdot U^\dagger = \mathbb{1}; \quad \det U = 1 \tag{B.7}$$

so that $\dim SU(n) = n^2 - 1$.

Let us now scetch a general scheme to create an atlas on these group manifolds. Consider infinitesimal operators Γ:

$$\Gamma_i = \left(\frac{\partial A}{\partial a^i}\right)_{|a=0}, \quad i = 1, 2, \ldots, \dim G, \tag{B.8}$$

which are the generators of a group G. In the vicinity of the unit element $\mathbb{1} \in G$ for small a_i we have in the linear approximation

$$A = \mathbb{1} + a^i \Gamma_i. \tag{B.9}$$

Then the matrix $a = a^i \Gamma_i$ one can regard as an element of a linear space with the basis $\{\Gamma_i\}$. More precisely, this is the tangent space at the point $\mathbb{1}$, endowed with a Lie algebra structure (for details see below). In order to construct an atlas on G we use the matrix exponent

$$\exp(a) = \exp(a^i \Gamma_i) = \sum_{n=o}^{\infty} \frac{a^n}{n} \tag{B.10}$$

which induces the coordinate system (local parametrization) in the vicinity of $\mathbb{1}$, thereby creating at this point the chart. Then via the transformations $g \in G$ one can attribute this chart to an arbitrary point of G thereby constructing the atlas, which would be definitely consistent with the group structure [for details see (Lyakhovskii and Bolotov, 1983), Chapter 3, Section 3]. This is only a general scheme showing how to construct the atlas on Lie groups, and its concrete realization is made by constructing one-dimensional subgroup for each generator. It is possible since by Lie's theorem any element of this group is either an element of its one-parametric subgroup or a product of such elements.

B.2 Tangent Spaces, Vector Fields and Lie Algebras

We feel now as qualified in determining "coordinate systems" on manifolds. The next step is to define functions and calculus machinery, such as gradients, directional derivatives and so on. It is clear that such notions should be determined locally, at first, and then extended by means of the atlas all over the manifold. To define *a function on the manifold* Φ means to suggest a rule by which to every point $\phi \in \Phi$ one relates a number. Let $(\phi^1, \ldots, \phi^n)$ be a local parametrization of a point $\phi \in U$ defined via a coordinate homeomorphism φ of U into a region $V \in \mathbb{R}^n$. Then the function $F(\phi)$ transforms into the function on $\mathbb{R}^n$: $F(x^1, \ldots, x^n)$, and if the latter is differentiable on $\mathbb{R}^n$, then $F(\phi)$ is a differentiable function on Φ.

A tangent vector at a point $\phi \in \Phi$ is the vector with components $\{\zeta^a\}_i$ in the local coordinate system $\{x^a\}_i$ so that i labels the chart. In the overlap region of two maps $U_i \cap U_k$ we have to impose the following count over rule

$$\zeta_i^a = \sum_{b=1}^{n} \left(\frac{\partial x_i^a}{\partial x_k^b}\right) \zeta_k^b. \tag{B.11}$$

One can easily note that we have got just a slight extention of the conventional notion of a vector in Euclidean space. The similar extention of directional derivative gives rise to the concept of **the vector field on a manifold,** which is defined as a set of directional derivatives $\{\zeta^a(x) \cdot \partial/\partial x^a\}$ associated to each point of Φ.

A tangent space $T_\phi(\Phi)$ is a n-dimensional vector space constituted with tangent vectors at a point $\phi \in \Phi$ to all smooth curves passing through this point. There is another equivalent definition of the tangent vector as the differentiation of smooth functions at a point $\phi \in \Phi$. In particular, partial differentiation $\partial/\partial x^a$ can be regarded as a tangent vector with components $(0, \ldots, 1, \ldots, 0)$ in the local coordinate system $(x^1, \ldots, x^n)$ in which unity is at the place labeled by a (Mitschenko and Fomenko, 1980; Schutz, 1982). Hence the tangent vectors $\{\partial/\partial x^a\}$ constitute the basis of $T_\phi(\Phi)$, so that any tangent vector can be presented as

$$\zeta = \sum_{a=1}^{n} \zeta^a \cdot \frac{\partial}{\partial x^a}.$$

The whole set of tangent vectors to the manifold Φ is written as a union $\bigcup_{\phi\in\Phi} T_\phi(\Phi)$, which is **the tangent fibre bundle** with $T_\phi(\Phi)$ as **the fibre**. Probably the most familiar pattern of the tangent space is the space of generalized velocities $\{\dot{q}^1, \ldots, \dot{q}^n\}$ in classical mechanics, where smooth manifolds appear as configuration spaces of mechanical systems, and generalized coordinates $\{q^1, \ldots, q^n\}$ define local parametrization of such manifolds.

Going back to vector fields on the manifold Φ, let us note that the set of smooth fields on Φ one can treat as smooth mappings $X : \Phi \to T(\Phi)$. In this respect one can regard them as elements of a vector space over $\mathbb{R}^n$, as far as the operations of summation, that of multiplication by a scalar $\alpha \in \mathbb{R}^1$ and by a smooth function f are well-defined

$$(X+Y)_\phi = X_\phi + Y_\phi; \quad (\alpha X)_\phi = \alpha X_\phi; \quad (fX)_\phi = f(\phi) X_\phi$$

locally at each point $\phi \in \Phi$. Moreover, these fields on Φ obey the structure of an infinite dimensional Lie algebra with respect to the commutator:

$$[X, Y] = XY - YX, \tag{B.12}$$

such that

$$[X, X] = 0; \quad [X, [Y, Z]] + [Y, [Z, X]] + [Z, [X, Y]] = 0. \tag{B.13}$$

In fact, representing the fields in the form

$$X = \zeta^a \frac{\partial}{\partial x^a}, \quad Y = \eta^b \frac{\partial}{\partial x^b}, \tag{B.14}$$

one convinces himself, that the commutator of fields (B.14)

$$[X, Y] = \left(\zeta^a \frac{\partial \eta^b}{\partial x^a} - \eta^a \frac{\partial \zeta^b}{\partial x^a} \right) \frac{\partial}{\partial x^b} \tag{B.15}$$

is again a vector field on Φ.

We underline here, that by use of algebraic structures the study of complicated geometrical objects can be significantly simplified. This becomes most essential in realizing the main idea of algebraic topology, i.e. to set a correspondence between the space of mappings (a powerful tull to describe rich spectrum of geometrical structures) and infinite series of homology groups (with their workable and relatively simple algebraic structure). Fruitfulness of the idea of "algebraization" can be demonstrated by an example of treating the manifolds of Lie groups by reducing them to much more simple algebraic structures, i.e. Lie algebras. Let us choose in this respect the intensively exploited in the main body example of the $SU(3)$ group.

The $SU(3)$ algebra is a linear subspace of 3×3 matrices with the commutator as defined by (B.12). Combining (B.7) and (B.9) we see that the $SU(3)$ algebra elements are traceless antihermitian matrices ($A = -A^\dagger$, $\operatorname{Tr} A = 0$). Basis elements are the $SU(3)$ group generators Γ_α, ($\alpha = \overline{1,8}$) given by (B.8) and obeying the equations

$$[\Gamma_\alpha, \Gamma_\beta] = \mathrm{i} f_{\alpha\beta\gamma} \Gamma_\gamma, \tag{B.16}$$

with $f_{\alpha\beta\gamma}$ being **the structural constants** (real-valued totally antisymmetric symbols). The representation $\underline{\mathbf{3}}$ with the generators $\Gamma_\alpha = \mathrm{i}\lambda_\alpha/2$ is the fundamental one. Here λ_α are the hermitian traceless Gell Mann matrices. By supplementing this set with the matrix $\lambda_0 = \sqrt{2/3}\,\mathbb{1}$ we come up to the basis of the Lie algebra for the $U(3)$ group: $\{\lambda_i\}, i = 0, 1, \ldots, 8$ with the following determining relations:

$$\begin{aligned} [\lambda_i, \lambda_j] = 2\mathrm{i}\, f_{ijk}\lambda_k; \quad [\lambda_i, \lambda_j]_+ = \tfrac{4}{3}\delta_{ij} + 2\, d_{ijk}\lambda_k; \\ \operatorname{Tr}(\lambda_i, \lambda_k) = 2\,\delta_{ij}; \quad \operatorname{Tr}(\lambda_i [\lambda_j, \lambda_k]) = 4\mathrm{i}\, f_{ijk}; \\ \operatorname{Tr}(\lambda_i [\lambda_j, \lambda_k]) = 4\mathrm{i}\, d_{ijk}; \quad \operatorname{Tr}(\lambda_\alpha) = 0. \end{aligned} \tag{B.17}$$

Here as before $[\ ,\ ]_+$ means anticommutator, d_{ijk} is a symmetric (with respect to any pair of indices) symbol [see e.g. (Huang, 1982), Table 2.3].

Now on the algebraic level we can easily solve the problem of the embedding of the $SU(2)$ group into the $SU(3)$ one, we refered to in Section 4.4. In fact, out of eight Gell Mann λ-matrices one can combine the three triads of generators:

$$\begin{aligned} \{\lambda_1, \lambda_2, \lambda_3\}; \quad \left\{\lambda_4, \lambda_5, \tfrac{1}{2}\left(\lambda_3 + \sqrt{3}\lambda_8\right)\right\}; \\ \left\{\lambda_6, \lambda_7, \tfrac{1}{2}\left(\sqrt{3}\lambda_8 - \lambda_3\right)\right\}, \end{aligned} \tag{B.18}$$

which are subject to the $SU(2)$ algebra commutation relations:

$$[\tau_a, \tau_b] = \mathrm{i}\epsilon_{abc}\tau_c; \quad a = 1, 2, 3. \tag{B.19}$$

Then the parametrization of a $SU(3)$ group element arises in a natural way through generalized Euler's angles and $SU(2)$ decomposition $U = V \cdot W \cdot V'$ with V, V' and W given by relations:

$$\begin{aligned}
V &= \exp\left[-\mathrm{i}\frac{\alpha}{2}\lambda_3\right] \exp\left[-\mathrm{i}\frac{\beta}{2}\lambda_2\right] \exp\left[-\mathrm{i}\frac{\gamma}{2}\lambda_3\right]; \\
W &= \exp\left[-\mathrm{i}\delta\lambda_4\right] \exp\left[-\mathrm{i}\frac{\nu}{\sqrt{3}}\lambda_8\right]; \\
V' &= V\left(\alpha, \beta, \gamma \rightarrow \alpha', \beta', \gamma'\right);
\end{aligned} \tag{B.20}$$

where

$$0 \le \alpha \le 2\pi;\ 0 \le \beta \le \pi;\ 0 \le \gamma \le 2\pi;\ 0 \le \delta \le \frac{\pi}{2};\ 0 \le \nu \le 3\pi.$$

When studying the Skyrme model we made use of the left chiral currents $\mathbf{L}_\mu$ which are, by definition, **left-invariant vector fields** on $G = SU(2)$. For the set of these fields $\mathcal{Z} = \{\mathbf{L}_\mu\}$ the following statements are valid [see e.g. (Warner, 1983)]:

1. $\mathcal{Z}$ is a vector space over $\mathbb{R}$. The mapping $\gamma : \mathcal{Z} \rightarrow G_I$ $[\gamma(\mathbf{L}_\mu) = \mathbf{L}_\mu(\mathbb{1})]$, with G_I being the connected component of the unit element $\mathbb{1} \in G$, gives the isomorphism between $\mathcal{Z}$ and the tangent space $T_I(G)$ associated to the unit element of G:

   $$\dim \mathcal{Z} = \dim G_I = \dim G.$$

2. Any invariant vector field is smooth.
3. Lie's brackets $[\mathbf{L}_\mu, \mathbf{L}_\nu] \in \mathcal{Z}$
4. $\mathcal{Z}$ is a Lie algebra with respect to Lie's brackets.

By use of the example of the relation between Lie algebras and groups let us explain more closely the concept of **the covering space** mentioned in Section 10.1. Each Lie algebra is known to generate several Lie groups, which might be different in their topological properties. The Lie algebra (B.19), in particular, generates both the $SU(2)$ group and the $SO(3)$ group, with dependence on the particular choice of the generators Γ_i. Both groups are compact, though the $SU(2)$ group is a simply connected one, whereas $SO(3)$ is a doubly-connected. In general case, among all these groups as generated by the same Lie algebra, there is only one simply connected group, which is called **covering** with respect to all the rest. The fact that the $SU(2)$ group is a simply connected one is readily seen from the equivalence $SU(2) \simeq \mathbf{S}^3$ and the fact that $\mathbf{S}^3$-manifold is a simply connected one. The latter follows from the definition (B.7): $U \in SU(2) : U^\dagger \cdot U = \mathbb{1}$, then

$$U = \begin{pmatrix} a & b \\ -b^* & a^* \end{pmatrix} : \det U = |a|^2 + |b|^2 = \sum_{i=1}^{4} x_i^2 = 1. \tag{B.21}$$

This example is a particular one of two-sheet **universal covering** given by the mapping $f : \mathbb{S}^n \to \mathbb{R}\mathbf{P}^n$, $n > 2$. This mapping identifies any two antipodal points of the sphere $\mathbb{S}^n$, producing the real projective space $\mathbb{R}\mathbf{P}^n$. In other words, the factorization $\mathbb{R}\mathbf{P}^n = \mathbb{S}^n/\mathbb{Z}_2$ takes place with $\mathbb{Z}_2 = \{\mathbb{1}, -\mathbb{1}\}$ being the center. Derivation of this type of relations proves to be essential for calculation of homotopy groups, as we shall see in Section B.5, by means of exact sequences method.

B.3 Differential Forms

In trying to extend calculus on manifolds to handle with tensor fields of arbitrary rank via convential fashion one encounters with great difficulties. The reason for this (among others) are:

1. In a well-defined calculus we need to compare tensor fields in neighboring points of the manifold, i.e. to compare elements of different tangent spaces. That is sometimes out of any sence, from the geometrical point of view.
2. It turned out to be impossible to determine in a consistent manner on manifolds some standard algebraic operations: addition, multiplication by a smooth function, tensor product, transposition of indices, convolution of indices, their symmetrization and alternation. A part of these operations do not preserve tensor character of fields under exchange of coordinates in a overlap region of two charts (for details consult with (Felsager, 1981, Chapter 7; Mitschenko and Fomenko, 1980, Chapter 5).

A solution to the problem was found through consideration of a special class of tensor fields, the so-called antisymmetric covariant tensor fields. The latter might be defined on a manifold Φ, which admits existence of the Riemannian metric, or, in a more general case, that of a nondegenerate tensor field of the type (0.2) or (2.0). We recall that the **Riemannian metric** on a smooth manifold Φ is given as a family of positive definite inner (scalar) products for each tangent space $T_\phi(\Phi)$:

$$(\boldsymbol{\xi}, \boldsymbol{\eta}) = \sum_{i,j} g_{ij}(\phi)\xi^i\eta^j, \tag{B.22}$$

where $g_{ij}(\phi)$ is a smooth function of local coordinates in any chart on Φ.

In the overlap regions the matrices $g_{ij}(\phi)$ are transformed according to the rule:

$$g_{ij} = \sum_{\alpha,\beta=1}^{n} g_{\alpha\beta}(\phi)\frac{\partial x^\alpha}{\partial y^i}\frac{\partial x^\beta}{\partial y^j}, \tag{B.23}$$

where $\{x^\alpha\}$ and $\{y^i\}$ are local coordinates of the overlapping charts. For the group manifolds $SO(n)$ and $SU(n)$ in particular we have respectively the scalar products

$$\langle \boldsymbol{A}, \boldsymbol{B} \rangle = \mathrm{Tr}\,(A \cdot B^T); \quad \langle U, V \rangle = \mathrm{Tr}\,(U \cdot V^\dagger). \tag{B.24}$$

Now a possibility arises to attribute the tangent space $T_\phi(\Phi)$, along with its dual vector space $T^*_\phi(\Phi)$, to each point $\phi \in \Phi$. The latter space $T^*_\phi(\Phi)$ is a space of covectors (1-forms) defined as linear functionals on vector fields. The basis $\{e^i\}$ of $T^*_\phi(\Phi)$ has to be dual to that of $T_\phi(\Phi)$, denoted by $\{E_i\}$, in other words they are to satisfy

$$\langle E_i, e^j \rangle = \delta^j_i, \tag{B.25}$$

where δ^j_i is the Kronecker's symbol. For the choice of natural basis $\{\partial/\partial x^i\}$ in $T_\phi(\Phi)$, the dual to it would be the basis $\{dx^i\}$ by virtue of (B.25). This allows one to represent any covector field as $\omega = \omega_i dx^i$.

The concept of the cotangent space is also very familiar for physicists, since along with the velocity space in classical mechanics it proves to be useful to consider the momentum space of covector fields $p_i dq^i$. This idea has been further developed and realized, for example, by consideration of the momentum representation in quantum field theory etc.

Differential forms are in general case just algebraic functions of differentials of local coordinates. But in what follows we will be interested in a special class of these forms, so called **external forms** of the type

$$\omega = \omega_{i_1,\dots i_p}(\phi^1, \dots \phi^n) d\phi^{i_1} \wedge \dots \wedge d\phi^{i_p}, \quad i_1 < i_2 < \dots < i_p, \tag{B.26}$$

where $d\phi^{i_1} \wedge \dots \wedge d\phi^{i_p}$ is the **exterior** or **wedge product** of differentials of local coordinates at a point $\phi \in \Phi$. This kind of product was introduced by Grassmann as an operation in the exterior algebra in particular aimed to generate multi-dimensional spaces out of one-dimensional ones. This operation is distributive and associative, and for any two forms ω_1 and ω_2 it holds to be true the following relation:

$$\omega_1 \wedge \omega_2 = (\ \ 1)^{\deg\omega_1 \deg\omega_2}\, \omega_2 \wedge \omega_1. \tag{B.27}$$

Here $\deg\omega$ is a rank of the form ω defined by a number of factors in the wedge product, in particular, the form (B.26) is a p-form. Coefficients $\omega_{i_1,\dots i_p}$ are supposed to be totally antisymmetric smooth functions of ϕ, and if it is the case they are in a one-to-one correspondence with antisymmetric tensor fields. The forms of an equal rank are usually combined into a set $\Lambda^p = \Lambda^p_\phi(\Phi)$ which is a vector space, and if $\dim\Phi = n$ we have

$$\dim \Lambda^p = \frac{n!}{p!(n-p)!}. \tag{B.28}$$

We denote by $\mathbf{C}^\infty(\Lambda^p)$ the space of smooth p-forms of the type (B.26), then it is straightforward to get the following list of such spaces, containing their dimension and a representative (Eguchi, Gilkey and Hanson, 1980):

$$\begin{aligned}
\mathbf{C}^\infty(\Lambda^0) &= \{\omega(x)\};\ \dim = 1;\\
\mathbf{C}^\infty(\Lambda^1) &= \{\omega_i(x)dx^i\};\ \dim = n;\\
\mathbf{C}^\infty(\Lambda^2) &= \{\omega_{ij}(x)\,dx^i \wedge dx^j\}; \dim = \frac{n(n-1)}{2!};\\
\mathbf{C}^\infty(\Lambda^3) &= \{\omega_{ijk}(x)\,dx^i \wedge dx^j \wedge dx^k\}; \dim = \frac{n(n-1)(n-2)}{3!};
\end{aligned} \tag{B.29}$$

$$\begin{aligned}
\mathbf{C}^\infty(\Lambda^{n-1}) &= \{\omega_{i_1\ldots i_{n-1}}(x)\,dx^{i_1} \wedge \ldots \wedge dx^{i_{n-1}}\}; \dim = n;\\
\mathbf{C}^\infty(\Lambda^n) &= \{\omega_{i_1\ldots i_n}(x)\,dx^{i_1} \wedge \ldots \wedge dx^{i_n}\}; \dim = 1.
\end{aligned}$$

Thus the 0-forms are usual c-number functions on Φ, with values in $\mathbb{R}^1$; the 1-forms are covectors $\Lambda^1_\phi(\Phi) = T^*_\phi(\Phi)$; the n-forms are the forms of maximal rank on Φ (the volume forms). Now one can construct the space

$$\Lambda^* = \Lambda^0 \oplus \Lambda^1 \oplus \ldots \oplus \Lambda^n, \tag{B.30}$$

which is closed with respect to the wedge product and is called the Cartan graded exterior algebra. The generators of this algebra are spaces Λ^0 and Λ^1, the algebra is associative one and the equation (B.27) proves to be the graded commutativity condition for this algebra.

Now we are in the position to list the differential operations, which one can define in analogy with conventional vector analysis.

1. The exterior derivative *d* transforms a p-form into a $(p+1)$-form

$$\begin{aligned}
d: \mathbf{C}^\infty(\Lambda^0) \to \mathbf{C}^\infty(\Lambda^1);\quad & d\big(\omega(x)\big) = \frac{\partial \omega}{\partial x^i}\,dx^i = d\omega;\\
d: \mathbf{C}^\infty(\Lambda^1) \to \mathbf{C}^\infty(\Lambda^2);\quad & d\big(\omega_i(x)dx^i\big) = \frac{\partial \omega_i}{\partial x^j}\,dx^i \wedge dx^j.
\end{aligned}$$

Here we adopted the convention that the new differential is always put as a left factor in the wedge product and therefore only antisymmetric parts of derivatives $(\partial\omega_i/\partial x^j)$ do contribute. In a general case of (B.26) we have

$$d\omega = d\omega_{i_1\ldots i_p} \wedge dx^{i_1} \wedge \ldots \wedge dx^{i_p} = f_{i_0 i_1\ldots i_p}\,dx^{i_0} \wedge dx^{i_1} \wedge \ldots \wedge dx^{i_p}, \tag{B.31}$$

where

$$f_{i_0 i_1 \ldots i_p} = \frac{1}{p+1}\sum_{s=0}^{p}(-1)^s \frac{\partial}{\partial x^{i_s}}\omega_{i_0\ldots i_{s-1} i_{s+1}\ldots i_p}. \tag{B.31a}$$

Important properties of thus defined exterior derivative are:

$$d\,(d\omega) \equiv 0 \quad \text{(the Poincaré lemma)}; \tag{B.32}$$

$$d\,(\omega_p \wedge \omega_q) = d\omega_p \wedge \omega_q + (-1)^p \omega_p \wedge d\omega_q \quad \text{(the Leibnitz rule)}. \tag{B.33}$$

As an illustration here we give some relations in three-dimensional space $\{x = (x^1, x^2, x^3)\}$, where

$$\omega^0 = f(x);$$
$$\omega^1 = \omega_i\, dx^i = \omega(x) \cdot d\boldsymbol{x};$$
$$\omega^2 = F_1 dx^2 \wedge dx^3 + F_2 dx^3 \wedge dx^1 + F_3 dx^1 \wedge dx^2 = \boldsymbol{F} \cdot d\boldsymbol{S};$$
$$\omega^3 = \rho\, dx^1 \wedge dx^2 \wedge dx^3 = \rho\, dV,$$

and we see that

$$\omega^1 \wedge \omega^2 = (\boldsymbol{\omega} \cdot \boldsymbol{F})\, dx^1 \wedge dx^2 \wedge dx^3;$$
$$d\omega^1 = \tfrac{1}{2}(\epsilon_{ijk}\partial_j\omega_k)\epsilon_{ilm}\, dx^l \wedge dx^m;$$
$$d\omega^2 = (\nabla \boldsymbol{F})\, dx^1 \wedge dx^2 \wedge dx^3 = \rho\, dV,$$

which are the convential operations of the 3-dimensional vector calculus, and (B.32) implies

$$\operatorname{rot}\operatorname{grad} f(x) = \operatorname{div}\operatorname{rot}\boldsymbol{\omega} = 0.$$

2. Hodge star ($*$) (dual conjugation operator). From (B.28) and (B.29) one can notice that the number of independent functions in $\mathbf{C}^\infty(\Lambda^p)$ is the same as that in $\mathbf{C}^\infty(\Lambda^{n-p})$, i.e. these two spaces are considered to be **dual** to each other with respect to a certain internal product, namely

$$* : \mathbf{C}^\infty(\Lambda^p) \to \mathbf{C}^\infty(\Lambda^{n-p}), \quad p < n. \tag{B.34}$$

We clarify this relation using as an example the p-form

$$\omega_p = \frac{1}{p!}\, \Omega_{i_1 \ldots i_p}\, dx^{i_1} \wedge \ldots \wedge dx^{i_p}, \tag{B.35}$$

then its dual $(n-p)$-form is

$$*\,\omega_p = \frac{1}{(n-p)!}\, \tilde{\Omega}_{j_1 \ldots j_{n-p}}\, dx^{j_1} \wedge \ldots \wedge dx^{j_{n-p}}, \tag{B.36}$$

with $\tilde{\Omega}_{j_1 \ldots j_{n-p}}$ being the tensor dual to $\Omega_{i_1 \ldots i_p}$ and defined by the formula

$$\tilde{\Omega}_{j_1 \ldots j_{n-p}} = \frac{1}{p!}\sqrt{g}\,\epsilon_{j_1 \ldots j_{n-p} i_1 \ldots i_p}\, \Omega_{i_1 \ldots i_p}. \tag{B.37}$$

Raising and lowering of indices of this tensor one performs with the help of the metric tensor g_{ij}, and then $\sqrt{g}\,\epsilon_{j_1, \ldots, j_p}$ is a coefficient function of the volume n-form on Φ. From the above expressions (B.34–37) it is straightforward that

$$*\,*\,\omega_p = (-1)^{p(n-p)}\omega_p, \tag{B.38}$$

and by means of the Hodge star $*$ it proves convinient to define the inner product of p-forms as

$$\langle \omega_p, \omega'_p \rangle = \int\limits_{\Phi} \omega_p \wedge *\omega'_p, \tag{B.39}$$

with the evident property

$$\langle \omega_p, \omega'_p \rangle = \langle \omega'_p, \omega_p \rangle,$$

which follows from the identity

$$\omega_p \wedge *\omega'_p = \omega'_p \wedge *\omega_p. \tag{B.40}$$

3. Exterior co-derivative δ (adjoint to exterior derivative). If the metric g_{ij} is available on Φ, then one can define the operation δ reducing the rank of a form by one unit, as given by the formula

$$\delta = (-1)^{n(p+1)+1} * d*, \tag{B.41}$$

so that

$$\delta : \mathbf{C}^\infty(\Lambda^p) \to \mathbf{C}^\infty(\Lambda^{p-1}).$$

The operators d and δ are adjoint, viz

$$\langle \omega_p, d\tau_{p-1} \rangle = \langle \delta\omega_p, \tau_{p-1} \rangle. \tag{B.42}$$

Moreover we also have for δ the analogue of the Poincaré lemma

$$\delta(\delta\omega) = \pm * d\, d * \omega \equiv 0 \tag{B.43}$$

as well as conventional tensor calculus operations. For example, when $n = 3$ and $p = 1$ we obtain

$$\delta(\boldsymbol{\omega} \cdot d\boldsymbol{x}) = - * (\nabla\boldsymbol{\omega})\, dx^1 \wedge dx^2 \wedge dx^3 = -\mathrm{div}\, \boldsymbol{\omega}.$$

4. Laplacian Δ (the Laplace-de Rham operator). This operator on a manifold Φ can be constructed once d and δ are defined (along with the metric) via the relation

$$\Delta = (d + \delta)^2 = d\delta + \delta d. \tag{B.44}$$

The Laplacian Δ is a self-adjoint operator, which sends p-forms to p-forms

$$\Delta : \mathbf{C}^\infty(\Lambda^p) \to \mathbf{C}^\infty(\Lambda^p),$$

and commutes with the above introduced operators d, $*$, and δ. Thus we have

$$\langle \omega_p, \Delta\omega_p \rangle = \langle \delta\omega_p, \delta\omega_p \rangle + \langle d\omega_p, d\omega_p \rangle \geq 0, \tag{B.45}$$

i.e. the operator Δ is a positive definite operator. All in all is sufficient to define some important classes of differential forms, namely:

Closed forms : $d\omega = 0$; Co-closed forms : $\delta\omega = 0$;
Exact forms : $\omega = d\alpha$; Co-exact forms : $\omega = \delta\alpha$;
Harmonic forms : $\Delta\omega = 0$.

Note that the harmonic forms are both closed and co-closed, as due to the equation

$$(\omega_p, \Delta\omega_p) = (\omega_p, d\delta\omega_p) + (\omega_p, \delta d\omega_p) = (\delta\omega_p, \delta\omega_p) + (d\omega_p, d\omega_p) \geq 0.$$

Theorem (Hodge). If Φ is a compact manifold without boundary, then any p-form ω_p can be uniquely decomposed into a sum:

$$\omega_p = d\alpha_{p-1} + \delta\beta_{p+1} + \gamma_p \tag{B.46}$$

as a sum of exact, co-exact and harmonic forms.

This completes the body of minimal knowledge, as required for realizing the ideas of integration on manifolds and these of construction of principal topological invariants such as homotopy groups, homology and co-homology groups, characteristic classes and so on.

B.4 Integration on Manifolds and De Rham Co-Homologies

The reader now realizes that in studying the properties of a manifold as a whole its local properties should be complemented with some analog of integral calculus. Here the basic idea to consider integrands as exterior forms belongs to Poincaré. This idea was realized in the construction of the first version of homology theory, the homologies of polyhedrons. In what follows we give only basic idea of this construction [for details see (Dubrovin, Novikov, and Fomenko, 1984; Felsager, 1981)].

As is well-known, one of the basic properties of integral sums in the Euclidean space is the independence of their limit upon the way of decomposition of the integration area into small cells. Poincaré chose oriented simplices as "building blocks" for integration of p-forms. These simplices are simplest convex polyhedrons in $\mathbb{R}^n$ (see Fig. B.3).

If Δ_p is a p-simplex, then its boundary $\partial\Delta_p$ is taken to be a formal sum of $(p-1)$-simplices with integer coefficients. In its turn, one can construct a p-chain, denoted by $C_p = \sum_i a_i \Delta_p^i$ with the boundary $\partial C_p = \sum_i a_i \partial\Delta_p^i$ such that $\partial(\partial C_p) = 0$. A set of arbitrary dimensional simplices forms a simplicial complex, and if a manifold M can be represented as a simplicial complex, then it is called a **triangulable** one. The latter notion one can regard as an appropriate generalization of the standard decomposition of the Euclidean plane, giving at the same time a method to study global properties of manifolds by algebraic means. We note that when $a_i \in \mathbb{R}$, $\mathbb{C}$, $\mathbb{Z}$ or take values in any other Abelian group, with respect to addition, then chains form an Abelian group too. One can define cycles z_p as chains with the trivial boundaries $(\partial C_p = 0)$, and two cycles are considered as equivalent – **homological** – when

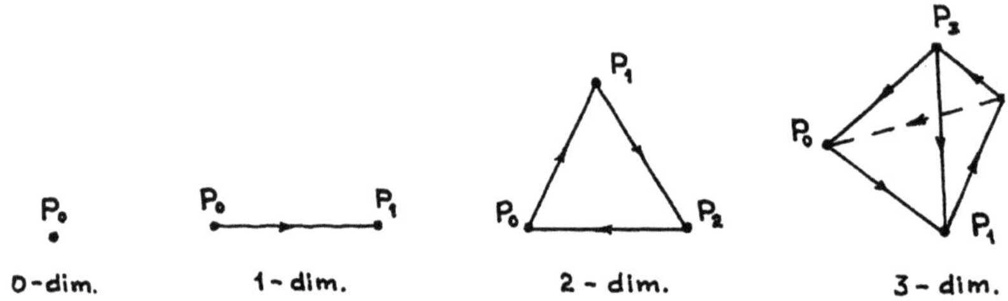

Fig. B.3. The simplest oriented simplices for low-dimensional Euclidean spaces: 0-simplex is a single point P_0; 1-simplex is a directed closed segment of a straight line and is completely determined by the ordered pair of vertices (P_0, P_1); 2-simplex is a closed triangle with the ordered triple of vertices (P_0, P_1, P_2); and 3-simplex is a tetrahedron with the ordered quadruple of vertices (P_0, P_1, P_2, P_3)

they are the same up to a boundary, i.e. $\partial C_{p+1} : z'_p = z_p + \partial C_{p+1}$. Families of cycles $Z_p = \{z_p\}$ and that of boundaries $B_p = \{\partial C_{p+1}\}$ also possess a group structure and the factor group

$$H_p(M; \mathbb{T}) = Z_p / B_p, \tag{B.47}$$

with equivalence classes of cycles as its elements, is called the **homology group** of M, where we denote by $\mathbb{T} = \mathbb{R}, \mathbb{C}, \mathbb{Z}$ the sets of real, complex and integer numbers, respectively. The relations among groups $H_p(M; \mathbf{T})$ with various choice of coefficients are given by the universal coefficient theorem

$$\begin{aligned} &H_p(M; \mathbb{R}) = H_p(M; \mathbb{Z}) \otimes \mathbb{R}, \\ &H_p(M; \mathbb{C}) = H_p(M; \mathbb{R}) \otimes \mathbb{C} = H_p(M; \mathbb{Z}) \otimes \mathbb{C}. \end{aligned} \tag{B.48}$$

Now we define **the de Rham cohomology groups** $H^p_{\mathrm{DR}}(M; \mathbf{T})$ as groups which consist with equivalence classes of closed modulo exact forms as elements:

$$\omega_p \sim \omega'_p : \quad \omega'_p = \omega_p + d\alpha_{p-1}. \tag{B.49}$$

Note that the group H^0_{DR} is a special one since there is no (-1) forms, hence no 0-forms can be given as exterior derivatives. Therefore H^0_{DR} is a space of *constant* functions and $\dim H^0_{\mathrm{DR}}$ equals the number of connected pieces of M. The following Poincaré theorem will be also of use: the de Rham cohomology of $\mathbb{R}^n$ is trivial:

$$\dim H^p_{\mathrm{DR}}(\mathbb{R}^n) = 0, \; p > 0; \quad \dim H^0_{\mathrm{DR}}(\mathbb{R}^n) = 1.$$

Any closed form ω_p in $\mathbb{R}^n$ can be expressed as the exterior derivative of a lower form ω_{p-1}. The latter means that only in the case when $\mathbb{R}^n$ coordinate neighbourhoods are patched together in a nontrivial manner we have a nontrivial de Rham cohomology.

The relations between $H_p(M; \mathbf{T})$ and $H^p(M; \mathbf{T})$ are established by the de Rham theorem: the inner product of a closed form $\omega_p \in Z^p$ and a cycle $C_p \in Z_p$ is given by

$$T(C_p, \omega_p) = \int\limits_{C_p} \omega_p, \tag{B.50}$$

where $T(C_p, \omega_p)$ is called a **period** of the form ω_p over the cycle C_p. The following **Stokes' theorem** defines the operation of integration: if M is a n-dimensional oriented manifold with a non-empty boundary ∂M then for any $(n-1)$-form ω_{n-1} we have

$$\int_M d\omega_{n-1} = \int_{\partial M} \omega_{n-1}. \tag{B.51}$$

In (B.51) the right-hand side has to be taken as an oriented sum (∂M is supplied with the induced orientation) and the left-hand side is given by

$$\int_M d\omega_{n-1} = \sum_\alpha \int_{U_\alpha} d\varphi_\alpha \, \omega_{n-1}, \tag{B.52}$$

with $\{\varphi_\alpha\}$ being decompositions of the unity, subject to the covering $\{U_\alpha\}$ of M.

A smooth mapping of manifolds $f : M_1 \to M_2$ induces a mapping of forms $f^* : \omega \to f^*(\omega)$, such that $d(f^*(\omega)) = f^*(d\omega)$ and closed forms render into closed, exact forms remain to be exact. By virtue of this homology classes are transformed into each other under f^* which then induces homomorphism between homology groups:

$$f^* : H^p(M_2; \mathbb{R}) \to H^p(M_1; \mathbb{R}). \tag{B.53}$$

Isham's algorithm (see Section 2.3) is based on the following assertion[3]: two homotopic smooth maps

$$f_1 : M_1 \to M_2; \quad f_2 : M_1 \to M_2, \tag{B.54}$$

induce two coincident maps of cohomology groups f_1^* and f_2^*, i.e.

$$f_1^* = f_2^* : \ H^p(M_2; \mathbb{R}) \to H^p(M_1; \mathbb{R}). \tag{B.55}$$

Two manifolds are said to be **homotopically equivalent** if there occur maps (B.54) (together with their inverse mappings), such that their superpositions $f_1 \cdot f_2^{-1} : \ M_2 \to M_2$ and $f_2 \cdot f_1^{-1} : \ M_1 \to M_1$ are homotopic to identity maps. Thereby the proposition is proved, in particular, **homotopic manifolds have the same cohomology groups.** The induced homomorphisms $(f_1 \cdot f_2)^* = f_2^* \cdot f_1^*$ and $(f_2 \cdot f_1)^* = f_1^* \cdot f_2^*$ are in other words identical homomorphisms of cohomology groups

$$\begin{aligned} \mathrm{Id} &= f_2^* \cdot f_1^* : \ H^p(M_2; \mathbb{R}) \to H^p(M_2; \mathbb{R}); \\ \mathrm{Id} &= f_1^* \cdot f_2^* : \ H^p(M_1; \mathbb{R}) \to H^p(M_1; \mathbb{R}), \end{aligned} \tag{B.56}$$

i.e. the homomorphisms f_1^* and f_2^* are in fact isomorphisms, such that $f_1^* = (f_2^*)^{-1}$.

[3] The rigorous proof one can find in (Isham, 1977).

All these facts simplify calculations of cohomology groups as well as construction of the explicit forms of topological invariants. We will be back to discuss these topic in a moment, and now we display some advantages of the technique exposed here. For example, by use of cohomologies one can get a knowledge on the conditions under which a smooth p-form ω is produced by a $(p-1)$-form α : $\omega = d\alpha$. In conventional way it requires to solve a system of partial differential equations, which is not an easy task even for 1-forms [see (Mitschenko and Fomenko, 1980), Chapter 6]. In cohomology terms the answer is given by the **existence theorem**

Theorem (de Rham). A smooth p-form ω_p is exact if and only if $d\omega_p = 0$ (the local condition) and its periods over all the cycles $T(\omega_p, C_p) = 0$ (the global condition), i.e. when cohomology class $[\omega_p] \notin H^p(M; \mathbb{R})$.

As it was pointed above, the group $H^p(M; \mathbb{R})$ is endowed with the structure of a vector space. Its dimensionality is called the p-th Betti number of M:

$$b_p = \dim H^p(M; \mathbb{R}), \tag{B.57}$$

which proves to be a topological invariant. Consider a manifold M with $\dim M = n$. Then we have $H^0(M; \mathbb{R}) = \mathbb{R}$, $b_0(M) = 1$, if M is simply connected, and $H^0(M; \mathbb{R}) = \mathbb{R}^k$, $b_0(M) = k$, if $M = \cup_{i=1}^k M_i$ (k-connected, with $M_i \cap M_j = \emptyset$, $i \neq j$); $H^n(M; \mathbb{R}) = \mathbb{R}$, if M is both connected and compact. For example, $H^0(\mathbb{R}^n; \mathbb{R}) = \mathbb{R}$, $H^p(\mathbb{R}^n; \mathbb{R}) = 0$ for $p > 0$.

In the case $M = \mathbb{S}^n$ we can use Hodge's theorem (B.46) in order to calculate $H^p(\mathbb{S}^n)$. For that we notice, that if ω is closed then in view of (B.46) $d\delta\beta = 0$ because $d^2\alpha = 0$ while the form γ is harmonic and hence closed, $d\gamma = 0$. It yields the closed forms to be defined by the gauge condition $\delta\beta = 0$, i.e. $\omega = d\alpha + \gamma$. In fact: from the equation

$$(\beta, d\delta\beta) = (\delta\beta, \delta\beta) \geq 0$$

it follows that $\delta\beta = 0$ if $d\delta\beta = 0$. Due to Eq. (B.46) one sees that each cohomology class in $H^p(M; \mathbb{R})$ contains exactly one harmonic form and there is an isomorphism between $H^p(M; \mathbb{R})$ and a set of harmonic p-forms $\mathrm{Harm}^p(M; \mathbb{R})$. Then, since on $\mathbb{S}^n$ only two harmonic forms are determined, namely: 0-form and n-form (the volume form), we have $H^0(\mathbb{S}^n; \mathbb{R}) = \mathbb{R}$, $H^n(\mathbb{S}^n; \mathbb{R}) = \mathbb{R}$ and $H^p(\mathbb{S}^n; \mathbb{R}) = 0$ for any $p \neq 0, n$.

The knowledge of homology and cohomology groups allows one to judge on the existence, as well as on number and properties of the independent solutions of the equation $\Delta\omega_p = 0$. Indeed, if M^n is a compact orientable Riemannian manifold, then

a. the Betti number $b_p(M)$ is finite and equals to the number of linearly independent soluitions to the equation $\Delta\omega_p = 0$;
b. it holds to be true **the Poincaré duality**:

$$H^p(M) = H^{n-p}(M); \quad b_p(M) = b_{n-p}(M). \tag{B.58}$$

Moreover, this information gives the possibility to evaluate the minimal number of critical points of functions and functionals depending on the manifold itself, where they are defined on, and so on [for more information consult with (Flanders, 1963)]. The relations between cohomology and homotopy are also used as a powerful tool to study homotopic invariants of manifolds.

B.5 Fundamental Groups, Homotopy Groups and Some Other Topological Invariants

In the main text we study at sufficient length topological properties of continuous mappings (physical fields) by means of homotopy theory (see Chapter 2). Here we intend mostly to make some involved concepts and statements more transparent.

Composition Law on the Set $\{[\mathbf{S}^1, \Phi]\}$. Let us consider $\phi : (\mathbf{S}^1, N) \to (\Phi, \phi_0)$, where N is the northen pole of $\mathbf{S}^1$ and ϕ_0 is a fixed point. The image of $\mathbf{S}^1$ on Φ is a closed path (loop) with the beginning and the end at the same point ϕ_0 (see Fig. B.4a).

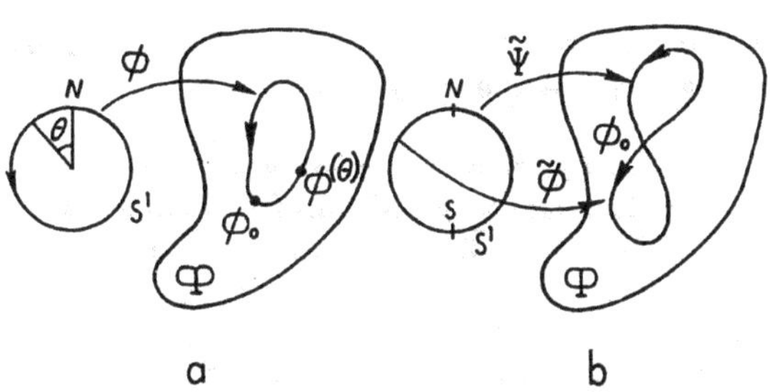

Fig. B.4 a, b. An explicit construction of the composition law on the set $\{[\mathbf{S}^1, \Phi]\}$, obtained by concentration of initial mappings ϕ and ψ of the whole circle $\mathbf{S}^1$ into Φ (as shown on picture a.) into the mappings $\tilde{\phi}$ and $\tilde{\psi}$ of the half of the circle into the same manifold (picture b.). The corresponding formulae are given as (B.59) and (B.60)

The map ϕ is homotopic to the map $\tilde{\phi}$, which is taken to be nontrivial on the semicircle $0 \le \theta \le \pi$ by virtue of the homotopy

$$F_\nu(\theta) = \left\{ \phi(\frac{1}{\nu}\theta) : 0 \le \theta \le 2\pi\nu; \quad \phi_0 : 2\pi\nu \le \theta \le 2\pi \right\}.$$

Let us transform $\phi = F_1(\theta)$, $(\nu = 1)$ into $\tilde{\phi} = F_{1/2}(\theta)$, $(\nu = 1/2)$ such that

$$\tilde{\phi} = \{\phi(2\theta) \text{ at } 0 \le \theta \le \pi; \quad \phi_0 \text{ at } \pi \le \theta \le 2\pi\}.$$

Now we consider the mapping $\psi : (\mathbf{S}^1, N) \to (\Phi, \phi_0)$, locate it on the semicircle $\pi \le \theta \le 2\pi$ and denote as $\tilde{\psi}$ (see Fig. B.4b) The composition of ϕ and ψ is given by the rule

$$\phi \cdot \psi(\theta) = \begin{cases} \tilde{\phi}(\theta), & \text{for} \quad 0 \le \theta \le \pi; \\ \tilde{\psi}(\theta), & \text{for} \quad \pi \le \theta \le 2\pi, \end{cases} \tag{B.59}$$

which means that we are running at doubled velocity first the ϕ-loop and then the ψ-one. One can easily notice that if $F_\nu(\theta)$ and $G_\nu(\theta)$ are the homotopies sanding ϕ and ψ respectively to $\tilde{\phi}$ and $\tilde{\psi}$, so will do their composition $G_\nu \star F_\nu(\theta)$, induced by equation (B.59). As a result we have got the composition law on the set of homotopy classes $\{[\phi]\}$:

$$[\phi \cdot \psi] = [\phi] \star [\psi], \tag{B.60}$$

which supplies the set $\{[\phi]\}$ with a group structure. The unity element of this group is the homotopy class of the constant map $\phi_0(\theta) : \mathbb{S}^1 \to \phi_0$, since

$$\phi_0 \cdot \phi = \tilde{\phi}(\sim \phi) \Rightarrow [\phi_0] \star [\phi] = [\phi].$$

The inverse element to $[\phi]$ would be the homotopy class of mappings $[\phi^{-1}(\theta)] = [\phi(2\pi - \theta)] = [\phi]^{-1}$, running the same ϕ-loop but in the opposite direction. We check that $\phi^{-1} \cdot \phi(\theta) = \phi_0(\theta)$ by presenting the homotopy

$$F_\nu(\theta) = \begin{cases} \phi_0, & \text{at} \quad 0 \le \theta \le \frac{\nu}{2};\ 2\pi - \frac{\nu}{2} \le \theta \le 2\pi; \\ \phi(2\theta - \nu), & \text{at} \quad \frac{\nu}{2} \le \theta \le \pi; \\ \phi(4\pi - \nu - 2\theta), & \text{at} \quad \pi \le \theta \le 2\pi - \frac{\nu}{2}; \end{cases} \tag{B.61}$$

which links $F_0(\theta) = \phi^{-1} \cdot \phi(\theta)$ to $F_{2\pi}(\theta) = \phi_0$. When $\nu = \pi$, $F_\pi(\theta)$ runs only over half of the ϕ-loop and then back to ϕ_0. All this follows from (B.61). The set $\{[\phi]\}$ with the composition law (B.60) is said to be the **fundamental group** of the manifold Φ with marked point ϕ_0 and is denoted as $\pi_1(\Phi, \phi_0)$. If Φ is a pathwise-connected, then fundamental groups at various points $\phi_0 \in \Phi$ are isomorphic modulo automorphism. If $\pi_1(\Phi, \phi_0)$ is an Abelian group such an automorphism is unique and all the groups $\pi_1(\Phi, \phi_0)$ are identified and denoted as $\pi_1(\Phi)$.

Let us consider the case $\Phi = \mathbb{S}^1$ in more detail by defining $\mathbb{S}^1$ via the equation $|z| = 1$ in the complex plane $\mathbb{C}$. The map $f : \mathbb{S}^1 \to \mathbb{S}^1$ is given by the smooth function $f(z) = z^n$, where n is an integer number. Any point $z_0 \in \mathbb{S}^1$ is a regular point under the map $f(z)$ since the pre-image $f^{-1}(z_0)$ consists of n points (the nth roots of z_0) and the f-mapping Jacobian has definitely the unit rank. The decomposition of $f(z)$ into the homotopy classes was discussed in detail in Section 2.2, therefore here we only clarify in what sense the number n evidently realizing the isomorphism $\pi_1(\mathbb{S}^1) = \mathbb{Z}$ stands for **the degree of mapping**.

The Degree of Mapping. In virtue of definition (Novikov and Fomenko, 1987) **the degree of mapping** or **the Brouwer degree** for smooth maps $f : M^n \to N^n$ between two connected oriented manifolds at a regular point $q \in N^n$, in which the total pre-image $f^{-1}(q)$ consists of a finite number of points p_i, is called the sum

$$\mathbf{Q} = \deg_q(f) = \sum_{p_i \in f^{-1}(q)} \operatorname{sgn}\det\left(\frac{\partial x^\alpha}{\partial y_i^\beta}\right). \tag{B.62}$$

Here $(x^1,\ldots,x^n)$, $(y_i^1,\ldots,y_i^n)$ are local coordinates in the neighbourhood of the point $q \in N^n$ and of the points $p_i \in f^{-1}(q) \subset M^n$, respectevely. The standard requirement for M^n to be a compact manifold one can avoid when considering automorphic mappings.

In practice to calculate the degree of mapping $\mathbf{Q}$ there are used more simple formulae than (B.62). If $\omega = \omega(x)\, dx^1 \wedge \ldots \wedge dx^n$ is an n-form on N^n, then

$$\int_M f^*(\omega) = \deg \mathrm{f} \int_N \omega, \tag{B.63}$$

or

$$\deg \mathrm{f} = \int_M f^*(\omega) \Big/ \int_N \omega, \tag{B.64}$$

where f^* is an induced mapping of forms, and one might impose standard normalization condition $\int_N \omega = 1$.

We turn back to the mapping $f : \mathbf{S}^1(\theta) \to \mathbf{S}^1(\varphi)$ given in the local (angular) parametrization. An appropriate function $\varphi = f(\theta)$ should satisfy the condition $f(\theta + 2\pi) = f(\theta) + 2\pi n$, which identifies the points differed by $2\pi n$. In particular, we consider $f(\theta) = n\theta$ and define the 1-form (here the 'volume' form) on $\mathbf{S}^1$ as $\omega = d\theta$, then

$$\begin{gathered} f^*(\omega) = d(f(\theta)) = n d\theta = n\omega, \\ \int_{\mathbf{S}^1} f^*(\omega) = n \int_{\mathbf{S}^1} \omega; \quad n = \int_{\mathbf{S}^1} f^*(\omega) \Big/ \int_{\mathbf{S}^1} \omega; \end{gathered} \tag{B.65}$$

i.e. $n \in \mathbb{Z}$ is the true degree of mapping $\mathbf{S}^1 \to \mathbf{S}^1$, and by virtue of (B.53) f^* realizes the homomorphism $H^1(\mathbf{S}^1) \to H^1(\mathbf{S}^1)$.

Notice, that any compact orienable manifold M possesses all necessary requirements for the n-cycle definition. The corresponding homology class $[M]$ generates the homology group $H_n(M)$ as far as for any connected M the homology class of n-cycles is a multiple of $[M]$. Hence, in terms of the homology theory the degree of mapping of $f : M_1 \to M_2$ can be defined as

$$f_*([M_1]) = (\deg f) \cdot [M_2]; \quad \deg f = f_*([M_1])/[M_2], \tag{B.66}$$

where f_* is the induced mapping of cycles.

Calculations of Higher Homotopy Groups. Consider a bit closer what is to calculate higher homotopy groups for the manifolds of the groups $SO(n)$ and $SU(n)$, those of spheres $\mathbf{S}^n$ and the coset spaces G/H. Higher homotopy groups $\pi_n(\Phi)$ are determined in the same manner as the fundamental groups above by giving a composition law on the set of the homotopy classes $\{[\mathbf{S}^n, \Phi]\}$.

If $\Phi = G$ is a Lie group, then the multiplication in G induces the desired composition law in $\pi_n(G)$. For topologically equivalent manifolds their homotopy groups are the same. A manifold Φ shows to be simply connected if $\pi_1(\Phi) = 0$, as far as any path on Φ one can shrink into a point (e.g. $\Phi = \mathbb{R}^n,\ \mathbb{S}^n\ (n \geq 2)$). However, the sphere $\mathbb{S}^1$ is not a simply connected manifold as $\pi_1(\mathbb{S}^1) = \mathbb{Z}$, the same is true for a torus $T^2 = \mathbb{S}^1 \otimes \mathbb{S}^1$, since $\pi_1(T^2) = \mathbb{Z} \otimes \mathbb{Z}$ and it means that T^2 is doubly connected. We also note, $\pi_1(SU(2)) = \pi_1(\mathbb{S}^3) = 0$, but $\pi_1(SO(3)) = \mathbb{Z}_2$. This confirms once more the fact of two-fold covering of $SO(3)$ by $SU(2)$.

Following (Olshanetskii, 1982) we give here an efficient method to express (instead to calculate) unknown groups $\pi_n(\Phi)$ through already known ones, since direct calculation of groups $\pi_n(\Phi)$ in general is not an easy task. Let a group G act on Φ, then through transformations from G the manifold Φ is decomposed into equivalence classes Φ/G. In its turn, each equivalence class is an orbit of the group G and the mapping Π, which relates to a point $\phi \in \Phi$ its orbit $\mathcal{O} \in \Phi/G$, is called the projection $\Pi : \Phi \to \Phi/G$ (Postnikov, 1982). One can also consider an embedding ι of a group G orbit into Φ, i.e. $\iota : \mathcal{O} \to \Phi$. Let $f : X \to \Phi$ be a continuous mapping, then there exist a homomorphism of homotopy groups

$$*f : \pi_n(X) \to \pi_n(\Phi), \tag{B.67}$$

and corresponding sequence of the homomorphisms

$$\pi_n(\mathcal{O}) \xrightarrow{*\iota} \pi_n(\Phi) \xrightarrow{*\Pi} \pi_n(\Phi/G). \tag{B.68}$$

If to extend this sequence by means of boundary homomorphism ∂, namely

$$\pi_n(\Phi/G) \xrightarrow{\partial} \pi_{n-1}(\mathcal{O}),$$

it gets to be an **exact sequence**. Recall, that the sequence of homomorphisms

$$A \xrightarrow{\mathrm{f}} B \xrightarrow{\mathrm{h}} C$$

is said to be exact in B if and only if $\operatorname{Im} f = \operatorname{Ker} h$, where $\operatorname{Im} f = f(A) \in B$ is an image of A in B and $\operatorname{Ker} h$ is a set of points in B, which are mapped into the zero-point of C (see illustration on Fig. B.5)

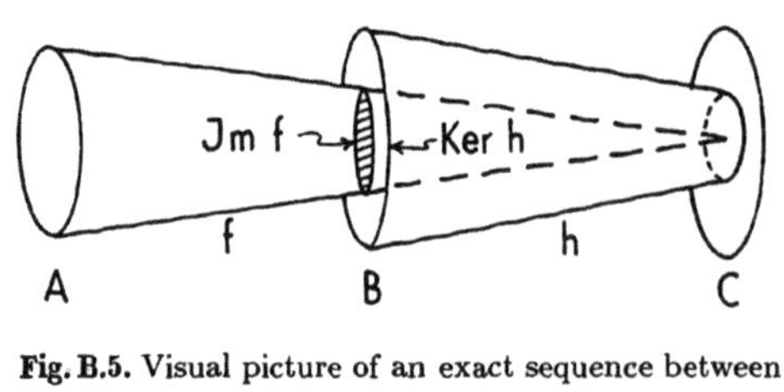

Fig. B.5. Visual picture of an exact sequence between three sets A, B and C

The following property of exact sequences will be important below:

Definition B.2. *The sequence* $e \to A \xrightarrow{f} B \xrightarrow{h} C \to e$ *is* **exact** *if and only if* $C = B/\mathrm{Im}\, f$. *When* $C = B/A$ *is a factor group, then* $e \to A \xrightarrow{f} B \xrightarrow{h} C \to e$ *is the exact sequence, where* e *are the identity elements.*

We give some examples of exact sequences of homomorphisms.

1. Let us start with well-known sequence:

$$e_1 \xrightarrow{f} \mathbb{Z}_2 \xrightarrow{g} SU(2) \xrightarrow{h} SO(3) \xrightarrow{k} e_2. \tag{B.69}$$

Here f is an embedding of the unity matrix $e_1 = \mathrm{diag}\,(1,1)$ in $\mathbb{Z}_2 = \{e_1, -e_1\}$; g is an embedding of the group $\mathbb{Z}_2$ in $SU(2)$; h is a stereographic projection into $SO(3)$ of the linear-fractional transformation in $\mathbb{C}^1$:

$$z \to \frac{\alpha z - \bar{\beta}}{\beta z + \alpha};$$

k is a mapping of elements of $SO(3)$ in $e_2 = \mathrm{diag}\,(1,1,1)$. As a consequence we obtain

$$SO(3) = SU(2)/\mathbb{Z}_2. \tag{B.70}$$

2. A sequence of unitary groups:

$$e_1 \xrightarrow{f} SU(n) \xrightarrow{g} U(n) \xrightarrow{h} U(1) \xrightarrow{k} e_1. \tag{B.71}$$

Here g is an embedding of $SU(n)$ in $U(n)$; h is a mapping of the type det : $\det g \in U(1)$, where $g \in U(n)$. As a consequence of (B.71) we have

$$U(1) = U(n)/SU(n). \tag{B.72}$$

Now it is high time to exploit an exact sequence, for example,

$$\xrightarrow{\partial} \pi_n(\mathcal{O}) \xrightarrow{*\iota} \pi_n(\Phi) \xrightarrow{*\Pi} \pi_n(\Phi/G) \xrightarrow{\partial} \pi_{n-1}(\mathcal{O}) \xrightarrow{*\iota} \tag{B.73}$$

in order to calculate $\pi_n(\Phi)$.

1. Since $\mathbb{S}^n = SO(n+1)/SO(n)$ then at $n > 2$ $\pi_1(\mathbb{S}^n) = \pi_2(\mathbb{S}^n) = 0$ and we have the sequence

$$0 = \pi_2(\mathbb{S}^n) \xrightarrow{\partial} \pi_1(SO(n)) \xrightarrow{*\iota} \pi_1(SO(n+1)) \xrightarrow{*\Pi} \pi_1(\mathbb{S}^n) = 0,$$

whereby

$$\pi_1(SO(n)) = \pi_1(SO(n+1)). \tag{B.74}$$

2. An odd-dimensional sphere $\mathbb{S}^{2n-1} = SU(n)/SU(n-1)$ for $n \geq 2$ gives the sequence

Table B.1. Homotopy groups for spheres and classical Lie groups

Φ	π_1	π_2	π_3	π_4	π_5	π_6	π_7	π_8	$\pi_k\ ,\ k>8$
S^1	Z	$\pi_k(S^1)=0 \quad k>1$							
S^2	0	Z	Z	Z_2	Z_2	Z_{12}	$\pi_k(S^2)=0\ ,\ k>6$		
S^3	0	0	Z	$\pi_k(S^3)=\pi_k(S^2)\ ,\ k>3$					
S^n $(n>3)$	$\pi_k(S^n)=0\ ,\ k<n,\quad \pi_n(S^n)=Z;\ \pi_{n+1}(S^n)=\pi_{n+2}(S^n)=Z_2$ $\pi_m(S^n)=\pi_{m+1}(S^{n+1}),\ m<2n-1;\ \pi_3(S^n)=Z_{24}$ $\pi_{n+4}(S^n)=0,\ n\geqslant 6;\ \pi_{n+5}(S^n)=0;\ n\geqslant 7;\ \pi_{4n-1}(S^{2n})=Z$									
$SO(3)$	$\pi_1(SO(n))=Z_2$, $n>1$	$\pi_2(SO(n))=0$, $n>1$	Z	Z_2	$\pi_k(SO(3))=\pi_k(S^3)\ k>1$				
$SO(4)$			$Z\oplus Z$	$Z_2\oplus Z_2$	$\pi_k(SO(4))=\pi_k(S^3)\oplus\pi_k(S^3),\ k>1$				
$SO(5)$			Z	Z_2	$\pi_3(SO(n))=Z,\ n\geqslant 5;\ \pi_4(SO(n))=0,\ n>5$				
$SO(n)$ $(n>5)$			$\pi_k(SO(n))\equiv\pi_k(SO)$ – STABLE GROUP, $1<k\leqslant n-2$						
SO	Z_2	0	Z	0	0	0	Z	Z	$\pi_{k+8}(SO)=\pi_k(SO)$
$U(1)$	$\pi_k(U(1))=0,\ k>1;\ \pi_2(U(1))=Z$								
$SU(2)$	0	0	Z	$\pi_k(SU(2))=\pi_k(S^3),\ k>1$					
$SU(3)$	0	0	Z	0	Z	$\pi_4(SU(n))=0,\ n\geqslant 3$			
$SU(n)$	$\pi_2(SU(n))=0,\ n\geqslant 1;\ \pi_2(SU(n))=0,\ n\geqslant 1$ $\pi_3(SU(n))=Z,\ n\geqslant 2;\ \pi_k(SU(n))=\pi_k(U(n)),\ k>1$								
$U(n)$	$\pi_1(U(n))=Z,\ n\geqslant 1$ $\pi_k(U(n))=\equiv\pi_k(U)$ – STABLE GROUP, $1\leqslant,k\leqslant 2n-1$								
U	Z	0	$\pi_{k+2}(U)=\pi_k(U)$						
$Sp(n)$	$\pi_1(Sp(n))=0,\ n\geqslant 1;\ \pi_3(Sp(n))=Z,\ n\geqslant 1$ $\pi_4(Sp(n))=Z\ ,n>1$ $\pi_k(Sp(n))=\pi(Sp)$ – STABLE GROUP, $\pi_{k+8}(Sp)=\pi_k(Sp)$								

$$0=\pi_2(\mathbb{S}^{2n-1})\xrightarrow{\partial}\pi_1(SU(n-1))\xrightarrow{*\iota}\pi_1(SU(n))\xrightarrow{*\Pi}\pi_1(\mathbb{S}^{2n-1})=0,$$

whence $\pi_1(SU(n))=0$, i.e. $SU(n)$ is simply connected.

3. From (B.72) it follows at $k>1$ the exact sequence

$$0=\pi_{k+1}(U(1))\xrightarrow{\partial}\pi_k(SU(n))\xrightarrow{*\iota}\pi_k(U(n))\xrightarrow{*\Pi}\pi_k(U(1))=0,$$

and then

$$\pi_k(SU(n))=\pi_k(U(n)). \qquad \text{(B.75)}$$

Other useful in physical applications results, which can be obtained in the same manner as above, are listed in Table B.1

We also give a few more facts from the cohomology theory of Lie groups, which had been used when constructing an explicit form of topological charges (see Section 2.3). There $SU(2)$-invariant form (2.26) was taken to be the generator of $H^3(SU(2); \mathbb{R})$ thanks to the general proposition that the cohomologies of homogeneous spaces of a compact connected Lie group might be expressed via invariant forms only (Dubrovin, Novikov and Fomenko, 1984; Section 1). Since any compact Lie group G appeares to be at the same time a symmetric space of the group $G \otimes G$, acting in accord with the rule

$$T_{(g,h)}(x) = g \cdot x \cdot h^{-1}, \tag{B.76}$$

the following assertions prove to be valid:

a. any closed form on G is cohomologic to the invariant form;
b. any invariant (non-zero) form is never cohomological to zero.

As far as there exists the invariant Cartan-Killing metric on G one can define a normalized invariant volume element $d\mu(g)$. Hence any closed form ω can be related to the invariant form $\tilde{\omega}$ by

$$\tilde{\omega} = \int_G T_g^* \, \omega d\mu(g). \tag{B.77}$$

For a compact group G the invariant forms $\tilde{\omega}$ are both left- and right-invariant forms, i.e. invariant with respect to left-hand-side and right-hand-side shifts. We made use of the fact that on G there exists the left-invariant 1-form $\omega = g^{-1} \cdot dg$, which takes its values in the Lie algebra of G. Under the right shifts by h^{-1} this form is transformed according to $\omega \to \omega' = h \cdot \omega h^{-1}$. Correspondingly, if the form ω is invariant with respect to (inner) automorphism, then it is double-hand-side invariant form. Such forms are in a one-to-one correspondence to antisymmetric poly-linear functions on the Lie algebra $\mathcal{G}$ of G, and the latter are invariant against inner automorphisms. In particular, as an example of such a function one can consider a 3-linear function on vector fields in the Lie algebra

$$\langle \boldsymbol{X}, (\boldsymbol{Y} \times \boldsymbol{Z}) \rangle = f_{\alpha\beta\gamma} \boldsymbol{X}^\alpha \boldsymbol{Y}^\beta \boldsymbol{Z}^\gamma, \tag{B.78}$$

constructed via the Cartan-Killing form on $\mathcal{G}$. For $G = SU(2)$ we come up to the exact form for topological charge (1.37) by taking into account (B.24), the appropriate normalization and the fact that the chiral currents $\mathbf{L}_\mu$ give a $SU(2)$ algebra realization.

Calculations of the Wess-Zumino Term. Now we are prepared to turn to the derivation of the formula (10.56), promised before. Since ω_5 as defined in Eq. (10.46) is a closed form ($d\omega_5 = 0$) on the disc D_5^+, to calculate it we use the representation $D_5^+ = \mathbb{S}^3 \otimes \mathbb{S}^1 \otimes [0,1]$ and extend the field $\mathbf{U}(\boldsymbol{x}, t) \to \mathbf{V}(\boldsymbol{x}, t, s)$ in such way, that $\mathbf{V}(\boldsymbol{x}, t, s = 1) \equiv \mathbf{U}(\boldsymbol{x}, t)$ on the Minkowski space boundary. Thanks to the fact that $SU(3)$ is simply connected ($\pi_1(SU(3)) = 0$) and $A(t)$

in (10.54) is a periodic function, a desired extention one can perform in a manner

$$\mathbf{V}(\boldsymbol{x},t,s) = A(t,s)\cdot \mathbf{U}(\boldsymbol{x})\cdot A^{\dagger}(t,s). \tag{B.79}$$

Dropping the terms which do not contribute into the Wess-Zumino term, we write down

$$\begin{aligned}\epsilon^{\mu\nu\lambda\rho\sigma}\mathrm{Tr}\left\{L_\mu L_\nu L_\lambda L_\rho L_\sigma\right\} &= 5\epsilon^{abc}\epsilon^{ij}\mathrm{Tr}\left\{L_i L_j L_a L_b L_c\right\}\\ &= 5\epsilon^{abc}\epsilon^{ij}\mathrm{Tr}\left\{L_i L_j A\cdot U^\dagger\partial_a U\cdot U^\dagger\partial_b U\cdot U^\dagger\partial_c U\cdot A^\dagger\right\},\end{aligned} \tag{B.80}$$

where a, b, c are spatial indices and i,j are indices related to the parameters t and s. Since ω_5 is invariant against the spatial rotations it is enough to calculate its values for any one direction (instead of all points on $\mathbf{S}^3$), for example, along the z-axis ($\boldsymbol{x} = (0,0,r)$). Then we have for the matrix (10.53)

$$\begin{aligned}&\mathbf{U}(\boldsymbol{x}) \to \mathbf{U}(r) = \cos\Theta(r) + \mathrm{i}\tau_3\sin\Theta(r),\\ &\mathbf{U}^\dagger\partial_{1,2}\mathbf{U} = \frac{\mathrm{i}}{r}\tau_{1,2}\sin\Theta;\quad \mathbf{U}^\dagger\partial_3\mathbf{U} = \mathrm{i}\tau_3\Theta',\end{aligned} \tag{B.81}$$

whence after some algebra one finds

$$\epsilon^{abc}\,U^\dagger\partial_a U\cdot U^\dagger\partial_b U\cdot U^\dagger\partial_c U \equiv F(r)\cdot \mathbb{1}. \tag{B.82}$$

By substituting the obtained result in (B.80) one gets

$$\frac{\mathrm{i}}{240\pi^2}\int_{D_5^+}\mathrm{Tr}\left[l^5\right] = \frac{\mathrm{i}\mathbf{Q}}{2}\int_{D_2} d^2x\epsilon^{ij}\,\mathrm{Tr}\left[\partial_i A^\dagger\partial_j A\right], \tag{B.83}$$

with $\mathbf{Q}$ being the topological charge. We parametrize the two-dimensional disc $D_2 = [0,2\pi]\otimes[0,1]$ through the 'polar' coordinates t and s, with s being the radius and t the angle variable, then

$$A(t,s) = \begin{pmatrix}1 & 0 & 0\\ 0 & s\exp(\mathrm{i}t) & \sqrt{1-s^2}\\ 0 & -\sqrt{1-s^2} & s\exp(-\mathrm{i}t)\end{pmatrix}. \tag{B.84}$$

Making use of this and (B.83) one finds

$$\epsilon^{ij}\,\mathrm{Tr}\left[\partial_i A^\dagger\partial_j A\right] = -2\mathrm{i}\,s,$$

and therefore

$$\frac{\mathrm{i}}{240\pi^2}\int_{D_5^+}\mathrm{Tr}\left[l^5\right] = \mathbf{Q}\int_0^1 s ds\int_0^{2\pi} dt = \pi\,\mathbf{Q}. \tag{B.85}$$

Appendix C

Methods of Reductions

Here we give outlines of the proofs of some propositions from Chapters 4 and 5, which allow one to narrow the area in search for possible substitutions, describing minimizers of chiral Hamiltonians in question. In particular, one can always limit oneself with the class of static fields for the functionals with quadratic dependence on time derivatives (see Proposition 4.1). We give simple calculations in proof of this proposition in Subsection C.1. Furthermore, if the functional is invariant under a group G, its minimum should be realized in the class of G-invariant fields by the Coleman-Palais theorem. In Subsection C.2 the idea of Palais' proof is presented together with a counterexample, which clarifies the statement of the theorem. One will find also a direct derivation of the structure of G_2-invariant field configurations, as it was discussed in Chapter 7 in relation with the minimal energy structures in higher homotopy classes. The spherical rearrangement technique is illustrated in the final subsection.

C.1 Reduction to Static Field Configurations

In order to prove the Proposition 4.1, that Inf $\mathbf{H}$ (if any) has to be realized on static fields (Sanyuk, 1979), we substitute Eqs. (1.32) and (1.33) into the functional (1.42) and after taking traces explicitly come up to the expression

$$\begin{aligned}\mathbf{H} = \int d^3x \Big\{ & \frac{1}{2\lambda^2}\left[(\partial_0\Theta)^2 + (\partial_i\Theta)^2 + \sin^2\Theta\left((\partial_0\underline{n})^2 + (\partial_i\underline{n})^2\right)\right] \\ & + \epsilon^2\sin^2\Theta\left[(\partial_0\underline{n})^2(\partial_i\Theta)^2 + (\partial_i\underline{n})^2(\partial_0\Theta)^2 + (\partial_i\underline{n})^2(\partial_k\Theta)^2\right] \\ & + \epsilon^2\sin^4\Theta\left[(\partial_0\underline{n})^2(\partial_i\underline{n})^2 - (\partial_0\underline{n}\cdot\partial_i\underline{n})^2\right] \\ & + \frac{\epsilon^2}{2}\sin^4\Theta\left[(\partial_i\underline{n})^2(\partial_k\underline{n})^2 - (\partial_i\underline{n}\cdot\partial_k\underline{n})^2\right] \\ & - 2\epsilon^2\sin^2\Theta\,\partial_0\Theta\,\partial_i\Theta(\partial_0\underline{n}\cdot\partial_i\underline{n}) - \epsilon^2\sin^2\Theta(\partial_i\Theta\partial_k\Theta)(\partial_i\underline{n}\cdot\partial_k\underline{n})\Big\}. \end{aligned} \tag{C.1}$$

The integrand in (C.1), being regarded as a function of $(\partial_0\underline{n})$ and $(\partial_0\Theta)$, one can minimize with respect to these variables and come up to the system:

$$\partial_0\underline{n}\left[\frac{1}{2\lambda^2} + \epsilon^2\,\underline{L}_i^2\right] = \epsilon^2\,\partial_i\underline{n}(\underline{L}_0\cdot\underline{L}_i); \tag{C.2a}$$

$$\partial_0\Theta\left[\frac{1}{2\,\lambda^2}+\epsilon^2\,\underline{L}_i^2\right]=\epsilon^2\,\partial_i\Theta(\underline{L}_0\cdot\underline{L}_i). \qquad \text{(C.2b)}$$

Here $\underline{L}_i$ are left chiral currents in accord with Eq. (1.33). Multiplying equation (C.2a) by $\partial_i\underline{n}\cdot\sin^2\Theta$, and correspondingly equation (C.2b) by $\partial_i\Theta$ and summing the obtained results, we come up to the necessary minimum condition:

$$(\underline{L}_0\cdot\underline{L}_i)=0. \qquad \text{(C.3)}$$

Then once again from (C.2a) and (C.2b) we deduce that either $\partial_0\underline{n}=\partial_0\Theta=0$ or $\underline{L}_0=0$, that proves Proposition 4.1. The same statement proves to be valid for any functional with quadratic dependence on time derivatives.

C.2 Reduction to *G*-Invariant Fields

To elucidate the idea of Palais' proof (Palais, 1979) of the Coleman-Palais theorem (see Section 4.3, Theorem 4.1), we start with introducing the notation $X\equiv\delta H/\delta\Phi[\phi_0]$ for the variational derivative, and write down the condition for a functional $F[\phi]$ to have an extremum in the class of invariant fields [or in the invariant set] Φ_0

$$\langle X,\delta\phi_0\rangle=O,\quad \text{for any}\quad \delta\phi_0\in\Phi_0, \qquad \text{(C.4)}$$

where the angular brackets have been used to denote a linear functional, taken as a scalar product. On the other hand, the G-invariance of the functional $F[\phi]$ means, that

$$\delta F[\phi]=\langle X,\delta\phi\rangle=\langle X_g,\delta\phi_g\rangle, \qquad \text{(C.5)}$$

where X_g and $\delta\phi_g$ denote the quantities X and $\delta\phi$, as transformed under the action of the element $g\in G$. At the same time

$$\delta F[\phi]=F[\phi_0+\delta\phi_g]-F[\phi_0]=\langle X,\delta\phi_g\rangle. \qquad \text{(C.6)}$$

Taking into account the arbirtrariness of $\delta\phi_g$ and compairing (C.6) and (C.5) we find that

$$X_g=X. \qquad \text{(C.7)}$$

Let us denote by $\tilde{\Phi}_0$ the subset of fields X, which satisfy the invariance condition (C.7). Let us stress that in general the set Φ_0 and its dual set $\tilde{\Phi}_0$ are different.

The condition (C.4) means that X simulteneously belongs to the annihilator of the set Φ_0, i.e. $X\in\Phi_0^*$. Now it is not difficult to conclude that ϕ_0 would be the true extremal, if

$$\tilde{\Phi}_0\cap\Phi_0^*=\emptyset, \qquad \text{(C.8)}$$

(where it was used the standard notation for the empty set) since in this case $X=0$ or $\delta F[\phi]=0$. This proves the statement of the Coleman-Palais theorem.

Illustrative Counterexample. The significance of condition (C.8) we illustrate by the example due to O. Ladyzhenskaya (Kapitanskii and Ladyzhenskaya, 1983). Consider on the plane $\mathbb{R}^2 = \{x^1, x^2\}$ a function $F[\phi] = f(x^2)$, which is invariant against the action of group G:

$$\phi = \{x^1, x^2\} \to g\phi = \{x^1 + \tau x^2, x^2\},$$

with a parameter $\tau \in \mathbb{R}^1$. The invariant set in this case is $\Phi_0 = \{x^1, 0\}$, so it coincides with the axis x^1. To construct the set $\tilde{\Phi}_0$ we have to find out the action of the group G in the dual set of vector fields $\{X_1, X_2\} \equiv \{\delta F/\delta x^1, \delta F/\delta x^2\}$. To achieve this we write the invariance condition (C.5)

$$X_1 \delta x^1 + X_2 \delta x^2 = X_1' \delta x'^1 + X_2' \delta x'^2,$$

where $\delta x'^1 = \delta x^1 + \tau \delta x^2$, $\delta x'^2 = \delta x^2$. Whence we deduce the transformation law

$$\{X_1, X_2\} \to \{X_1, X_2 - \tau X_1\}$$

and the structure of the set $\tilde{\Phi}_0 = \{0, X_2\}$. Finally, the annihilator Φ_0^* one finds from the condition (C.4):

$$\delta F_{|\Phi_0} = (X_1, \delta x^1) = 0,$$

so that $\Phi_0^* = \{0, X_2\} = \tilde{\Phi}_0$. Thus the condition (C.8) appeares to be broken and the Coleman principle does not work in this case. Indeed, if we put $f'(0) \neq 0$, then $df(0) = f'(0)dx^2 \neq 0$, although on the invariant set $\{x^1, 0\}$ we have $df = \partial_1 f dx^1 \equiv 0$.

Let us also perform a direct derivation of the structure of G_2-invariant fields, announced in Section 7.1.1 and given by Eq. (7.5). In the manner similar to that in Section 4.4 we just write down the condition (4.23), but now for G_2-invariant fields it is better to write it in terms of spherical coordinates (r, ϑ, α):

$$-\mathrm{i}\partial_\alpha \mathbf{U} + \frac{k}{2}[\tau_3, \mathbf{U}] = 0, \tag{C.9}$$

where k is an integer number, with its value determined from the requirement on $\mathbf{U}$ to be a periodic function in α. Taking the trace of (C.9) we find $\partial_\alpha \Theta = 0$, or $\Theta = \Theta(r, \vartheta)$. As a result the equation (C.9) is simplified:

$$-\mathrm{i}\partial_\alpha(\underline{n} \cdot \underline{\tau}) + \frac{k}{2}[\tau_3, (\underline{n} \cdot \underline{\tau})] = 0. \tag{C.10}$$

Making use of the relation $[\tau_i, \tau_k] = 2\,\mathrm{i}\,\epsilon_{ikj}\tau_j$, from (C.10) we derive the equations on n_i:

$$\partial_\alpha n_3 = 0, \quad \partial_\alpha n_1 = -kn_2, \quad \partial_\alpha n_2 = kn_1.$$

Introducing the polar coordinates β, γ of the vector $\underline{\mathbf{n}}$ by setting

$$n_3 = \cos\beta, \quad n_1 + \mathrm{i}n_2 = \sin\beta e^{\mathrm{i}\gamma},$$

we get the structure (7.5) for the G_2-invariant chiral fields:

$$\Theta = \Theta(r,\vartheta); \quad \beta = \beta(r,\vartheta); \quad \gamma = k\alpha; \quad k \in \mathbb{Z}.$$

C.3 Spherical Rearrangement Technique: An Illustration

We demonstrate how this method works by proving, as an illustrative example, the inequality (5.14), with $f(\boldsymbol{x})$ being a non-negative and decreasing function. Using the notations

$$F(M) = \int_{f>M} |\nabla f|^2 \, d^3x; \quad F_R(M) = \int_{f_R>M} |\nabla_r f_R|^2 \, d^3x, \tag{C.11}$$

we rewrite the condition (5.10) as

$$\int_{f>M} d^3x = \int_{f_R>M} d^3x \overset{\text{def}}{=} V. \tag{C.12}$$

Let dS_M and dS_{MR} be the area elements of the equilevel surfaces $f(\boldsymbol{x}) = M$ and $f_R(\boldsymbol{x}) = M$, respectively, ∇f and $\nabla_r f_R$ their normals, correspondingly

$$dV = \int dS_M \cdot dh(\boldsymbol{x})$$

would be the volume element, and

$$dM = |\nabla f| \, dh(\boldsymbol{x}) = \text{const}$$

is the level increment (see Fig. C.1).

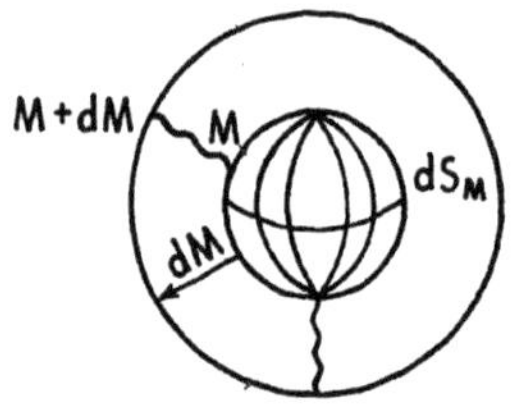

Fig. C.1. An equilevel surface M with the area element dS_M, and the increment dM

In terms of the above notations we can write down the relation

$$\left|\frac{dV}{dM}\right| = \oint \frac{dS_M dh}{|\nabla f| \, dh} = \oint_{S_M} \frac{dS_M}{|\nabla f|} = \oint_{S_{MR}} \frac{dS_{MR}}{|\nabla_r f_R|}. \tag{C.13}$$

Analogously from (C.11) one gets

$$\left|\frac{dF(M)}{dM}\right| = \oint |\nabla f| dS_M; \quad \left|\frac{dF_R(M)}{dM}\right| = \oint |\nabla_r f_R| \, dS_{MR}. \tag{C.14}$$

Multiplying equations (C.13) with (C.14) we have by virtue of the Shwartz inequality

$$\left|\frac{dV}{dM}\right| \cdot \left|\frac{dF(M)}{dM}\right| = \oint_{S_M} \frac{dS_M}{|\nabla f|} \cdot \oint |\nabla f| dS_M \geq \left|\oint dS_M\right|^2;$$

$$\left|\frac{dV}{dM}\right| \cdot \left|\frac{dF_R(M)}{dM}\right| = \oint_{S_{MR}} \frac{dS_{MR}}{|\nabla_r f_R|} \cdot \oint |\nabla_r f_R| dS_{MR} = \left|\oint dS_{MR}\right|^2 . \quad (C.15)$$

The equality sign in (C.15) follows from the fact, that $|\nabla_r f_R| = \text{const}$ at the surface $f_R = M = \text{const}$ by the construction. The equilevel surfaces $f = M$ and $f_R = M$ confine via (C.12) the same volume and, hence, the sphere S_{MR} has a smaller area:

$$\oint dS_M \geq \oint dS_{MR}.$$

Whence

$$\left|\frac{dV}{dM}\right| \cdot \left|\frac{dF(M)}{dM}\right| \geq \left|\frac{dV}{dM}\right| \cdot \left|\frac{dF_R(M)}{dM}\right|; \quad \left|\frac{dF(M)}{dM}\right| \geq \left|\frac{dF_R(M)}{dM}\right|.$$

Integrating the latter inequality in the limits from $M = 0$ to $M = M_{\max}$, one gets the inequality (5.14), we sought about. For further information on this subject we refer the reader to (Glaser, Martin, Gross, and Thirring, 1976).

Appendix D

Proofs of Stability and Existence Theorems

D.1 Proof of Generalized Hobart-Derrick Theorem

Let us start with an equivalent formulation of Theorem 3.2 from Chapter 3:

Theorem. The second variation of an additive Lyapunov functional of the type (3.27) in a neighbourhood of a stationary soliton solution is a sign changing one for all dimensions $D \geq 2$.

Proof. We consider for simplicity the case $D = 3$ and assume, following (Rybakov and Chakrabarti, 1984; Rybakov 1988, 1991), that the field $\phi = u(\boldsymbol{r}) : \mathbb{R}^3 \to \mathbb{R}^n$ has solitonic behaviour, i.e for $r \to \infty$

$$|\nabla u| = \mathcal{O}\left[r^{-(3/2+\alpha)}\right], \quad \alpha > 0, \tag{D.1}$$

and the function $u(\boldsymbol{r})$ is the critical point of the functional (3.27) i.e. it satisfies the Euler-Lagrange equations

$$F_s - \partial_i F^i_s = 0; \quad s = \overline{1,n}; \; i = 1,2,3, \tag{D.2}$$

where it is denoted

$$F_s = \frac{\partial F}{\partial \phi^s}; \quad F^i_s = \frac{\partial F}{\partial(\partial_i \phi^s)}.$$

Let us write the second variation of the functional

$$\delta^2 \Phi = \int d^3x \left(F_{sr}\xi^s\xi^r + F^{ik}_{sr}\partial_i\xi^s\partial_k\xi^r + 2\,F^i_{sr}\partial_i\xi^s\xi^r\right),$$

where $\xi^s = \delta\phi^s$, and take a particular perturbations in the form $\xi^s = f^j(\boldsymbol{r})\partial_j u^s$. Then we shall find

$$\delta^2 \Phi = \int d^3x \left[\partial_i f^l A^{ik}_{lj} \partial_k f^j + (\partial_i f^l \cdot f^j - \partial_i f^j \cdot f^l) B^i_{jl}\right], \tag{D.3}$$

where it is denoted

$$A^{ik}_{lj} = \partial_l u^r F^{ik}_{rs} \partial_j u^s, \; 2\,B^i_{jl} = -2\,B^i_{lj} = \partial_{[j} F^i_r \partial_{l]} u^r. \tag{D.4}$$

Note, that the second term in (D.3) is sign changing, and by virtue of the Eq. (D.2) the following equality

$$\partial_i B^i_{jl} = 0$$

holds, and it implies

$$2\, B^i_{jl} = \epsilon^{ikm} \partial_k a_{mjl}, \tag{D.5}$$

where, in turn,

$$a_{mjl} = \frac{1}{2\pi} \epsilon_{mki} \partial^k \int d^3x' B^i_{jl}(\boldsymbol{r}') |\boldsymbol{r} - \boldsymbol{r}'|^{-1}. \tag{D.6}$$

Substituting (D.5) into (D.3) after integration by parts one finds

$$\delta^2 \Phi = \int d^3x\, \partial_i f^l \big(A^{ik}_{lj} + \epsilon^{ikm} a_{mjl}\big) \partial_k f^j. \tag{D.7}$$

To study the integrand sign-definiteness in (D.7) consider the asymptotical region $r \to \infty$. Then from (D.1) and (D.4) it follows, that

$$A^{ik}_{lj} = \mathcal{O}\left[r^{-(3+2\alpha)}\right]; \quad B^i_{jl} = \mathcal{O}\left[r^{-(3+2\alpha)}\right], \tag{D.8}$$

and from (D.6) it is easy to deduce the estimate

$$a_{mjl} = \mathcal{O}(r^{-\nu}),$$

where $\nu = 3$ for $\alpha > 1/2$ and $\nu = 2(1+\alpha)$ for $\alpha < 1/2$. Indeed, from (D.2), (D.4) and (D.5) via the integration by parts one obtains the identity

$$\int d^3x\, B^i_{jl} = 0,$$

which allows to transform the equality (D.6) into the form

$$a_{mjl} = \frac{1}{2\pi} \epsilon_{mki} \partial^k \int d^3x' B^i_{jl}(\boldsymbol{r}') \left(\frac{1}{|\boldsymbol{r} - \boldsymbol{r}'|} - \frac{1}{r}\right). \tag{D.6a}$$

From the assumption (D.1) it follows that for some finite value of a takes place the estimate

$$\left|\partial_k \int\limits_{r'>\alpha} d^3x' B^i_{jl}(\boldsymbol{r}')\left(\frac{1}{|\boldsymbol{r} - \boldsymbol{r}'|} - \frac{1}{r}\right)\right|$$
$$< C \int\limits_{r'>\alpha} d^3x' (r')^{(3+2\alpha)} \left|\partial_k \left(\frac{1}{|\boldsymbol{r} - \boldsymbol{r}'|} - -\frac{1}{r}\right)\right| \equiv M_k(r). \tag{D.9}$$

Now either by direct integration of the right-hand side of (D.9), or by the mean-value theorem after some analysis one obtains

$$|a_{mjl}| < C_1 \left|\frac{\boldsymbol{r} - \boldsymbol{r}'}{|\boldsymbol{r} - \boldsymbol{r}'|^3} - \frac{\boldsymbol{r}}{r^3}\right| = \mathcal{O}(r^{-3}), \tag{D.10}$$

what is the required estimate for the behaviour of a_{mjl}. By comparing (D.10) with (D.8) we notice that the second term in (D.7) would be greater than the first one. Thus the quadratic form in (D.7) is sign-changing. □

The proof of the theorem can be easily extended to solitons in $D = N+2 \geq 2$ dimensional models, as in this case instead of (D.5) and (D.6) one will have

$$2\, B^i_{jl} = \epsilon^{ik\alpha_1 \dots \alpha_N}\, \partial_k a_{\alpha_1 \dots \alpha_N jl};$$

$$a_{\alpha_1 \dots \alpha_N jl} = \frac{2}{N!}\, \epsilon_{ki\alpha_1 \dots \alpha_N}\, \partial^k \int d^{N+2}x'\, G\,(\boldsymbol{r} - \boldsymbol{r}') B^i_{jl}(\boldsymbol{r}'),$$

where $G\,(\boldsymbol{r} - \boldsymbol{r}')$ is the Green function of the D-dimensional Laplace equation.

Thus, it has been proved, that the functional of the type (3.27) has the sign changing second variation in the neighbourhood of a soliton solution, and it proves to be valid for any case of space dimension $D \geq 2$, as well as for soliton solutions with harmonic dependence on time. The obtained result one can regard as a generalization of the well-known Hobart-Derrick criterion (see Section 3.6) and it proves that only conditionally stable solitons are possible, i.e. they can be stable under some constraints on admissible perturbations. For instance, such constraints emerge in the case of chiral solitons, endowed with the topological charge $\mathbf{Q}$.

D.2 Existence of the Axially-Symmetric Solutions

The proof of existence of the axially-symmetric configurations, which are minimizers of the Skyrme model energy functional in higher homotopy classes as it was shown in Chapter 7, can be performed almostly along the lines of the proof of the Skyrmion existence (see Section 6.3). Once again we introduce the notations (5.7), dimensionless coordinates $\boldsymbol{r} \to \epsilon\lambda\boldsymbol{r}'$ and investigate the shifted functional

$$4\pi\, \frac{\epsilon}{\lambda}\, \mathcal{I} \stackrel{\text{def}}{=} \mathbf{H} - 6\sqrt{2}\pi^2 \frac{\epsilon}{\lambda}\, |\mathbf{Q}|, \tag{D.11}$$

taking it in the form (5.8). For the axisymmetric ansatz (7.5):

$$\Theta = \Theta(r, \vartheta), \quad \beta = \beta(r, \vartheta), \quad \gamma = m\alpha, \quad m \in \mathbb{Z}, \tag{D.12}$$

where (r, ϑ, α) are spherical coordinates, one can introduce orts $\boldsymbol{e}_1 = \boldsymbol{e}_r$, $\boldsymbol{e}_2 = \boldsymbol{e}_\vartheta$, $\boldsymbol{e}_3 = \boldsymbol{e}_\alpha$, so that the components of vectors $\boldsymbol{X}, \boldsymbol{Y}, \boldsymbol{Z}$ from Eqs. (5.7) would take the form:

$$\begin{aligned} X_1 &= \partial_r \Theta; \quad X_2 = r^{-1}\partial_\vartheta \Theta; \\ Y_1 &= \partial_r \beta; \quad Y_2 = r^{-1}\partial_\vartheta \beta; \\ Z_3 &\equiv Z = \frac{m}{r \sin\vartheta}\, \sin\Theta \sin\beta. \end{aligned} \tag{D.13}$$

In terms of (D.13) the functional $\mathcal{I}$ from (D.11) will read

$$\mathcal{I} = \tfrac{1}{2}\int_0^\infty r^2 dr \int_0^\pi \sin\vartheta d\vartheta \Big\{ \tfrac{1}{2} Z^2 + 3\sqrt{2}Z\,(X_1Y_2 - X_2Y_1)$$
$$+ (X_1^2 + X_2^2 + Y_1^2 + Y_2^2)\,(\tfrac{1}{2} + Z^2) + (X_1Y_2 - X_2Y_1)^2 \Big\}. \qquad \text{(D.14)}$$

To proceed we derive the second variation $F \equiv \delta^2\mathcal{I}$ of the integrand in (D.14) with respect to $\delta\boldsymbol{X} = \boldsymbol{x}$, $\delta\boldsymbol{Y} = \boldsymbol{y}$ and obtain

$$F = A\,(\boldsymbol{x}^2 + \boldsymbol{y}^2) + B(\boldsymbol{x}\times\boldsymbol{y})^2 + (x_1Y_2 - x_2Y_1 + y_2X_1 - y_1X_2)^2,$$

where it was denoted

$$A = (\tfrac{1}{2} + Z^2)\,; \quad B = X_1Y_2 - X_2Y_1 + \tfrac{3}{2}\,Z.$$

The necessary condition for the functional $\mathcal{I}$ to attain the minimum on the axisymmetric configuration (D.12), when keeping fixed the value of the topological charge $|\mathbf{Q}| \geq 2$, i.e. for $|m| \geq 2$, is the fulfilment of the inequality $F \geq 0$, which proves to hold if $|B| \leq A$, or in detail

$$-\frac{1}{2}\,Z^2 - \frac{3}{\sqrt{2}}\,Z \leq X_1Y_2 - X_2Y_1 \leq \frac{1}{2}\,Z^2 - \frac{3}{\sqrt{2}}\,Z. \qquad \text{(D.15)}$$

Taking a positive number δ_1, from (D.15) it is readily seen, that in the region $r\sin\vartheta > \delta_1 > 0$ the gradients $\nabla\Theta$, $\nabla\beta$ appear to be bounded. Therefore we choose a spatial domain $\Omega \subset \mathbb{R}^3$, defined by restrictions

$$r\sin\vartheta > \delta_1, \quad \sin\Theta > \delta, \quad r < R < \infty,$$

and consider the functional $\mathcal{I}(\Omega)$, as restricted on Ω.

Next we construct a minimizing sequence of functions $\{\Theta_n, \beta_n\}$ from the Sobolev space $W^1_\infty(\Omega)$. Due to the Tikhonov-Alaoglu theorem such a sequence would be a weakly* convergent one, i.e. convergent with respect to the weak topology $\sigma\big(W^1_\infty, W^1_1\big)$ [definition of the latter concept one can find in any available modern textbook on functional analysis, e.g. (Kolmogorov and Fomin, 1972)]. In their turn the sequences $\{\sin\Theta_n\}$ and $\{\sin\beta_n\}$ converge strongly in $\mathbf{C}[\Omega]$, and sequences $\{\boldsymbol{X}_n, \boldsymbol{Y}_n\}$ are weakly* convergent in $L_\infty(\Omega)$.

Now we make use of the Mazur lemma (Ekeland and Temam, 1976), which allows one to construct a convex combination of sequences

$$\mathrm{Co}\boldsymbol{Y}_n = \sum_{k=n}^{N} \lambda_k \boldsymbol{Y}_k, \quad \sum_{k=n}^{N} \lambda_k = 1, \quad \lambda_k \geq 0, \qquad \text{(D.16)}$$

which is strongly convergent in $L_2(\Omega)$. Since the functional $\mathcal{I}$ is definitely convex with respect to $\boldsymbol{Y}$, then the sequence (D.16) remains to be the minimizing one. Finally, one can represent the term $K = X_1Y_2 - X_2Y_1$, in the functional $\mathcal{I}$ as the product of two sequences $(\boldsymbol{X}_n \times \mathrm{Co}\boldsymbol{Y}_n)$, one of them $\{\boldsymbol{X}_n\}$ is weakly* convergent in $L_\infty(\Omega)$, and another one $\{\mathrm{Co}\boldsymbol{Y}_n\}$ is strongly convergent in $L_2(\Omega)$. Therefore the sequence K_n weakly converges in $L_2(\Omega)$. By virtue of the fact , that a squared norm is weakly semicontinuous from below, the functional $\mathcal{I}$ also possesses this property. Thus we come to the conclusion that the minimum of the functional $\mathcal{I}(\Omega)$ is attainable on some limit functions. It only remains to prove that this result holds true when $\Omega \to \mathbb{R}^3$, i.e. when we put $\delta_1 = \delta = 0$, $R = \infty$, with a preliminary proof that $\mathcal{I}(\Omega)$ is continuously dependent on δ_1, δ, R [for more details consult with (Rybakov, 1991; Makhankov, Rybakov and Sanyuk, 1992)].

D.3 Existence of a Nonsingular Matrix

As it was promised in Section 11.2.1 we expose the proof of existence of a nonsingular matrix, denoted by $N_{\alpha\beta}$, with thereby introduced "angular momentum" variables in (11.33). Note that this problem is of general type and arises whenever elements of a group G appear as dynamical variables (Balachandran, Marmo, Skagerstam and Stern, 1983). In particular, this is the case when one sets up the canonical formalism in the framework of collective coordinate quantization.

To be precise, instead of general case, we address the collective variable $A(t) \in SU(3)$, which can be parametrized via

$$A[\zeta(\rho)] = \exp(\mathrm{i}\lambda_\alpha \rho_\alpha) A[\zeta]; \quad \zeta(0) = \zeta, \tag{D.17}$$

where $\zeta(\rho) = \{\zeta_1(\rho), \dots \zeta_8(\rho)\}$ and $\rho = (\rho_1, \dots, \rho_8)$ are sets of functions and parameters, respectively; λ_α, $\alpha = \overline{1,8}$, are $SU(3)$ generators, which are known to form a basis for the Lie algebra. Differentiating (D.17) with respect to ρ_α and setting $\rho = 0$ at the final stage, we find

$$\mathrm{i}\lambda_\alpha A[\zeta] = \frac{\partial A[\zeta]}{\partial \zeta_\beta} \cdot N_{\alpha\beta}(\zeta); \tag{D.18}$$

so that

$$N_{\alpha\beta}(\zeta) = \frac{\partial \zeta_\alpha}{\partial \rho}\Big|_{\rho=0}.$$

Let as assume the opposite, i.e. that thus obtained matrix $N_{\alpha\beta}(\zeta)$ is a singular one, or equivalently that $\det N = 0$. Then there exists a set of variables $\{\chi_\alpha\}$, not all equal to zero, such that the linear combination

$$\chi_\alpha \cdot N_{\alpha\beta} = 0. \tag{D.19}$$

Multiplying (D.18) with χ_α we have

$$\mathrm{i}\chi_\alpha \lambda_\alpha A[\zeta] = \chi_\alpha N_{\alpha\beta}(\zeta) \cdot \frac{\partial A[\zeta]}{\partial \zeta_\beta} = 0,$$

and hence $\chi_\alpha \lambda_\alpha = 0$. But the latter statement contradicts the linear independence of the basis elements λ_α, or holds true only if all $\chi_\alpha = 0$. Thus we come up to a contradiction with the previous assumption (D.19), which proves that $N_{\alpha\beta}$ should be a nonsingular matrix.

Appendix E

Finkelstein-Williams' Spinor Structures

Following Williams (1970) we first show how to reduce calculations of the fundamental group $\pi_1(\Sigma_0)$ (with notations given in Section 10.1) to that of $\pi_4(\Phi)$, i.e. to establish isomorphism between elements of these groups. Consider as a representative of the homotopic class Σ_0 the mapping

$$\eta : \mathbb{R}^1 \to \Sigma_0; \quad \eta(-\infty) = \eta(+\infty) = \sigma_0, \tag{E.1}$$

where $\sigma_0 \in \Sigma_0$ is a fixed point, so that the path $\eta \in \Sigma_0$ starts and ends at this point. Next we define the map

$$F : \mathbb{R}^4 \to \Phi; \quad F(\boldsymbol{x}, s) = [\eta(s)](\boldsymbol{x}). \tag{E.2}$$

Note that $F(\boldsymbol{x}, s) \to \phi_0 \in \Phi$, whenever either $|\boldsymbol{x}|$ or s tends to infinity. Thereby the relation between the paths (E.1) and the mappings (E.2) is established. However, the map (E.2) belongs to one of homotopic classes, which is an element of the group $\pi_4(\Phi)$. Since the triviality or non-triviality of the path η in Σ_0 now depends on that of the map F, one can say that groups $\pi_1(\Sigma_0)$ and $\pi_4(\Phi)$ are isomorphic:

$$\pi_1(\Sigma_0) \simeq \pi_4(\Phi) \tag{E.3}$$

in support to the statement (10.8). By virtue of the Whitehead theorem (10.4) from (E.3) it follows that

$$\pi_1(\Sigma_1) \simeq \pi_4(\Phi), \tag{E.4}$$

and it allows one to perform the aforementioned reduction. Recall, that we call the maps **trivial**, if they are homotopically equivalent to constant maps given by (10.7).

Now we display Williams' argumentation in proof that nontrivial paths (among those starting and ending at a point $\phi_1 \in \Sigma_1$) are paths, which correspond to rotations of a soliton through the angle 2π [see also exposition of this result in (Pak and Tze, 1979)]. For that we should choose a map from Σ_1 $\{\varphi_1 : \mathbb{R}^3 \to \mathbb{S}^3;\ \deg(\varphi_1) = 1\}$ by putting e.g. $\varphi_1(\boldsymbol{x}) = (\phi_0, \phi_1, \phi_2, \phi_3)$ with (ϕ_0, ϕ^a) being the components of stereographic projection (10.10) at $R^{ab}(\lambda) = \mathbb{1}$. For simplicity consider a rotation about z-axis: $\boldsymbol{x} \in \mathbb{R}^3$

$$x_i \to x'_i = R_{ij}(\lambda)x_j; \quad \|R_{ij}(\lambda)\| = \begin{pmatrix} \cos 2\pi\lambda & \sin 2\pi\lambda & 0 \\ -\sin 2\pi\lambda & \cos 2\pi\lambda & 0 \\ 0 & 0 & 1 \end{pmatrix}. \tag{E.5}$$

As the parameter λ varies between 0 and 1 one obtains a continuous sequence of mappings (10.9), (10.10), which determines a closed path (a loop) in Σ_1, To check whether this path is trivial or not, it proves to be convenient to present it as a mapping between two spheres: $\mathbb{S}^3 \to \mathbb{S}^3$ topologically equivalent to the identical one: $\tilde{\varphi}_1 : \mathbb{S}^3 \to \mathbb{S}^3$, with the latter defined as

$$\phi_a = \psi_a, \ a = \overline{1,3}; \ \phi_0 = \psi_0.$$

Here (ψ_0, ψ^a) is a point of the pre-image sphere, and (ϕ_0, ϕ^a) is the corresponding point on the image sphere. In this way, the path on the image-sphere $\mathbb{S}^3$ might be obtained by mapping the point (ψ_0, ψ^a) through $\tilde{\varphi}_1$ with the subsequent rotation of the pre-image sphere via the matrix $R_{ij}(\lambda)$. While doing this we have to be sure that the desired correspondence of image and pre-image points is fulfilled, namely

$$\phi_a = R_{ab}(\lambda)\psi_b; \quad \phi_0 = \psi_0; \quad a, b = \overline{1,3}. \tag{E.6}$$

The space of mappings $\mathrm{Map}(\mathbb{S}^3, \mathbb{S}^3) \equiv \{\Sigma\} = \cup_i \Sigma_i$, when endowed with compact-open topology, has a structure of a topological group, since the range (the image-sphere) $\mathbb{S}^3$ definitely has the group structure of $SU(2)$. In virtue of this fact (Montgomery and Zippin, 1955) the pathwise-connected components of a topological group $\{\Sigma\}$ show to be homeomorphic, so homotopy classes Σ_i are all homeomorphic to each other as well.

The path (E.6) in Σ_1 one can write as an unitary matrix

$$\begin{aligned} U_1(\lambda) &= \begin{pmatrix} \phi_0 + \mathrm{i}\phi_3 & \phi_2 + \mathrm{i}\phi_1 \\ -\phi_2 + \mathrm{i}\phi_1 & \phi_0 - \mathrm{i}\phi_3 \end{pmatrix} \\ &= \begin{pmatrix} \psi_0 + \mathrm{i}R_{3b}\psi_b & R_{2b}\psi_b + \mathrm{i}R_{1b}\psi_b \\ -R_{2b}\psi_b + \mathrm{i}R_{1b}\psi_b & \psi_0 - \mathrm{i}R_{3b}\psi_b \end{pmatrix}. \end{aligned} \tag{E.7}$$

Via an appropriate homeomorphism the path in Σ_1 can be transformed to a path in Σ_0. The simplest way to construct such a homeomorphism comes from a clear physical idea. Since U_1 'creates' a soliton, then U_1^{-1} has to 'create' an antisoliton, which annihilates U_1. So $U_1^{-1} \cdot U_1$ is an automorphism which transfers a path from the 1-soliton sector (Σ_1) to the 0-soliton sector (Σ_0). Thus a path in Σ_0 might be constructed via that in Σ_1 as

$$\begin{aligned} U_0(\lambda) &= U_1^{-1}(0) \cdot U_1(\lambda) \\ &= \begin{pmatrix} \psi_0 - \mathrm{i}\psi_3 & -\psi_2 - \mathrm{i}\psi_1 \\ \psi_2 - \mathrm{i}\psi_1 & \psi_0 + \mathrm{i}\psi_3 \end{pmatrix} \begin{pmatrix} \phi_0 + \mathrm{i}\phi_3 & \phi_2 + \mathrm{i}\phi_1 \\ -\phi_2 + \mathrm{i}\phi_1 & \phi_0 - \mathrm{i}\phi_3 \end{pmatrix}. \end{aligned} \tag{E.8}$$

Correspondingly the path in Σ_0 with identified starting ($\lambda = 0$) and ending points ($\lambda = 1$) would be

$$U_0(0) = U_1^{-1}(0) \cdot U_1(0) = \mathbb{1} = \begin{pmatrix} 1 & 0 \\ 0 & 1 \end{pmatrix}, \tag{E.9}$$

which corresponds to the constant map $\phi_0 = 1,\ \phi_a = 0$. Taking (E.7) in the explicit form one obtains

$$\begin{aligned}
\phi_1 &= \psi_0[\psi_1(\cos 2\pi\lambda - 1) + \psi_2 \sin 2\pi\lambda] \\
&\quad - \psi_3[-\psi_1 \sin 2\pi\lambda + \psi_2(\cos 2\pi\lambda - 1)]; \\
\phi_2 &= \psi_0[\psi_2(\cos 2\pi\lambda - 1) - \psi_1 \sin 2\pi\lambda] \\
&\quad + \psi_3[\psi_2 \sin 2\pi\lambda + \psi_1(\cos 2\pi\lambda - 1)]; \\
\phi_3 &= -(\psi_1^2 + \psi_2^2) \sin 2\pi\lambda; \\
\phi_0 &= 1 + (\psi_1^2 + \psi_2^2)(\cos 2\pi\lambda - 1).
\end{aligned} \tag{E.10}$$

In accord with (E.2) the path (E.10) can be associated to the map

$$f : \mathbb{S}^4 \to \mathbb{S}^3, \tag{E.11}$$

with the parameter λ being an extra variable, describing in a sense splitting of $\mathbb{S}^4$ into spheres $\mathbb{S}^3$. The path (E.10) in Σ_0 is nontrivial if and only if the mapping (E.11) is nontrivial. Since $\pi_4(\mathbb{S}^3) = \mathbb{Z}_2$, it indicates that there are only two homotopy classes $\Sigma_0{}'$ and $\Sigma_1{}'$, so that all the nontrivial maps have to belong to $\Sigma_1{}'$. The essential point of Williams' proof is the observation that for $\lambda = 1/2$ one finds from (E.10)

$$\begin{aligned}
\phi_1 &= -2(\psi_0\psi_1 - \psi_3\psi_2); \\
\phi_2 &= -2(\psi_0\psi_2 + \psi_3\psi_1); \\
\phi_3 &= 1 - 2(\psi_1^2 + \psi_2^2),
\end{aligned} \tag{E.12}$$

which is the celebrated Hopf map $\kappa : \mathbb{S}^3 \to \mathbb{S}^2,\quad \pi_3(\mathbb{S}^2) = \mathbb{Z}$, [see e.g. (Steenrod, 1951)]. Hence, f from (E.11) being a suspension of the Hopf map is definitely nontrivial. This completes the proof.

References

Introduction

Adkins, G.S. (1987): "Static Properties of Skyrmions," in *Chiral Solitons,* ed. by K.F. Liu (World Scientific, Singapure), pp. 99–170

Balachandran, A.P. (1986): "Skyrmions," in: *High Energy Physics 1985,* Proc. of the Yale Theoretical Advanced Study Institute, ed. by M.J. Bowick, F. Gürsey (World Scientific, Singapure), Vol. I, pp. 1–82

Brown, G.E., Rho, M. (1986): "Towards a Basis in QCD for Nuclear Physics," *Comm. Nucl. Part. Phys.* ***XV*** pp. 245–267

Chodos, A., Hadjimichael, E., Tze Ch., eds. (1984): *Solitons in Nuclear and Elementary Particle Physics,* Proc. of the Lewes Workshop, (World Scientific, Singapure)

Dalitz, R.N. (1988): "An Outline of Life and Work of Tony Hilton Royle Skyrme (1922–1987)," *Int. J. Mod. Phys.* ***A3*** pp. 2719–2744

Elliott, J.P., Skyrme, T.H.R. (1956): "Center of Mass Effects in the Nuclear Shell Model," *Proc. Roy. Soc. London* ***A232*** pp. 561–566

Holzwarth, G., Schwesinger, B. (1986): "Baryons in the Skyrme Model," *Rep. Progr. Phys.* **49** pp. 825–871

Jezabek, M., Praszalowicz, M., eds. (1987): *Skyrmions and Anomalies,* Proc. of the Krakow Workshop, (World Scientific, Singapure)

Kelvin, W.T. (1904): *Baltimore Lectures on Molecular Dynamics and the Wave Theory of Light,* (Clay&Son, Baltimore)

Kelvin, W.T. (1910): *Mathematical and Physical Papers,* vol. IV Hydrodynamics and General Dynamics, (Cambridge Univ. Press, Cambridge) pp. 1–69

Makhankov, V.G., Rybakov, Yu.P., Sanyuk, V.I. (1989): *The Skyrme Model and Solitons in Hadron Physics,* Lecture Notes for Young Scientists Series, No. 55 (JINR Publ., Dubna), (in Russian) 172 pp.

Makhankov, V.G., Rybakov, Yu.P., Sanyuk, V.I. (1992): "The Skyrme Model and Strong Interactions," *Usp. Fiz. Nauk* **162** pp. 1–61 (in Russian), [Engl. transl. Sov. Phys. Usp. **35** (1992) pp. 55–84]

Meissner, U.-G. (1988): "Low-Energy Hadron Physics from Effective Chiral Lagrangians with Vector Mesons," *Phys. Reports* **161** pp. 213–362

Meissner, U.-G., Zahed, I. (1986): "Skyrmions in Nuclear Physics," in *Adv. Nucl. Phys* **17** ed. by J.W. Negele, E. Vogt (Plenum Press, New York - London) pp. 143–222

Pak, N.K., Tze, H.C. (1979): "Chiral Solitons and Current Algebra," *Ann. Phys.* **117** pp. 164–194

Sanyuk, V.I. (1992): "Genesis and Evolution of the Skyrme Model from 1954 till the Present," *Int. J. Mod. Phys.* ***A7*** pp. 1–40

Skyrme, T.H.R. (1954): "A New Model for Nuclear Matter," *Proc. Roy. Soc. London* ***A226*** pp. 521–530

Skyrme, T.H.R. (1955): "Meson Theory and Nuclear Matter," *Proc. Roy. Soc. London* ***A230*** pp. 277–286

Skyrme, T.H.R. (1958): "A Nonlinear Theory of Strong Interactions," *Proc. Roy. Soc. London* ***A247*** pp. 260–278

Skyrme, T.H.R. (1959): "The Effective Nuclear Potential," *Nucl. Phys.* **9** pp. 615–634

Skyrme, T.H.R. (1961): "A Nonlinear Field Theory," *Proc. Roy. Soc. London* **A260** pp. 127–138

Skyrme, T.H.R. (1962): "A Unified Field Theory of Mesons and Baryons," *Nucl. Phys.* **31** pp. 556–569

Skyrme, T.H.R. (1988): "The Origins of Skyrmions," *Int. J. Mod. Phys.* **A3** pp. 2745–2751

Zahed, I., Brown, G.E. (1986): "The Skyrme Model," *Phys. Reports* **142** pp. 1–102

Chapter 1

Ablowitz, M.J., Kaup, D.J., Newell, A.C., Segur, H. (1973): "Method for Solving the sine-Gordon Equation," *Phys. Rev. Lett.* **30** pp. 1262–1265

Balachandran, A.P., Gomm, H., Sorkin, R.D. (1987): "Quantum Symmetries From Quantum Phases, Fermions from Bosons, a Z_2 Anomaly and Galilean Invariance," *Nucl. Phys.* **B281** pp. 573–612

Canright, G.S., Johnson, M.D. (1990): "Anyons and Condensed Matter: Why There is More Room in Two Dimensions?" *Comm. Cond. Matt. Phys.* **XV** pp. 77–97

Caudrey, P.J., Gibbon, J.D., Eilbeck, J.C., Bullough, R.K. (1973): "Exact Multisoliton Solutions of the Self-Induced Transparency and sine-Gordon Equations," *Phys. Rev. Lett.* **30** pp. 237–238

Chew, G.F. (1961): *S-Martix Theory of Strong Interactions,* (W.A. Benjamin Inc., New York, Amsterdam)

Coleman, S. (1975): "Quantum sine-Gordon Equations as the Massive Thirring Model," *Phys. Rev.* **D11** pp. 2088–2097

Dalitz, R.N. (1988): "An Outline of Life and Work of Tony Hilton Royle Skyrme (1922–1987)," Int. J. Mod. Phys. **A3** pp. 2719–2744

Faddeev, L.D. (1975): "Hadrons from Leptons?," *ZhETF Pis'ma* **21** pp. 141–144 (in Russian), [Engl. transl. *JETP Lett.* **21** (1975) pp. 64–65]

Finkelstein, D. (1966): "Kinks," *J. Math. Phys.* **7** pp. 1218–1225

Finkelstein, D., Misner, C. (1959): "Some New Conservation Laws," *Ann. Phys. (N.Y)* **6** pp. 230–243

Flowers, B. H. (1952): "Magnetic Moments and Quadrupole Moments of Odd-mass Nuclei," *Phil. Mag.* **43** pp. 1330–1334

Gell-Mann, M., Lévy, M. (1960): "The Axial Vector Current in Beta Decay," *Nuovo Cim.* **16** pp. 705–726

Gürsey, F. (1960): "On the Symmetries of Strong and Weak Interactions," *Nuovo Cim.* **16** pp. 230–240

Gürsey, F. (1961): "On the Structure and Parity of Weak Interaction Currents," *Ann. Phys. (N.Y)* **12** pp. 91–117

Hirota, R. (1972): "Exact Solutions of the sine-Gordon Equations for Multiple Collisions of Solitons," *J. Phys. Soc. Jap.* **33** pp. 1459–1463

t'Hooft, G. (1974): "A Planar Diagram Theory for Strong Interactions," *Nucl. Phys.* **B72** pp. 461–473

Jacob, M., Chew, G.F. (1964): *Strong-Interaction Physics,* (W.A. Benjamin Inc., New York, Amsterdam)

Kelvin, W.T. (1904): Baltimore Lectures on Molecular Dynamics and the Wave Theory of Light (Clay & Son, Baltimore)

Mackenzie, R., Wilczek, F. (1988): "Peculiar Spin and Statistics in Two Space Dimensions," *Int. J. Mod. Phys.* **A3** pp. 2827–2853

Mandelstam, S. (1975): "Soliton Operators for the Quantized sine-Gordon Equation," *Phys. Rev.* **D11** pp. 3026–3030

Pais, A. (1953): "Isotopic Spin and Mass Quantization," *Physica* **19** pp. 869–887

Pais, A. (1954): "On the Program of Systematization of Particles and Interactions," *Proc. Nat. Acad. Sci. (USA)* **40** pp. 484–492

Pauli, W. (1946): *Meson Theory of Nuclear Forces,* (Interscience, New York)

Perring, J.K., Skyrme, T.H.R. (1962): "A Model Unified Field Equations," *Nucl. Phys.* **31** pp. 550–555

Rajaraman, R. (1982): *Solitons and Instantons – An Introduction to Solitons and Instantons in Quantum Field Theory,* (North-Holland Publ. Co., Amsterdam)

Sanyuk, V.I. (1990a): "The Skyrme Model and Hadron Structure in QCD," in *"Solitons and Applications,"* Proc. of IV Int. Workshop, Dubna, ed. by V.G. Makhankov, V.K. Fedyanin, O.K. Pashaev (World Scientific, Singapure), pp. 374–378

Sanyuk, V.I. (1990b): "The Vortex Nature of the Skyrmions," in *"Problems of High Energy Physics and Field Theory,"* Proc. of the XII Workshop, Protvino, ed. by V.A. Petrov (Nauka, Moscow), pp. 254–259

Schwinger, J. (1957): "A Theory of the Fundamental Interactions," *Ann. Phys. (N.Y)* **2** pp. 407–434

Skyrme, T.H.R. (1954): "A New Model for Nuclear Matter," *Proc. Roy. Soc. London* **A226** pp. 521–530

Skyrme, T.H.R. (1955): "Meson Theory and Nuclear Matter," *Proc. Roy. Soc. London* **A230** pp. 277–286

Skyrme, T.H.R. (1958): "A Nonlinear Theory of Strong Interactions," *Proc. Roy. Soc. London* **A247** pp. 260–278

Skyrme, T.H.R. (1959): "A Unified Model of K- and π-Mesons," *Proc. Roy. Soc. London* **A252** pp. 236–245

Skyrme, T.H.R. (1961a): "A Nonlinear Field Theory," *Proc. Roy. Soc. London* **A260** pp. 127–138

Skyrme, T.H.R. (1961b): "Particle States of Quantized Meson Fields," *Proc. Roy. Soc. London* **A262** pp. 237–245

Skyrme, T.H.R. (1962): "A Unified Field Theory of Mesons and Baryons," *Nucl. Phys.* **31** pp. 556–569

Skyrme, T.H.R. (1971): "Kinks and the Dirac Equations," *J. Math. Phys.* **12** pp. 1755–1764

Skyrme, T.H.R. (1988): "The Origins of Skyrmions," *Int. J. Mod. Phys.* **A3** pp. 2745–2751

Sommerfield, C. (1968): "Currents as Dynamical Variables," *Phys. Rev.* **176** pp. 2019–2025

Sugawara, H. (1968): "A Field Theory of Currents," *Phys. Rev.* **170** pp. 1659–1662

Thirring, W.E. (1958): "A Soluble Relativistic Field Theory," *Ann. Phys. (N.Y)* **3** pp. 91–112

Tze, Ch.-H. (1987): "Skyrmions and Current Algebra," in *Chiral Solitons,* ed. by K.F. Liu (World Scientific, Singapure), pp. 1–63

Williams, J. (1970): "Topological Analysis of Nonlinear Field Theory," *J. Math. Phys.* **11** pp. 2611–2616

Witten, E. (1979): "Baryons in the 1/N Expansion," *Nucl. Phys.* **B160** pp. 57–115

Zakharov, V.E., Takhtadzhian, L.A., Faddeev, L.D. (1974): "Complete Description of the "sine-Gordon" equation," *Dokl. Akad. Nauk SSSR* **219** pp. 1334–1337 (in Russian), [Engl. transl. *Soviet Physics Doklady* **19** (1975) pp. 824–826]

Chapter 2

Barnes, K.J., Ketley, I.J., Nicole, D.A., O'Donnel, P.J. (1977): "Nonlinear Chiral Models and Many Dimensional Solitons," *Phys. Rev.* **D16** pp. 511–516

Bott, R., Tu, L.W. (1982): *Differential Forms in Algebraic Topology,* (Springer-Verlag, New York, Heidelberg, Berlin)

Boya, L.J., Cariñena, J.F., Mateos J. (1978): "Homotopy and Solitons," *Fortschr. Phys.* **26** pp. 175–214

Chang, P., Gürsey, F. (1967): "Unified Formulation of Effective Nonlinear Pion-Nucleon Lagrangians," *Phys. Rev.* **164** pp. 1752–1761

Choquet-Bruhat, Y., De Witt-Morette, C., Dillard-Bleick, M. (1977): *Analysis, Manifolds and Physics,* (North-Holland Publ. Co., Amsterdam, New York, Oxford)

Coleman, S. (1977): "Classical Lumps and Their Quantum Descendants," in *New Phenomena in Sub-Nuclear Physics*, ed. by A. Zichichi (Plenum Press, New York) pp. 297–421

Dubrovin, B.A., Fomenko, A.T., Novikov, S.P. (1984a): *Modern Geometry – Methods and Applications*, Part I The Geometry of Surfaces, Transformation Groups and Fields (Springer-Verlag, New York, Berlin, Heidelberg, Tokyo)

Dubrovin, B.A., Novikov, S.P., Fomenko, A.T. (1984b): *Sovremennaja Geometria: Metody Teorii Gomologii*, (Nauka, Moskva) (in Russian)

Dubrovin, B.A., Fomenko, A.T., Novikov, S.P. (1985): *Modern Geometry – Methods and Applications*, Part II The Geometry and Topology of Manifolds (Springer-Verlag, New York, Berlin, Heidelberg, Tokyo)

Ebrahim, A., Gürsey, F. (1974): "General Chiral $SU_2 \times SU_2$ Lagrangian and Representation Mixing," *Nuovo Cim. Lett.* **9** pp. 9–14

Ebrahim, A., Savci, M. (1984): "General Nonlinear Realization of Chiral $SU(2) \otimes SU(2)$ Symmetry in Curved Isospin Space," *J. Math. Phys.* **25** pp. 1746–1751

Eguchi, T., Gilkey, P.B., Hanson, A.J. (1980): "Gravitation, Gauge Theories and Differential Geometry," *Phys. Reports* **66** pp. 213–393

Faddeev, L.D. (1976a): "Some Comments on the Many-Dimensional Solitons," *Lett. Math. Phys.* **1** pp. 289–296

Faddeev, L.D. (1976b): "In Search for Multidimensional Solitons," in *Nonlocal, Nonlinear and Non-Renormolizable Field Theories*, (JINR, Dubna), pp. 207–223 (in Russian)

Felsager, B. (1981): *Geometry, Particles and Fields*, (Odense Univ. Press)

Finkelstein, D. (1966): "Kinks," *J. Math. Phys.* **7** pp. 1218–1225

Finkelstein, D., Misner, C. (1959): "Some New Conservation Laws," *Ann. Phys.* **6** pp. 230–243

Gasiorovicz, S., Geffen, D.A. (1969): "Effective Lagrangians and Field Algebras with Chiral Symmetry," *Rev. Mod. Phys.* **41** pp. 531–573

Goddard, P., Mansfield, P. (1986): "Topological Structures in Field Theory," *Rep. Prog. Phys.* **49** pp. 725–781

Gürsey, F. (1960): "On the Symmetries of Strong and Weak Interactions," *Nuovo Cim.* **16** pp. 230–240

Gürsey, F. (1961): "On the Structure and Parity of Weak Interaction Currents," *Ann. Phys. (N.Y)* **12** pp. 91–117

Hu, S.T. (1959): *Homotopy Theory*, (Academic Press, New York)

Isham, C.J. (1977): "Topological Currents for Arbitrary Chiral Groups in Three Space Dimensions," *J. Phys.* **A10** pp. 1397–1407

Jackson, A., Jackson, A.D., Goldhaber, A.S., Brown, G.E., Castillejo L.S. (1985): "A Modified Skyrmion," *Phys. Lett.*, **B154** pp. 101–106

Kundu, A., Rybakov, Yu.P., Sanyuk, V.I. (1979): "Topological Solitons in the Gauge Skyrme Model," *Indian J. Pure & Appl. Phys.* **17** pp. 673–677

Marleau, L. (1990): "The Skyrme Model and Higher Order Terms," *Phys. Lett.* **B235** pp. 141–146

Meetz, K. (1969): "Realization of Chiral Symmetry in Curved Isospin Space," *J. Math. Phys.* **10** pp. 589–593

Nash, Ch., Sen, S. (1983): *Topology and Geometry for Physicists*, (Academic Press, London, Orlando)

Olshanetskii, M.A. (1982): "A Short Guide to Modern Geometry for Physicists," *Uspekhi Fizicheskikh Nauk* **136** pp. 421–433 (in Russian) [Engl. transl. *Soviet Physics – Uspekhi* **25** (1982)]

Rybakov, Yu.P., Sanyuk, V.I. (1981): "Topological Skyrmeons," Niels Bohr Inst. preprint, NBI-HE-81-49 36 pp.

Skyrme, T.H.R. (1958): "A Nonlinear Theory of Strong Interactions," *Proc. Roy. Soc. London* **A247** pp. 260–278

Skyrme, T.H.R. (1961): "A Nonlinear Field Theory," *Proc. Roy. Soc. London* **A260** pp. 127–138

Steenrod, N. (1951): *The Topology of Fibre Bundles,* (Princeton Univ. Press, Princeton, N.J.)
Weinberg, S. (1967): "Dynamical Approach to Current Algebra," *Phys. Rev. Lett.* **18** pp. 188–191
Weinberg, S. (1969): "Phenomenological Lagrangians," *Physica,* **96** pp. 327–340

Chapter 3

Bogomol'nyi, E.B. (1976): "Stability of Classical Solutions," *Yadernaya Fizika,* **24** pp. 861–870 (in Russian), [Engl. transl. *Sov. J. Nucl. Phys.* **24** (1976) pp. 449–454]
Christ, N.H., Lee, T.D. (1975): "Quantum Expansion of Soliton Solutions," *Phys. Rev.* **D12** pp. 1606–1627
Courant, R., Hilbert, D. (1953): *Methods in Mathematical Physics,* Vol. 1, Chap. VI. (Interscience Publ., New York)
Duff, M.J., Isham, C.J. (1976): "Solitons and Vortex Type Solutions in Nonlinear Chiral Theories," *Nucl. Phys.* **B108** pp. 130–140
Derrick, G.H. (1964): "Comments on Nonlinear Wave Equations as Models for Elementary Particles," *J. Math. Phys.* **5** pp. 1252–1254
Friedberg, R., Lee, T.D., Sirlin, A (1976) "Class of Scalar-Field Soliton Solutions in Three Space Dimensions," *Phys. Rev.* **D13** pp. 2739–2761
Gelfand, I.M., Fomin, S.V. (1963): *Calculus of Variations,* Chap. 5. (Prentice-Hall, Englewood Cliffs, N.J.)
Goldstone, J., Jackiw, R. (1975): "Quantization of Nonlinear Waves," *Phys. Rev.* **D11** pp. 1486–1498
Hietarinta, J. (1977): "On the Existence of Localized Solutions in Nonlinear Chiral Models," *J. Math. Phys.* **18** pp. 65–71
Hobart, R.H. (1963): "On the Instability of a Class of Unitary Field Models," *Proc. Phys. Soc. London* **82** pp. 201–203
Hobart, R.H. (1964): "Nonlinear Field Equilibria," *Proc. Phys.Soc. London* **85** pp. 610–611
Kumar, A., Nisichenko, V.P., Rybakov, Yu.P. (1979): "Stability of Charged Solitons," *Int. J. Theor. Phys.* **18** pp. 425–432
Makhankov, V.G. (1977): "On Stability of Charged Solitons in Klein-Gordon Equation with Saturable Nonlinearity," Preprint JINR P2-10362, Dubna (in Russian) [Engl. transl. *Algorithm and Programs for Solving Certain Physical Problems,* KFKI, Budapest, issue 2, pp. 113–123]
Makhankov, V.G. (1978): "Dynamics of Classical Solitons (in Nonintegrable Systems)," *Phys. Reports* **35** pp. 1–128
Makhankov, V.G. (1980): "Computer Experiments in Soliton Theory," *Comp. Phys. Comm.,* **21** pp. 1–49
Makhankov, V.G. (1983): "Solitons and Numerical Experiment," *Fiz. Elem. Chastits At. Yadra* **14** pp. 123–180 (in Russian) [Engl. transl. *Sov. J. Part. Nucl.* **14** (1983) pp. 50–75]
Makhankov, V.G. (1990): *Soliton Phenomenology,* (Kluwer Academic Publ., Dordrecht)
Malkin, I.G. (1962): "Certain Questions on the Theory of the Stability in the sense of Lyapunov," in *Stability and Dynamics Systems,* Transl. Amer. Math. Soc., Ser. 1, Vol. 5. (Providence, Rhode Island), pp. 1–175
Movchan, A.A. (1960): "Stability of Processes with Respect to two Metrics," *Prikladnaya Matematika i Mechanika* **24** pp. 988–1001 (in Russian)
Rybakov, Yu.P. (1966): "Stability Problems in Nonlinear Field Theory," in *Problems of Gravitation and Elementary Particles Theory,* (Atomizdat, Moscow), pp. 68–97 (in Russian)
Rybakov, Yu.P. (1979): "On Conditional Stability of Regular Solutions in Nonlinear Field Theory," in *Problems of Gravitation and Elementary Particles Theory,* Vol. 10. (Atomizdat, Moscow), pp. 194–202 (in Russian)

Rybakov, Yu.P. (1984): "On Stability of Multiple-Charged Solitons," in *Problems of Gravitation and Elementary Particles Theory*, Vol. 14. (Energoatomizdat, Moscow), pp. 161–169 (in Russian)
Rybakov, Yu.P. (1985a): "Stability of Solitons," in *Problems of Gravitation and Elementary Particles Theory*, Vol. 16. (Energoatomizdat, Moscow), pp. 174–189 (in Russian)
Rybakov, Yu.P. (1985b): *Particles Structure in Nonlinear Field Theory*, (People's Friendship Univ. Press, Moscow) (in Russian)
Rybakov, Yu.P. (1991): "Stability of Multi-Dimensional Solitons in Chiral Models and Gravitation," in *Advances of Science and Technology*, Ser. Classical Field Theory and Theory of Gravitation Vol. 2 (VINITI AN SSSR, Moscow) pp. 56–111 (in Russian)
Rybakov, Yu.P., Chakrabarti, S. (1984): "Conditional Stability of Multiple-Charged Solitons," *Int. J. Theor. Phys.* **23** pp. 325–333
Synge, J.L. (1961): "On a Certain Non-Linear Differential Equations," *Proc. of Royal Irish Acad. Sci.* **62A** pp. 17–41
Tricomi, F.G. (1961): *Differential Equations*, Chap. 3. (Blackie and Son Ltd., London)
Vakhitov, N.G., Kolokolov, A.A. (1973): "Stationary Solutions of the Wave Equation in a Medium with Nonlinear Saturation," *Izvestiya Vuzov, Radiofizika* **16** pp. 1020–1028 (in Russian) [Engl. transl. *Radiophysics and Quantum Electronics* **16** (1973) pp. 783–789]
Zastavenko, L.G. (1965): "Particle-like Solutions for a Nonlinear Wave Equation," *Prikladnaya Matematika i Mechanika* **29** pp. 430–439 (in Russian)
Zubov, V.I. (1957): *Lyapunov's Methods and Their Applications*, (Leningrad State Univ. Press, Leningrad) (in Russian)

Chapter 4

Balachandran, A.P. (1986): "Skyrmions," in *High Energy Physics 1985*, Proc. of the Yale Theoretical Advanced Study Institute, ed. by M.J. Bowick, F. Gürsey (World Scientific, Singapure), Vol. I pp. 1–82
Balachandran, A.P., Barducci, A., Lizzi, F., Rogers, V.G.J., Stern, A. (1984): "Doubly Strange Dibaryon in the Chiral Model," *Phys. Rev. Lett.* **52** pp. 887–890
Coleman, S. (1977): "Classical Lumps and Their Quantum Descendants," in *New Phenomena in Sub-Nuclear Physics*, ed. by A. Zichichi (Plenum Press, New York), pp. 297–421
Dubrovin, B.A., Novikov, S.P., Fomenko, A.T. (1979): *Sovremennaja Geometria*, (Nauka, Moskva) (in Russian)
Ekeland, I., Temam, R. (1976): *Convex Analysis and Variational Problems*, (North-Holland Pub. Co., Amsterdam, Oxford)
Esteban, M.J. (1986): "A Direct Variational Approach to Skyrme's Model for Meson Fields," *Commun. Math. Phys.* **105** pp. 571–591
Faddeev, L.D. (1976): "In Search for Multidimensional Solitons," in *Nonlocal, Nonlinear and Non-Renormolizable Field Theories*, (JINR, Dubna), pp. 207–223 (in Russian)
Ioffe, A.D., Tikhomirov, V.M. (1979): *Theory of Extremal Problems*, (North-Holland Pub. Co., Amsterdam, New York, Oxford)
Jaffe, R.L., Manohar, A. (1990): "The g_1 Problem: Deep Inelastic Electron Scattering and the Spin of the Proton," *Nucl. Phys.* **B337** pp. 509–546
Kapitanskii L.V., Ladyzhenskaya, O.A. (1983): "Coleman's Principle for the Determination of the Stationary Points of Invariant Functionals," *Zapiski Nauchnykh Seminarov Leningradskogo Otdeleniya Matematicheskogo Instituta im. V.A. Steklova AN SSSR* Vol. 127 (Nauka, Leningrad), pp. 84–102 (in Russian), [Engl. transl. *Journ. Sov. Math.* **27** (1984) pp. 2606–2616]
Kundu, A., Rybakov, Yu.P., Sanyuk, V.I. (1979): "Topological Solitons in the Gauge Skyrme Model," *Indian J. Pure & Appl. Phys.* **17** pp. 673–677
Lions, P.L. (1984): "The Concentration-Compactness Principle in the Calculus of Variations," *Ann. Inst. Henri Poincaré, Ann. Non. Lin.* **1** Part 1 – pp. 109–145; Part 2 – pp. 223–283

Lions, P.L. (1985): "The Concentration-Compactness Principle in the Calculus of Variations. The Limit Case," *Rev. Math. Iber.* **1** Part 1 – pp. 145–200; **2** Part 2 – pp. 45–121

Makhankov, V.G., Rybakov, Yu.P., Sanyuk, V.I. (1989): *The Skyrme Model and Solitons in Hadron Physics,* Lecture Notes for Young Scientists Series, No. 55 (JINR Publ., Dubna), (in Russian)

Makhankov, V.G., Rybakov, Yu.P., Sanyuk, V.I. (1992): "The Skyrme Model and Strong Interactions," *Usp. Fiz. Nauk* **162** pp. 1–61 (in Russian) [Engl. transl. Sov. Phys. Usp. **35** (1992) pp. 55–84]

Mandula, J. (1990): "Lattice Simulations of the Anomalous Gluon Contribution to the Proton Spin," *Phys. Rev. Lett.* **65** pp. 1403–1406

Manton, N.S. (1987): "Geometry of Skyrmions," *Commun. Math. Phys.* **111** pp. 469–478

Manton, N.S., Ruback, P.J. (1986): "Skyrmions on Flat Space and Curved Space," *Phys. Lett.* ***B*181** pp. 137–140

Palais, R.S. (1979): "The Principle of Symmetric Criticality," *Commun. Math. Phys.* **69** pp. 19–30

Romanov, V.N., Frolov, I.V., Schwarz A.S. (1978): "Spherically Symmetric Solitons," *Teor. Mat. Fizika* **37** pp. 305–319 (in Russian), [Engl. transl. *Teor. & Math. Physics* **37** (1978) pp. 1038–1046]

Rybakov, Yu.P. (1988): "On Critical Points of Invariant Functionals," in *Actual Problems of Quantum Mechanics and Statistical Physics,* (People's Friendship Univ. Press, Moscow), pp. 71–77 (in Russian)

Waterhouse, W.G. (1983): "Do Symmetric Problems Have Symmetric Solutions?," *Amer. Math. Monthly,* **90** pp. 378–387

Weyl, H. (1951): *Space-Time-Matter,* (Dover, New York)

Chapter 5

Coleman, S., Glaser, V., Martin, A. (1978): "Action Minima Among Solutions of Euclidean Scalar Field Equations," *Commun. Math. Phys.* **58** pp. 211–221

Glaser, V., Martin, A., Grosse, H., Thirring, W. (1976): "A Family of Optimal Conditions for the Absence of Bound States in a Potential," in *Studies in Mathematical Physics,* ed. by E.H. Lieb, B. Simon, A.S. Wightman (Princenton, N.J.) pp. 169–194

Luttinger, J.M. (1973): "Generalized Isoperimetric Inequalities," *J. Math. Phys.* **14** pp. 586–593

Pólya, G., Szegö, G. (1951): *Isoperimetric Inequalities in Mathematical Physics,* (Princeton Univ. Press, Princeton, N.J.)

Rybakov, Yu.P. (1982): "A Search for the Absolute Minimum of Energy in Chiral Models," in *Problems of Gravitation and Elementary Particles Theory,* Vol. 13. (Energoatomizdat, Moscow), pp. 187–192 (in Russian)

Rybakov, Yu.P., Sanyuk, V.I. (1981): "Topological Skyrmeons," Niels Bohr Inst. preprint, NBI-HE-81-49 36 pp.

Rybakov, Yu.P., Sanyuk, V.I. (1991): "Methods to Study (3+1) Localized Structures: Skyrmion as the Absolute Minimum of Energy," Syracuse University preprint SU-4228-473, 38 pp.

Rybakov, Yu.P., Sanyuk, V.I. (1992): "Methods for Studying 3+1 Localized Structures: the Skyrmion as the Absolute Minimizer of Energy," Int. J. Mod. Phys. ***A*7** pp. 3235–3264

Chapter 6

Akhiezer, N.I., Glazman, I.M. (1961–1963): *Theory of Linear Operators in Hilbert Space,* (F.Ungar Pub. Co., New York) Vol. 1 – (1961); Vol. 2 – (1963)

Adkins, G.S., Nappi, C.R., Witten, E. (1983): "Static Properties of Nucleons in the Skyrme Model," *Nucl. Phys.* ***B*228** pp. 552–566

Bogomol'nyi, E.B., Fateev, V.A. (1983): "Asymptotic Calculation of Soliton Masses," *Yadernaya Fizika,* **37** pp. 228–241 (in Russian), [Engl. transl. *Sov. J. Nucl. Phys.* **37** (1983) pp. 134–141]

Gelfand, I.M., Fomin, S.V. (1963): *Calculus of Variations,* Chap. 8. (Prentice-Hall, Englewood Cliffs, N.J.)

Jackson, A.D., Rho, M. (1983): "Baryons as Chiral Solitons," *Phys. Rev. Lett.* **51** pp. 751–754

Nisichenko, V.P. (1981): *Studies of Regular Solutions to Nonlinear Equations in Some Field Models,* Ph.D. Thesis, People's Friendship Univ. (Moscow) unpublished (in Russian)

Romanov, V.N., Frolov, I.V., Schwarz A.S. (1978): "Spherically Symmetric Solitons," *Teor. Mat. Fizika* **37** pp. 305–319 (in Russian), [Engl. transl. *Teor. & Math. Physics* **37** (1978) pp. 1038–1046]

Rybakov, Yu.P. (1979): "On Conditional Stability of Regular Solutions in Nonlinear Field Theory," in *Problems of Gravitation and Elementary Particles Theory,* Vol. 10. (Atomizdat, Moscow), pp. 194–202 (in Russian)

Rybakov, Yu.P., Sanyuk, V.I. (1981): "Topological Skyrmeons," Niels Bohr Inst. preprint, NBI-HE-81-49 36 pp.

Rybakov, Yu.P. (1991): "Stability of Multi-Dimensional Solitons in Chiral Models and Gravitation," in *Advances of Science and Technology,* Ser. Classical Field Theory and Theory of Gravitation Vol. 2 (VINITI AN SSSR, Moscow) pp. 56–111 (in Russian)

Rybakov, Yu.P., Sanyuk, V.I. (1991): "Methods to Study (3+1) Localized Structures: Skyrmion as the Absolute Minimum of Energy," Syracuse University preprint SU-4228-473, 38 pp.

Rybakov, Yu.D., Sanyuk, V.I. (1992): "Methods for Studying 3+1 Localized Structures: the Skyrmion as the Absolute Minimizer of Energy," Int. J. Mod. Phys. **A7** pp. 3235–3264

Sanyuk, V.I. (1979): *Studies of Nonlinear Classical Field Models with Topological Charges,* Ph.D. Thesis, People's Friendship Univ. (Moscow) unpublished (in Russian)

Sobolev, S.L. (1963): *Applications of Functional Analysis in Mathematical Physics,* (Amer. Math. Soc, Providence, Rhode Island)

Skyrme, T.H.R. (1962): "A Unified Field Theory of Mesons and Baryons," *Nucl. Phys.* **31** pp. 556–569

Vainberg, M.M. (1964): *Variational Methods for the Study of Nonlinear Operators,* (Holden-Day Inc., San Francisco, London, Amsterdam)

Chapter 7

Adkins, G.S., Nappi, C.R., Witten, E. (1983): "Static Properties of Nucleons in the Skyrme Model," *Nucl. Phys.* **B228** pp. 552–566

Balachandran, A.P., Barducci, A., Lizzi, F., Rogers, V.G.J., Stern, A. (1984): "Doubly Strange Dibaryon in the Chiral Model," *Phys. Rev. Lett.* **52** pp. 887–890

Bander, M., Hayot, F. (1984): "Instability of Rotating Chiral Solitons," *Phys. Rev.* **D30** pp. 1837–1838

Braaten, E. (1984): "Centrifugal Effects in Skyrmions," in *Solitons in Nuclear and Elementary Particles Physics,* Proc. of the Lewes Workshop, ed. by A. Chodos, E. Hadjimichael, Ch. Tze (World Scientific, Singapure), pp. 57–72

Braaten, E., Carson, L. (1986a): "Deuteron as a Soliton in the Skyrme Model," *Phys. Rev. Lett.* **56** pp. 1897–1900

Braaten, E., Carson, L. (1986b): "Nuclei as Multiskyrmions," in *Workshop on Nuclear Chromodynamics,* ed. by S. Brodsky, E. Moniz (World Scientific, Singapore), pp. 454–462

Braaten, E., Carson, L. (1986c): "Nuclei in the Skyrme Model," in *Relativistic Dynamics and Quark-Nuclear Physics,* ed. by M.B. Johnson, A. Picklesimer (Wiley, New York), pp. 854–861

Braaten, E., Carson, L. (1988): "Deuteron as a Toroidal Skyrmion," *Phys. Rev.* ***D*38** pp. 3525–3539

Braaten, E., Carson, L. (1989): "Deuteron as a Toroidal Skyrmion: Electromagnetic Form Factors," *Phys. Rev.* ***D*39** pp. 838–843

Braaten, E., Carson, L., Townsend, S. (1990): "Novel Structure of Static Multisoliton Solutions in the Skyrme Model," *Phys. Lett.* ***B*235** pp. 147–152

Kopeliovich, V.B., Stern, B.E. (1987): "Exotic Solitons in the Skyrme Model," *ZhETP Pis'ma* **45** pp. 165–168 (in Russian), [Engl. transl. "Exotic Skyrmions," *JETP Lett.* **45** (1987) pp. 203–207]

Kozhevnikov, I.R., Rybakov, Yu.P., Fomin, M.B. (1988): "Structure of Topological Solitons in the Skyrme Model," *Teor. Mat. Fizika* **75** pp. 353–360 (in Russian), [Engl. transl. *Teor. & Math. Physics* **75** (1988) pp. 575–580]

Kudrjavtsev, A.E., Martem'yanov, B.V. (1985): "Does the Rotating Chiral Soliton Exist?," ITEP preprint ITEP-9, 10 pp.

Makhankov, V.G., Rybakov, Yu.P., Sanyuk, V.I. (1989): *The Skyrme Model and Solitons in Hadron Physics*, Lecture Notes for Young Scientists Series, No. 55 (JINR Publ., Dubna), (in Russian)

Makhankov, V.G., Rybakov, Yu.P., Sanyuk, V.I. (1992): "The Skyrme Model and Strong Interactions," *Usp. Fiz. Nauk* **162** pp. 1–61 (in Russian) [Engl. transl. Sov. Phys. Usp. **35** (1992) pp. 55–84]

Potapov, M.V., Rybakov, Yu.P. (1989): "On the Structure of the Rotating Skyrmions," in *"Problems of High Energy Physics and Field Theory,"* Proc. of the XI Workshop, Protvino, ed. by V.A. Petrov (Nauka, Moscow), pp. 236–239 (in Russian)

Rybakov, Yu.P. (1984): "Bound States of Two Skyrmions," in *Problems of Gravitation and Elementary Particles Theory*, Vol. 14. (Atomizdat, Moscow), pp. 166–170 (in Russian)

Rybakov, Yu.P. (1988): "On Critical Points of Invariant Functionals," in *Actual Problems of Quantum Mechanics and Statistical Physics*, (People's Friendship Univ. Press, Moscow), pp. 71–77 (in Russian)

Rybakov, Yu.P., Sutantio, V. (1985): "The Search for an Optimal Configuration of Two Bounded Skyrmions," in *Problems of Gravitation and Elementary Particles Theory*, Vol. 15. (Atomizdat, Moscow), pp. 137–141 (in Russian)

Sanyuk, V.I. (1992): "Genesis and Evolution of the Skyrme Model from 1954 till the Present," *Int. J. Mod. Phys.* ***A*7** pp. 1–40

Schramm, A.J., Dothan, L.C., Biedenharn, L.C. (1988): "A Calculation of the Deuteron as a Biskyrmion," *Phys. Lett.* ***B*205** pp. 151–155

Verbaarshot, J.J.M. (1987): "Axial Symmetry of Bound Baryon Number – Two Solution of the Skyrme Model," *Phys. Lett.* ***B*195** pp. 235–239

Verbaarshot, J.J.M., Walhout, T.S., Wambach, J., Wyld, H.W. (1987): "Symmetry and Quantization of the Two-Skyrmions System. The Case of Deuteron," *Nucl. Phys.* ***A*468** pp. 520–538

Chapter 8

Adkins, G.S. (1987): "Static Properties of Skyrmions," in *Chiral Solitons*, ed. by K.F. Liu (World Scientific, Singapure), pp. 99–170

Adkins, G.S., Nappi, C.R., Witten, E. (1983): "Static Properties of Nucleons in the Skyrme Model," *Nucl. Phys.* ***B*228** pp. 552–566

Balachandran, A.P. (1986a): "Solitons in $He^3 - B$," *Nucl. Phys.* ***B*271** pp. 227–252

Balachandran, A.P. (1986b): "Skyrmions," in: *High Energy Physics 1985*, Proc. of the Yale Theoretical Advanced Study Institute, ed. by M.J. Bowick, F. Gürsey (World Scientific, Singapure), Vol. I, pp. 1–82

Balachandran, A.P., Lizzi, F., Rodgers, V.G.J., Stern, A. (1985): "Dibaryons as Chiral Solitons," *Nucl. Phys.* ***B*256** pp. 525–556

Balachandran, A.P., Marmo, G., Skagerstam, B.S., Stern, A. (1991): *Classical Topology and Quantum States*, (World Scientific, Singapore) in press

Blaizot J.P., Ripka, G. (1988): "Nonlinear Distortions of Rotating Chiral Solitons," *Phys. Rev.* ***D38*** pp. 1556–1570

Bogolubov, N.N. (1950): "On a New Form of the Adiabatic Perturbation Theory in the Problem of Interaction Between a Particle and Quantum Field," *Ukr. Mat. Zh.* **2** pp. 3–24 (in Russian)

Braaten, E., Ralston, J.P. (1985): "Limitations of a Semiclassical Treatment of the Skyrme Soliton," *Phys. Rev.* ***D31*** pp. 598–602

Braaten, E., Carson, L. (1988): "Deuteron as a Toroidal Skyrmion," *Phys. Rev.* ***D38*** pp. 3525–3539

Chemtob, M. (1985): "Three Flavour Skyrme Model Supplied to Baryon Octet and Decimet," *Nuovo Cim.* ***A89*** pp. 381–435

D'yakonov D.I., Mirlin A.D. (1988): "Is a Condensed State of Nuclear Matter Possible?," *Yad. Fiz.* **47** pp. 662–672 [Engl. transl. *Sov. J. Nucl. Phys.* **47** pp. 421–427

Edmonds, A.R. (1957): *Angular Momentum in Quantum Mechanics*, (Princeton University Press, Princeton NJ)

Jain S., Wadia, S.R. (1985): "Large-N Baryons: Collective Coordinates of the Topological Soliton in the $SU(3)$ Chiral Model," *Nucl. Phys.* ***B258*** pp. 713–725

Konoplich, R.V., Kudrjavtsev, A.E., Martemyanov, B.V., Rubin, S.G., (1988): "On the Quantum Correction to the Skyrmion Mass," *Hadronic J.* **11** pp. 271-277

Kopeliovich, V.B. (1988): "Quantization of the Rotations of Axially Symmetric Systems in the Skyrme Model," *Yad. Fiz.* **47** pp. 1495–1503 [Engl. transl. *Sov. J. Nucl. Phys.* **47** pp. 949–953]

Khrustalev, O.A., Razumov, A.V., Taranov, A.Yu. (1980): "Collective Coordinate Method in the Canonical Formalizm: Bogolubov's Transformations," *Nucl. Phys.* ***B172*** pp. 44–58

Landau, L.D., Lifshitz, E.M. (1977): *Quantum Mechanics (Non-Relativistic Theory)*, (Pergamon, New York)

Li, B.A., Liu, K.F., Zhang, M.M. (1987): "Semiclassical Skyrmion Equation of Motion," *Phys. Rev.* ***D35*** pp. 1693–1697

Makhankov, V.G., Rybakov, Yu.P., Sanyuk, V.I. (1989): *The Skyrme Model and Solitons in Hadron Physics*, Lecture Notes for Young Scientists Series, No. 55 (JINR Publ., Dubna), (in Russian)

Meissner, U.-G., Zahed, I. (1986): "Skyrmions in Nuclear Physics," in *Adv. Nucl. Phys* **17** ed. by J.W. Negele, E. Vogt (Plenum Press, New York - London) pp. 143–222

Potapov, M.V., Rybakov, Yu.P. (1991): "Deformed Solitons in the Skyrme Model," in *"Problems of High Energy Physics and Field Theory,"* Proc. of the XIII Workshop, Protvino, ed. by V.A. Petrov (Nauka, Moscow), pp. 236–239 (in Russian)

Parmentola, A.J. (1984): "The Skyrme Model and the Strong-Coupling Approximation," *Phys. Rev.* ***D30*** pp. 685–687

Rajaraman, R. (1982): *Solitons and Instantons – An Introduction to Solitons and Instantons in Quantum Field Theory*, (North-Holland Publ. Co., Amsterdam)

Verbaarshot, J.J.M., Walhout, T.S. Wambach, J., Wyld, H.W. (1987): "Symmetry and Quantization of the Two-Skyrmion System. The Case of Deuteron," *Nucl. Phys.* ***A468*** pp. 520–538

Wambach, J., Wyld, H.W., Sommermann, H.M. (1987): "Axially Deformed Skyrmions in a Constrained Variational Approach," *Phys. Lett.* ***B*** pp. 272–275

Wudka, J. (1985): "Canonical Quantization of the Skyrme Model," *Nucl. Phys.* ***B259*** pp. 689–701

Chapter 9

Adler, S. (1969): "Axial-Vector Vertex in Spinor Electrodynamics," *Phys. Rev.* **177** pp. 2426–2438

Andrianov, A.A. (1985): "Bosonization in Four Dimensions Due to Anomalies and an Effective Lagrangian for Pseudoscalar Mesons," *Phys. Lett.* **157*B*** pp. 425–429

Andrianov, A.A., Novozhilov, Yu.V. (1985): "Chiral Bosonization in non Abelian Gauge Theories," *Phys. Lett.* **153*B*** pp. 422–426

Andrianov, A.A., Andrianov, V.A., Novozhilov, V.Yu., Novozhilov, Yu.V. (1986): "Spectral Asymmetry, Condensates and the QCD Chiral Lagrangian," *Lett. Math. Phys.* **11** pp. 217-224

Balachandran, A.P. (1986): "Skyrmions," in: *High Energy Physics 1985*, Proc. of the Yale Theoretical Advanced Study Institute, ed. by M.J. Bowick, F. Gürsey (World Scientific, Singapure), Vol. I pp. 1–82

Balachandran, A.P., Marmo, G., Skagerstam, B.S., Stern, A. (1991): *Classical Topology and Quantum States*, (World Scientific, Singapore) in press

Balachandran, A.P., Nair, V.P., Rajeev, S.G., Stern, A. (1982): "Exotic Levels from Topology in the Quantum Chromodynamics Effective Lagrangian," *Phys. Rev. Lett.* **49** pp. 1124–1127

Bando, M., Kugo, T., Yamawaki, K. (1985): "Composite Gauge Bosons and "Low Energy Theorems" of Hidden Local Symmetries," *Prog. Theor. Phys.* **73** pp. 1541–1559

Bando, M., Kugo, T., Yamawaki, K. (1988): "Nonlinear Realization and Hidden Local Symmetries," *Phys. Reports* **164** pp. 217–314

Bardeen, W.A. (1969): "Anomalous Ward Identities in Spinor Field Theories," *Phys. Rev.* **184** pp. 1848

Berlin, T.H., Kac, M. (1952): "The Spherical Model of Ferromagnet," *Phys. Rev.* **86** pp. 821–835

Bell, J.S., Jackiw, R. (1969): "A PCAC Puzzle $\pi^0 \to \gamma\gamma$ in the σ-model," *Nuovo Cim.* **A60** pp. 47–60

Bogolubov, N.N., Struminskii, B.V., Tavkhelidze, A.N. (1965): "On the Problem of Composite Models in Elementary Particle Theory," Preprint JINR D-1968, Dubna, 13 pp. (in Russian)

Cahill, R.T. (1989): "Hadronisation of QCD," *Aust. J. Phys.* **42** pp. 171–186

Chatterjee, A. (1990): "Large-N Expansions in Quantum Mechanics, Atomic Physics and Some $O(N)$ Invariant Systems," *Phys. Reports* **186** pp. 249–370

D'yakonov, D.I., Petrov, V.Yu. (1984): "Baryons as Solitons," Preprint LINF - 967, Leningrad, 59 pp. (in Russian)

Ebert, D., Pervushin, V.N. (1978): "Diquark Spectrum in Two-Dimensional Quantum Chromodynamics," *Teor. Mat. Fizika* **36** pp. 313–323 (in Russian)

Ebert, D., Reinhardt, H. (1986): "Effective Chiral Hadron Lagrangian with Anomalies and Skyrme Terms from Quark Flavour Dynamics," *Nucl. Phys.* **B271** pp. 188–226

Faddeev, L.D., Slavnov, A.A. (1980): *Gauge Fields: Introduction to Quantum Theory*, (W.A. Benjamin/Cummings, Reading MA)

Fritszch, H., Gell Mann, M., Leutwyler, H. (1973): "Advantages of the Color Octet Gluon Picture," *Phys. Lett.* **B47** pp. 365–368

Gasiorovicz, S., Geffen, D.A. (1969): "Effective Lagrangians and Field Algebras with Chiral Symmetry," *Rev. Mod. Phys.* **41** pp. 531–573

Gell Mann, M. (1985): "From Renormalizability to Calculability?," in *Shelter Island II* Proc. of the 1983 Shelter Island Conference, ed. by R. Jackiw, N.N. Khuri, S. Weinberg, E. Witten (The MIT Press, Cambridge Mass.) pp. 3–23

Gell Mann, M., Neeman, Y., eds. (1964): *The Eightfold Way*, (W.A. Benjamin, Inc., New York)

Gipson, J.M. (1984): "Quasi-Solitons in the Strongly Coupled Higgs Sector of the Standard Model," *Nucl. Phys.* **B231** pp. 365–385

Goldstone, J., Wilczek, F. (1981): "Fractional Quantum Numbers on Solitons," *Phys. Rev. Lett.* **47** pp. 986–989

Gonzales, D., Redlich, A.N. (1984): "Non-Abelian Bosonization in Two Dimensions using Path Integrals," *Phys. Lett.* ***B*147** pp. 150–156

Greenberg, O.W. (1964): "Spin and Unitary-Spin Independence in a Paraquark Model of Baryons and Mesons," *Phys. Rev. Lett.* **13** pp. 598–602

Gross, D.J., Wilczek, F. (1973a): "Ultraviolet Behavior of non Abelian Gauge Theories," *Phys. Rev. Lett.* **30** pp. 1343–1346;

Gross, D.J., Wilczek, F. (1973b): "Asymptotically Free Gauge Theories, I," *Phys. Rev.* ***D*8** pp. 3633–3652

Gross, D.J., Wilczek, F. (1974): "Asymptotically Free Gauge Theories, II," *Phys. Rev.* ***D*9** pp. 980–993

Han, M., Nambu, Y. (1965): "Three-Triplet Model with Double $SU(3)$ Symmetry," *Phys. Rev.* ***B*139** pp. 1006–1010

t'Hooft, G. (1974a): "A Planar Diagram Theory for Strong Interactions," *Nucl. Phys.* ***B*72** pp. 461–473

t'Hooft, G. (1974b): "A Two-Dimensional Model for Mesons," *Nucl. Phys.* ***B*75** pp. 461–470

Karchev, N.I., Slavnov, A.A. (1985): "Effective Chiral Lagrangian from the $SU(\infty)$ Quantum Chromodynamics," *Teor. Mat. Fizika* **65** pp. 192–201 (in Russian)

Kaymakcalan, O, Rajeev, S., Schechter, J. (1984): "Non-Abelian Anomaly and Vector-Meson Decay," *Phys. Rev.* ***D*30** pp. 594–602

Marciano, W., Pagels, H. (1978): "Quantum Chromodynamics," *Phys. Reports* **36*C*** pp. 137–276

Meissner, U.-G. (1988): "Low-Energy Hadron Physics from Effective Chiral Lagrangians with Vector Mesons," *Phys. Reports* **161** pp. 213–362

Politzer, H.D. (1973): "Reliable Perturbative Results for Strong Interactions?," *Phys. Rev. Lett.* **30** pp. 1346–1349

Rajaraman, R. (1982): *Solitons and Instantons – An Introduction to Solitons and Instantons in Quantum Field Theory,* (North-Holland Publ. Co., Amsterdam)

Roberts, C.D., Cahill, R.T. (1987): "A Bosonisation of QCD and Realization of Chiral Symmetry," *Aust. J. Phys.* **40** pp. 449–517

Weinberg, S. (1973): "Non-Abelian Gauge Theories of the Strong Interactions," *Phys. Rev. Lett.* **31** pp. 494–497

Witten, E. (1979): "Baryons in the 1/N Expansion," *Nucl. Phys.* ***B*160** pp. 57–115

Witten, E. (1980): "Quarks, Atoms and 1/N Expansion," *Physics Today* **33** pp. 38–43

Witten, E. (1984): "Skyrmions and QCD," in *Solitons in Nuclear and Elementary Particle Physics,* Proc. of the Lewes Workshop, ed. by A. Chodos, E. Hadjimichael, Ch. Tze (World Scientific, Singapure), pp. 306–312

Wrzecionko, J. (1988): "Direct Derivation of the Skyrme-like Lagrangians from QCD," *J. Phys.* ***G*14** pp. 411–424

Yang, C.N., Mills, R.L. (1954): "Conservation of Isotopic Spin and Isotopic Gauge Invariance," *Phys. Rev.* **96** pp. 191–195

Yndurâin, F.J. (1983): *Quantum Chromodynamics,* (Springer-Verlag, Berlin)

Zahed, I., Brown, G.E. (1986): "The Skyrme Model," *Phys. Reports* **142** pp. 1–102

Chapter 10

Arovas, D.P., Schrieffer, R., Wilczek, F., Zee, A. (1985): "Statistical Mechanics of Anyons," *Nucl. Phys.* ***B*251** pp. 117–126

Arovas, D.P., Wilczek, F., Schrieffer, R. (1984): "Fractional Statistics and the Quantum Hall Effect," *Phys. Rev. Lett.* **53** pp. 722–723

Azcárraga, de J.A., Izquierdo, J.M., Macfarlane, A.J. (1990): "Current Algebra and Wess-Zumino Terms: a Unified Geometric Treatment," *Ann. Phys.* **202** pp. 1–21

Balachandran, A.P., Gomm, H., Sorkin, R.D. (1987): "Quantum Symmetries From Quantum Phases, Fermions from Bosons, a Z_2 Anomaly and Galilean Invariance," *Nucl. Phys.* ***B*281** pp. 573–612

Balachandran, A.P., Marmo, G., Skagerstam, B.S., Stern, A. (1991): *Classical Topology and Quantum States,* (World Scientific, Singapore) in press

Dirac, P.A.M. (1931): "Quantized Singularities in the Electromagnetic Field," *Proc. Roy. Soc. London* ***A*133** pp. 60–72

Dirac, P.A.M. (1948): "The Theory of Magnetic Poles," *Phys. Rev.* **74** pp. 817–830

Finkelstein, D. (1966): "Kinks," *J. Math. Phys.* **7** pp. 1218–1225

Finkelstein, D., Rubinstein, J. (1968): "Connection Between Spin, Statistics and Kinks," *J. Math. Phys.* **9** pp. 1762–1779

Friedman, J.L., Sorkin, R.D. (1980): "Spin 1/2 from Gravity," *Phys. Rev. Lett.* **44** pp. 1100–1103

Fröhlich, J., Marchetti, P.A. (1988): "Quantum Field Theory of Anyons," *Lett. Math. Phys.* **16** pp. 348–358

Goddard, P., Mansfield, P. (1986): "Topological Structures in Field Theory," *Rep. Prog. Phys.* **49** pp. 725–781

Jackiw, R. (1984a): "Quantization of Physical Parameters," *Commun. Nucl. and Part. Phys.* **13** pp. 141–156

Jackiw, R. (1984b): "Topological Investigations of Quantized Gauge Theories," in *Relativity, Groups and Topology II,* Les Houches 1983, ed. by B.S. De Witt, R. Stora (North-Holland, Amsterdam)

Jain, S., Wadia, S.R. (1985): "Large-N Baryons: Collective Coordinates of the Topological Soliton in the $SU(3)$ Chiral Model," Nucl. Phys. ***B*258** pp. 713–725

Krichever, I.M., Olshanetsky, M.A., Perelomov, A.M. (1986): "Wess-Zumino Lagrangians in Chiral Models and Quantization of their constants," *Nucl. Phys.* ***B*264** pp. 415–422

Laughlin, R.B. (1988): "The Relationship Between High-Temperature Superconductivity and the Fractional Hall Effect," *Science* **242** pp. 525–533

Mackenzie, R., Wilczek, F. (1988): "Peculiar Spin and Statistics in Two Space Dimensions," *Int. J. Mod. Phys.* ***A*3** pp. 2827–2853

Novikov, S.P. (1981): "Multivalued Functions and Functionals. Analogue of Morse Theory," *Dokl. Akad. Nauk* **260** pp. 31–34 (in Russian)

Novikov, S.P. (1982): "The Hamiltonian Formalizm and the Multivalued Analogue of Morse Theory," *Usp. Mat. Nauk* **37** pp. 3–49 [Engl. transl *Russian Math. Surveys* **37** pp. 1–56]

Ravenel, D.C., Zee, A. (1985): "Spin and Statistics of Solitons and J-Homomorphism," *Commun. Math. Phys.* **98** pp. 239–243

Schwarz, A.S. (1989): "Statistics of Skyrmions," *Mod. Phys. Lett.* ***A*4** pp. 403–407

Sorkin, R.D. (1986): "Introduction to Topological Geons," in *Topological Properties and Global Structure of Spacetime,* ed. by P.G. Bergmann, V. de Sabata (Plenum Press, New York)

Sorkin, R.D. (1988): "A General Relation Between Kink-Exchange and Kink-Rotation," *Commun. Math. Phys.* **115** pp. 421–434

Tamm, Ig. (1931): "Die Verallgemeinerten Kugelfunctionen und die Wellenfunctionen eines Electrons im Felde eines Magnetopoles," *Z. Phys.* **71** pp. 141–150

Wess, J., Zumino, B. (1971): "Consequences of Anomalous Ward Identities," Phys. Lett. ***B*37** pp. 95–97

Whitehead, G.W. (1966): *Homotopy Theory,* (MIT Press, Cambridge MA)

Wilczek, F. (1982): "Quantum Mechanics of Fractional-Spin Particles," *Phys. Rev. Lett.* **49** pp. 957–959

Wilczek, F. (1990): "Lectures on Fractional Statistics and Anyon Superconductivity," in *Anomalies, Phases and Defects,* Proc. of the Second Ferrara School, Ferrara 1989, ed. by M. Bregola, G. Marmo, G. Morandi (Bibliopolis, Naples)

Williams, J. (1970): "Topological Analysis of Nonlinear Field Theory," *J. Math. Phys.* **11** pp. 2611–2616

Witten, E. (1983a): "Global Aspects of Current Algebra," *Nucl. Phys.* ***B*223** pp. 422–432

Witten, E. (1983b): "Current Algebra, Baryons and Quark Confinement," *Nucl. Phys.* **B223** pp. 433–444

Zahed, I., Brown, G.E. (1986): "The Skyrme Model," Phys. Reports **142** pp. 1–102

Chapter 11

Adkins, G.S., Nappi, C.R. (1984): "The Skyrme Model with Pion Masses," *Nucl. Phys.* **B233** pp. 109–115

Adkins, G.S., Nappi, C.R. (1985): "Model-Independent Relations for Baryons as Solitons in Mesonic Theories," *Nucl. Phys.* **B249** pp. 507–518

Balachandran, A.P. (1986): "Skyrmions," in: *High Energy Physics 1985,* Proc. of the Yale Theoretical Advanced Study Institute, ed. by M.J. Bowick, F. Gürsey (World Scientific, Singapure), Vol. I, pp. 1–82

Balachandran, A.P., Lizzi, F., Rodgers, V.G.J., Stern, A. (1985): "Dibaryons as Chiral Solitons," *Nucl. Phys.* **B256** pp. 525–556

Balachandran, A.P., Marmo, G., Skagerstam, B.S., Stern, A. (1983): *Gauge Theories and Fibre Bundles. Applications to Particle Dynamics,* Lect. Notes in Phys., Vol. 188 (Springer-Verlag, Berlin, Heidelberg)

Balachandran, A.P., Marmo, G., Skagerstam, B.S., Stern, A. (1991): *Classical Topology and Quantum States,* (World Scientific, Singapore) in press

Braaten, E., Cai, D. (1989): "Evidence for K Symmetry in Nucleon-Nucleon Scattering," *Phys. Rev. Lett.* **62** pp. 1243–1246

Callan, C.G., Klebanov, I. (1985): "Bound-State Approach to Strangeness in the Skyrme Model," *Nucl. Phys.* **B262** pp. 365–382

Callan, C.G., Hornbostel, K., Klebanov, I. (1988): "Baryon Masses in the Bound-State Approach to Strangeness in the Skyrme Model," *Phys. Lett.* **202B** pp. 269–272

Castillejo, L., Jones, P.S.J., Jackson, A.D., Verbaarschot, J.J.M., Jackson, A. (1989): "Dense Skyrmion Systems," *Nucl. Phys.* **A501** pp. 801–812

Chemtob, M. (1985): "Skyrme Model of Baryon Octet and Decouplet," *Nucl. Phys.* **B256** pp. 600–608

Gomm, H., Jain, P., Johnson, R., Schechter, J. (1986a): "Scale Anomaly and the Scalars," *Phys. Rev.* **D33** pp. 801–812

Gomm, H., Jain, P., Johnson, R., Schechter, J. (1986b): "Bag Formation in the Chiral Model," *Phys. Rev.* **D33** pp. 3476–3478

Guadagnini, E. (1984): "Baryons as Solitons and Mass Formulae," *Nucl. Phys.* **B236** pp. 35–47

Holzwarth, G., Schwesinger, B. (1986): "Baryons in the Skyrme Model," *Rep. Progr. Phys.* **49** pp. 825–871

Jackson, A., Jackson, A.D., Pasquier, V. (1985): "The Skyrmion-Skyrmion Interaction," *Nucl. Phys.* **A432** pp. 567–609

Jain S., Wadia, S.R. (1985): "Large-N Baryons: Collective Coordinates of the Topological Soliton in the $SU(3)$ Chiral Model," *Nucl. Phys.* **B258** pp. 713–725

Karliner, M., Mattis, M.P. (1986): "πN, KN and $\bar{K}N$ Scattering: Skyrme Model versus experiment," *Phys. Rev.* **D34** pp. 1991–2024

Kanazawa, A. (1987): "Static Properties of Baryons in the $SU(3)$ Skyrme model," *Progr. Theor. Phys.* **77** pp. 1240–1252

Kutschera, M., Pethick, C.J. (1985): "The Short-Range Interaction Between Skyrmions," *Nucl. Phys.* **A440** pp. 670–684

Landau, L.D., Lifshitz, E.M. (1977): *Quantum Mechanics (Non-Relativistic Theory),* (Pergamon, New York)

Loss, M. (1987): "The Skyrme Model on Riemannian Manifold," *Lett. Math. Phys.* **14** pp. 149–156

Makhankov, V.G., Rybakov, Yu.P., Sanyuk, V.I. (1992): "The Skyrme Model and Strong Interactions," *Usp. Fiz. Nauk* **162** pp. 1–61 (in Russian) [Engl. transl. Sov. Phys. Usp. **35** (1992) pp. 55–84]

Manton, N.S. (1987): "Geometry of Skyrmions," *Commun. Math. Phys.* **111** pp. 469–478

Manton, N.S., Ruback, P.J. (1986): "Skyrmions in Flat Space and Curved Space," *Phys. Lett.* **181*B*** pp. 137–140

Manton, N.S., Samols, T.M. (1990): "Skyrmions on S^3 and H^3 from Instantons," J. Phys. ***A*23** pp. 3749–3759

Mattis, M.P. (1986): "Skyrmions and Vector Mesons," *Phys. Rev. Lett.* **56** pp. 1103–1106

Mattis, M.P., Braaten, E. (1989): "Hadron Scattering in the Large N_c Limit as a Problem in Linear Algebra," *Phys. Rev.* ***D*39** pp. 2737–2750

Mattis, M.P., Karliner, M. (1985): "Baryon Spectrum of the Skyrme Model," *Phys. Rev.* ***D*31** pp. 2833–2848

Mattis, M.P., Peskin, M.E. (1985): "Systematics of πN Scattering in Chiral Soliton Models," *Phys. Rev.* ***D*32** pp. 58–73

Mattis, M.P., Mukerjee, M. (1988): "$I_t = J_t$ rule: A New Large-N_c Selection Rule for Meson-Baryon Scattering," *Phys. Rev. Lett.* **61** pp. 1344–47

Nowak, M.A., Rho, M., Scoccola, N. (1988): "The Hyperons as Skyrmions with Vector Mesons," *Phys. Lett* **201*B*** pp. 425–43

Nyman, E.M., Riska, D.O. (1990): "Low-Energy Properties of Baryons in the Skyrme Model," *Rep. Progr. Phys.* **53** pp. 1137–1180

Pethick, C.J. (1985): "Dense Matter in Stellar Collapse and Neutron Stars," Nucl. Phys. ***A*434** pp. 537–604

Praszalowicz, M. (1985): "A Comment on the Phenomenology of the $SU(3)$ Skyrme Model," *Phys. Lett.* **158*B*** pp. 264–269

Praszalowicz, M. (1989): "Baryon Properties in Chiral Models," *Nuovo Cim.* **102*A*** pp. 39–62

Rabinovici, E., Schwimmer, A., Yankielowicz, S., (1984): "Quantization in the Presence of Wess-Zumino Term," *Nucl. Phys.* ***B*248** pp. 523–535

Riska, D.O., Schwesinger, B. (1989): "The Isospin Independent Spin-Orbit Interaction in the Skyrme Model," Phys. Lett. ***B*229** pp. 339–342

Schnitzer, H.J. (1984): "The Soft-Pion Skyrmion Lagrangian and Strong CP-violation," *Phys. Lett.* **139*B*** pp. 217–222

Skyrme, T.H.R. (1962): "A Unified Field Theory of Mesons and Baryons," *Nucl. Phys.* **31** pp. 556–569

Sudarshan, E.C.G., Mukunda, N. (1974): *Classical Dynamics: A Modern Perspective*, (Wiley, New York)

Yabu, H., Ando, K., (1988): "A New Approach to the SU(3) Skyrme Model. Exact treatment of the SU(3) symmetry breaking," *Nucl. Phys.* ***B*301** pp. 601–626

Yabu, H., Schwesinger, B., Holzwarth, G. (1989): "The Intermediate-Range Attraction of the Nucleon-Nucleon Force in a Skyrme Model with Explicit σ-mesons," *Phys. Lett.* **224*B*** pp. 25–28

Walhout, T.S. (1990): "The Equation of State of Dense Skyrmion Matter," *Nucl. Phys.* ***A*519** pp. 816–830

Weinberg, S. (1967): "Dynamical Approach to Current Algebra," *Phys. Rev. Lett.* **18** pp. 188–191

Wirzba, A., Bang, H. (1990): "The Mode Spectrum and the Stability Analysis of Skyrmions on a 3-sphere," *Nucl. Phys.* ***A*515** pp. 571–598

Witten, E. (1983): "Current Algebra, Baryons and Quark Confinement," *Nucl. Phys.* ***B*223** pp. 433–444

Zahed, I., Brown, G.E. (1986): "The Skyrme Model," *Phys. Reports* **142** pp. 1–102

Appendix A

Barnes, K.J., Dondi, P.H., Sarcar, S.G. (1972): "Nonlinear Realization of Chiral Algebras and Phenomenological Lagrangians," *Proc. Roy. Soc. London,* **A330** pp. 389–415

Barnes, K.J., Isham, C.J. (1970): "Chiral Metric Meson Lagrangians," *Nuclear Physics,* **15** pp. 333–349

Barnes, K.J., Ketley, I.J., Nicole, D.A., O'Donnel, P.J. (1977): "Nonlinear Chiral Models and Many Dimensional Solitons," *Phys. Rev.,* **D16** pp. 511–516

Callan, C., Coleman, S., Wess, J., Zumino, B. (1969): "Structure of Phenomenological Lagrangians," *Phys. Rev.,* **177** pp. 2247–2250

Cartan, É. (1929): "Groupes Simples closet Ouverts et Géométrie Riemannienne," *J. Math. Pures Appl.,* **8** pp. 1–33

Chang, F., Gürsey, F. (1967): "Unified Formulation of Effective Nonlinear Pion-Nucleon Lagrangians," *Phys. Rev.,* **164** pp. 1752–1761

Coleman, S., Wess, J., Zumino, B. (1969): "Structure of Phenomenological Lagrangians," *Phys. Rev.,* **177** pp. 2239–2247

De Alfaro, V., Fubini, S., Furlan, , C. Rossetti, G. (1973): *Currents in Hadron Physics,* (North-Holland, Amsterdam-London)

Feynman, R., Gell-Mann, M. (1958): "Theory of Fermi Interaction," *Phys. Rev.,* **109** pp. 193–198

Gasiorovicz, S., Geffen, D.A. (1969): "Effective Lagrangians and Field Algebras with Chiral Symmetry," *Rev. Mod. Phys.* **41** pp. 531–573

Isham, C.J. (1969): "A Group-Theoretical Approach to Chiral Transformations," *Nuovo Cim.* **59*A*** pp. 356–375

Lee, B.W. (1971): *Chiral Dynamics,* (Gordon and Breach, New York)

Meetz, K. (1969): "Realization of Chiral Symmetry in Curved Isospin Space," *J. Math. Phys.* **10** pp. 589–593

Pais, A. (1953): "Isotopic Spin and Mass Quantization," *Phisica* **19** pp. 869–887

Pais, A. (1954): "On the Program of Systematization of Particles and Interactions," *Proc. Nat. Acad. Sci. (USA)* **40** pp. 484–492

Schwartz, J.T. (1968): *Differential Geometry and Topology,* (New York)

Slavnov, A.A., Faddeev, L.D. (1971): "Invariant Perturbation Theory for Nonlinear Chiral Lagrangians," *Teor. Mat. Fizika* **8** pp. 297–307 (in Russian)

Touschek, B.F. (1957): "Parity Conservation and the Mass of the Neutrino," *Nuovo Cim.* **5** pp. 754–755

Treiman, S.B., Jackiw, R., Gross, D.J. (1972): *Lectures on Current Algebra and its Applications,* (Princenton Univ. Press, Princenton NJ)

Volkov, M.K., Pervushin, V.N. (1978): *Essentially Nonlinear Quantum Theories, Dynamical Symmetries and Meson Physics,* (Atomizdat, Moscow) (in Russian)

Watanabe, S. (1957): "Chirality of K-particle," *Phys. Rev.* **106** pp. 1306–1315

Weinberg, S. (1967): "Dynamical Approach to Current Algebra," *Phys. Rev. Lett.* **18** pp. 188–191

Weinberg, S. (1968): "Nonlinear Realization of Chiral Symmetry," *Phys. Rev.* **166** pp. 1568–1577

Zhelobenko, D.P. (1970): *Compact Lie Groups and their Representations,* (Moscow, Nauka) (in Russian)

Appendix B

Dubrovin, B.A., Novikov, S.P., Fomenko, A.T. (1984): *Sovremennaja Geometria: Metody Teorii Gomologii,* (Nauka, Moskva) (in Russian)

Eguchi, T., Gilkey, P.B., Hanson, A.J. (1980): "Gravitation, Gauge Theories and Differential Geometry," Phys. Reports **66** pp. 213–393

Felsager, B. (1981): *Geometry, Particles and Fields,* (Odense Univ. Press)

Finkelstein, D., Misner, C. (1959): "Some New Conservation Laws," *Ann. Phys.* **6** pp. 230–243
Flanders, H. (1963): *Differential Forms with Applications to the Physical Sciences,* (Academic Press, New York)
Huang, K. (1980): *Quarks, Leptons and Gauge Fields,* (World Scientific, Singapore)
Isham, C.J. (1977): "Topological Currents for Arbitrary Chiral Groups in Three Space Dimensions," *J. Phys.* **A10** pp. 1397–1407
Lyakhovskii, V.D., Bolotov, A.A. (1983): *Symmetry Groups and Elementary Particles,* (Leningrad, Leningrad Univ. Press) (in Russian)
Mitschenko, A.S., Fomenko, A.T. (1980): *Course in Differential Geometry and Topology,* (Moscow, Moscow Univ. Press) (in Russian)
Novikov, S.P., Fomenko, A.T. (1987): *Elements of Differential Geometry and Topology,* (Nauka, Moskva) (in Russian)
Olshanetskii, M.A. (1982): "A Short Guide to Modern Geometry for Physicists," *Uspekhi Fizicheskikh Nauk* **136** pp. 421–433 (in Russian) [Engl. transl. *Soviet Physics - Uspekhi* **25** (1982)]
Postnikov, M.M. (1982): *Lie Groups and Algebras,* Lecture Notes in Geometry, Semester V (Moscow, Nauka) (in Russian)
Schutz, B.F. (1982): *Geometrical Methods of Mathematical Physics,* (Cambridge Univ. Press, Cambridge)
Schwarz, A.S. (1989): *Quantum Field Theory and Topology,* (Moscow, Nauka) (in Russian)
Warner, F.W. (1983): *Foundations of Differentiable Manifolds and Lie Groups,* (Springer-Verlag, New York-Berlin-Heidelberg)

Appendix C

Glaser, V., Martin, A., Grosse, H., Thirring, W. (1976): "A Family of Optimal Conditions for the Absence of Bound States in a Potential," in *Studies in Mathematical Physics,* ed. by E.H. Lieb, B. Simon, A.S. Whightman (Princenton, N.J.) pp. 169–194
Kapitanskii L.V., Ladyzhenskaya, O.A. (1983): "Coleman's Principle for the Determination of the Stationary Points of Invariant Functionals," *Zapiski Nauchnykh Seminarov Leningradskogo Otdeleniya Matematicheskogo Instituta im. V.A. Steklova AN SSSR* Vol. 127 (Nauka, Leningrad), pp. 84–102 (in Russian), [Engl. transl. *Journ. Sov. Math.* **27** (1984) pp. 2606–2616]
Palais, R.S. (1979): "The Principle of Symmetric Criticality," *Commun. Math. Phys.* **69** pp. 19–30
Sanyuk, V.I. (1979): *Studies of Nonlinear Classical Field Models with Topological Charges,* Ph.D. Thesis, People's Friendship Univ. (Moscow) unpublished (in Russian)

Appendix D

Balachandran, A.P., Marmo, G., Skagerstam, B.S., Stern, A. (1983): *Gauge Theories and Fibre Bundles. Applications to Particle Dynamics,* Lect. Notes in Phys., Vol. 188 (Springer-Verlag, Berlin, Heidelberg)
Ekeland, I., Temam, R. (1976): *Convex Analysis and Variational Problems,* (North-Holland Pub. Co., Amsterdam, Oxford)
Kolmogorov, A.N., Fomin, S.V. (1972): *Elements of Theory of Functions and Functional Analysis,* (Nauka, Moscow) (in Russian)
Makhankov, V.G., Rybakov, Yu.P., Sanyuk, V.I. (1992): "The Skyrme Model and Strong Interactions," *Usp. Fiz. Nauk* **162** pp. 1–61 (in Russian) [Engl. transl. Sov. Phys. Usp. **35** (1990) pp. 55–84]
Rybakov, Yu.P. (1988): "On Critical Points of Invariant Functionals," in *Actual Problems of Quantum Mechanics and Statistical Physics,* (People's Friendship Univ. Press, Moscow), pp. 71–77 (in Russian)

Rybakov, Yu.P. (1991): "Stability of Multi-Dimensional Solitons in Chiral Models and Gravitation," in *Advances of Science and Technology,* Ser. Classical Field Theory and Theory of Gravitation Vol. 2 (VINITI AN SSSR, Moscow) pp. 56–111 (in Russian)
Rybakov, Yu.P., Chakrabarti, S. (1984): "Conditional Stability of Multiple-Charged Solitons," *Int. J. Theor. Phys.* **23** pp. 325–333

Appendix E

Montgomery, D., Zippin, L. (1955): *Topological Transformation Groups,* (Interscience, New York)
Pak, N.K., Tze, H.C. (1979): "Chiral Solitons and Current Algebra," *Ann. Phys.* **117** pp. 164–194
Steenrod, N. (1951): *The Topology of Fibre Bundles,* (Princeton Univ. Press, Princeton, N.J.)
Williams, J. (1970): "Topological Analysis of Nonlinear Field Theory," *J. Math. Phys.* **11** pp. 2611–2616

Subject Index

Springer Series in
Nuclear
and **Particle Physics**

Enquiries and manuscripts should be addressed to the editor working on your subject (or to the managing editor)

Springer

MIX
Papier aus verantwortungsvollen Quellen
Paper from responsible sources
FSC® C105338
FSC
www.fsc.org

If you have any concerns about our products,
you can contact us on
ProductSafety@springernature.com

In case Publisher is established outside the EU,
the EU authorized representative is:
Springer Nature Customer Service Center GmbH
Europaplatz 3, 69115 Heidelberg, Germany

Printed by Libri Plureos GmbH
in Hamburg, Germany